国外优秀物理著作
原版系列

量子世界中的蝴蝶
——最迷人的量子分形故事

Butterfly in the Quantum World
—The story of the most fascinating quantum fractal

［印］金杜·萨蒂亚（Indubala I. Satija） 著

哈尔滨工业大学出版社
HARBIN INSTITUTE OF TECHNOLOGY PRESS

黑版贸审字 08-2019-043 号

图书在版编目(CIP)数据

量子世界中的蝴蝶：最迷人的量子分形故事：英文/(印)金杜·萨蒂亚(Indubala I. Satija)著. —哈尔滨：哈尔滨工业大学出版社,2020.7

书名原文：Butterfly in the Quantum World：The story of the most fascinating quantum fractal

ISBN 978-7-5603-8870-0

Ⅰ.①量… Ⅱ.①金… Ⅲ.①量子论-普及读物-英文 Ⅳ.①O413-49

中国版本图书馆 CIP 数据核字(2020)第 104932 号

Copyright © 2016 by Morgan & Claypool Publishers
All rights reserved.
The English reprint rights arranged through Rightol Media（本书英文影印版权经由锐拓传媒取得 Email：copyright@rightol.com）

策划编辑	刘培杰
责任编辑	杜莹雪　宋　淼
封面设计	孙茵艾
出版发行	哈尔滨工业大学出版社
社　　址	哈尔滨市南岗区复华四道街10号　邮编150006
传　　真	0451-86414749
网　　址	http://hitpress.hit.edu.cn
印　　刷	哈尔滨博奇印刷有限公司
开　　本	787mm×1092mm　1/16　印张22.25　字数428千字
版　　次	2020年7月第1版　2020年7月第1次印刷
书　　号	ISBN 978-7-5603-8870-0
定　　价	118.00元

（如因印装质量问题影响阅读，我社负责调换）

To my Father

Hofstadter Butterfly

Contents

Summary	vii
About the author	viii
Preface	ix
Prologue	xv
Prelude	xlv

Part I The butterfly fractal

0   Kiss precise 0-1
0.1 Apollonian gaskets and integer wonderlands 0-5
     Appendix: An Apollonian sand painting—the world's largest artwork 0-7
     References 0-7

1   The fractal family 1-1
1.1 The Mandelbrot set 1-3
1.2 The Feigenbaum set 1-7
     1.2.1 Scaling and universality 1-8
     1.2.2 Self-similarity 1-9
1.3 Classic fractals 1-10
     1.3.1 The Cantor set 1-11
     1.3.2 The Sierpinski gasket 1-12
     1.3.3 Integral Apollonian gaskets 1-13
1.4 The Hofstadter set 1-14
     1.4.1 Gaps in the butterfly 1-17
     1.4.2 Hofstadter meets Mandelbrot 1-17
     1.4.3 Concluding remarks: A mathematical, physical, and poetic magπ 1-18
     Appendix: Harper's equation as an iterative mapping 1-22
     References 1-23

2   Geometry, number theory, and the butterfly: Friendly numbers and kissing circles 2-1
2.1 Ford circles, the Farey tree, and the butterfly 2-3
     2.1.1 Ford circles 2-3
     2.1.2 Farey tree 2-4

	2.1.3	The saga of even-denominator and odd-denominator fractions	2-6
	2.1.4	The sizes of butterflies	2-9
2.2	A butterfly at every scale—butterfly recursions	2-9
2.3	Scaling and universality	2-13
	2.3.1	Flux scaling	2-13
	2.3.2	Energy scaling	2-14
	2.3.3	Universality	2-15
2.4	The butterfly and a hidden trefoil symmetry	2-17
2.5	Closing words: Physics and number theory	2-17
	Appendix A: Hofstadter recursions and butterfly generations	2-18
	Appendix B: Some theorems of number theory	2-21
	Appendix C: Continued-fraction expansions	2-22
	Appendix D: Nearest-integer continued fraction expansion	2-23
	Appendix E: Farey paths and some comments on universality	2-23
	References	2-25

3   The Apollonian–butterfly connection ($\mathcal{ABC}$)   3-1

3.1	Integral Apollonian gaskets ($\mathcal{IAG}$) and the butterfly	3-3
	3.1.1 A duality transformation	3-3
	3.1.2 Illustrating the Apollonian–butterfly connection	3-5
3.2	The kaleidoscopic effect and trefoil symmetry	3-6
	3.2.1 Seeing an Apollonian gasket as a kaleidoscope	3-6
	3.2.2 How nested butterflies are related to kaleidoscopes	3-8
	3.2.3 $\mathcal{ABC}$ and trefoil symmetry	3-9
3.3	Beyond Ford Apollonian gaskets and fountain butterflies	3-13
	Appendix: Quadratic Diophantine equations and $\mathcal{IAG}$s	3-15
	References	3-16

4   Quasiperiodic patterns and the butterfly   4-1

4.1	A tale of three irrationals	4-2
4.2	Self-similar butterfly hierarchies	4-5
4.3	The diamond, golden, and silver hierarchies, and Hofstadter recursions	4-9
4.4	Symmetries and quasiperiodicities	4-12
	Appendix: Quasicrystals	4-13
	A.1 One-dimensional quasicrystals	4-14
	A.2 Two-dimensional quasicrystals: Quasiperiodic tiles	4-15
	A.3 A brief history of the discovery of quasicrystals	4-17

A.4	Excerpts from the ceremony of the Nobel Prize in chemistry in 2011	4-17
	References	4-19

Part II Butterfly in the quantum world

5   The quantum world   5-1

5.1	Wave or particle—what is it?	5-3
	5.1.1 Matter waves	5-4
5.2	Quantization	5-5
5.3	What is waving?—The Schrödinger picture	5-8
5.4	Quintessentially quantum	5-10
	5.4.1 The double-slit experiment, first hypothesized and finally realized	5-11
	5.4.2 The Ehrenberg–Siday–Aharonov–Bohm effect (ESAB)	5-13
5.5	Quantum effects in the macroscopic world	5-17
	5.5.1 Central concepts of condensed-matter physics	5-18
	5.5.2 Summary	5-23
	References	5-23

6   A quantum-mechanical marriage and its unruly child   6-1

6.1	Two physical situations joined in a quantum-mechanical marriage	6-2
6.2	The marvelous pure number ϕ	6-2
6.3	Harper's equation, describing Bloch electrons in a magnetic field	6-4
6.4	Harper's equation as a recursion relation	6-11
6.5	On the key role of inexplicable artistic intuitions in physics	6-12
6.6	Discovering the strange eigenvalue spectrum of Harper's equation	6-13
6.7	Continued fractions and the looming nightmare of discontinuity	6-16
6.8	Polynomials that dance on several levels at once	6-18
6.9	A short digression on INT and on perception of visual patterns	6-23
6.10	The spectrum belonging to irrational values of ϕ and the "ten-martini problem"	6-25
6.11	In which continuity (of a sort) is finally established	6-28
6.12	Infinitely recursively scalloped wave functions: Cherries on the doctoral sundae	6-31
6.13	Closing words	6-33
	Appendix: Supplementary material on Harper's equation	6-33
	References	6-35

Part III Topology and the butterfly

7 A different kind of quantization: The quantum Hall effect   7-1

7.1 What is the Hall effect? Classical and quantum answers   7-2
7.2 A charged particle in a magnetic field: Cyclotron orbits and their quantization   7-4
  7.2.1 Classical picture   7-4
  7.2.2 Quantum picture   7-5
  7.2.3 Semiclassical picture   7-7
7.3 Landau levels in the Hofstadter butterfly   7-9
7.4 Topological insulators   7-11
  Appendix A: Excerpts from the 1985 Nobel Prize press release   7-12
  Appendix B: Quantum mechanics of electrons in a magnetic field   7-13
  Appendix C: Quantization of the Hall conductivity   7-14
  References   7-14

8 Topology and topological invariants: Preamble to the topological aspects of the quantum Hall effect   8-1

8.1 A puzzle: The precision and the quantization of Hall conductivity   8-2
8.2 Topological invariants   8-3
  8.2.1 Platonic solids   8-4
  8.2.2 Two-dimensional surfaces   8-5
  8.2.3 The Gauss–Bonnet theorem   8-7
8.3 Anholonomy: Parallel transport and the Foucault pendulum   8-8
8.4 Geometrization of the Foucault pendulum   8-10
8.5 Berry magnetism—effective vector potential and monopoles   8-13
8.6 The ESAB effect as an example of anholonomy   8-18
  Appendix: Classical parallel transport and magnetic monopoles   8-19
  References   8-20

9 The Berry phase and the quantum Hall effect   9-1

9.1 The Berry phase   9-2
9.2 Examples of Berry phase   9-6
9.3 Chern numbers in two-dimensional electron gases   9-10
9.4 Conclusion: the quantization of Hall conductivity   9-11
9.5 Closing words: Topology and physical phenomena   9-13

Appendix A: Berry magnetism and the Berry phase — 9-14
Appendix B: The Berry phase and 2 × 2 matrices — 9-16
Appendix C: What causes Berry curvature? Dirac strings, vortices, and magnetic monopoles — 9-17
Appendix D: The two-band lattice model for the quantum Hall effect — 9-19
References — 9-20

10 The kiss precise and precise quantization — 10-1

10.1 Diophantus gives us two numbers for each swath in the butterfly — 10-3
    10.1.1 Quantum labels for swaths when ϕ is irrational — 10-7
10.2 Chern labels not just for swaths but also for bands — 10-7
10.3 A topological map of the butterfly — 10-8
10.4 Apollonian–butterfly connection: Where are the Chern numbers? — 10-10
10.5 A topological landscape that has trefoil symmetry — 10-12
10.6 Chern-dressed wave functions — 10-14
10.7 Summary and outlook — 10-14
References — 10-17

Part IV Catching the butterfly

11 The art of tinkering — 11-1

11.1 The most beautiful physics experiments — 11-3
References — 11-4

12 The butterfly in the laboratory — 12-1

12.1 Two-dimensional electron gases, superlattices, and the butterfly revealed — 12-7
12.2 Magical carbon: A new net for the Hofstadter butterfly — 12-12
12.3 A potentially sizzling hot topic in ultracold atom laboratories — 12-16
Appendix: Excerpts from the 2010 Physics Nobel Prize press release — 12-20
References — 12-20

13 The butterfly gallery: Variations on a theme of Philip G Harper — 13-1

14 Divertimento — 14-1

15	Gratitude	15-1
16	Poetic Math&Science	16-1
17	Coda	17-1
18	Selected bibliography	18-1

编辑手记   E-1

Summary

In 1976, several years before fractals became well-known, Douglas Hofstadter, then a physics graduate student at the University of Oregon, was trying to understand the quantum behavior of an electron in a crystal in the presence of a magnetic field. As he carried out his explorations by graphing the allowed energies of the electron as a function of the magnetic field, which he had theoretically calculated, he discovered that the graph resembled a butterfly with a highly intricate recursive structure that nobody had anticipated. It turned out to consist of nothing but copies of itself, nested infinitely deeply. Originally dubbed "Gplot"—a "picture of God"—the graph is now fondly known to physicists and mathematicians as the "Hofstadter butterfly".

The butterfly graph is a rare quantum fractal exhibiting some parallels with other well-known fractals, such as the Mandelbrot set, and it also turns out to be intimately related to Apollonian gaskets, which are mesmerizing mathematical kaleidoscopes—magical structures in which the images of tangent circles are reflected again and again through an infinite collection of curved mirrors. Apollonian gaskets can be decorated by integers at all levels, and similarly, the butterfly can be decorated by integers at all levels that describe the quantization of resistance, which is itself an exotic physical phenomenon called the quantum Hall effect.

This book narrates the story of the butterfly and its connection to the quantum Hall effect. It reveals how the secret behind the astonishingly precise quantization of Hall resistance is encoded in the branch of mathematics called topology. Topology reveals that there are hidden numerical quantities that unite a sphere and a cube while distinguishing them both from a doughnut and a coffee cup. The deep topological phenomenon underlying the quantum Hall effect is an abstract version of the physics that underlies the daily precession of a Foucault pendulum; it can be thought of as a quantum cousin to that precession, and it is known as the Berry phase.

The book begins by remembering the ancient Greek mathematician Apollonius who, around 300 BC, coined the terms "ellipse" and "hyperbola", and who explored the beautiful phenomenon of mutually tangent circles, ultimately leading to Apollonian gaskets. This problem was rediscovered by the French philosopher René Descartes, and then again by the chemistry Nobel laureate Frederick Soddy, who glorified it in a poem titled *The Kiss Precise*. Using a few concepts of quantum mechanics, the book mostly takes a geometrical approach, ultimately linking the "kiss precise" of Apollonius, Descartes, and Soddy to the "precise quantization" of Hall resistance and to the Hofstadter butterfly, which, when it is color-coded to reflect the topological integers lurking in the quantum Hall effect, stunningly displays the marvelous nature of that mysterious physical phenomenon.

May this exotic butterfly, today familiar to just a tiny community of physicists, spread its colorful wings and fly to ever more distant and unknown territories!

About the author

Born in Amritsar, India, Indu Satija grew up in Bombay. After graduating with a Masters degree in physics from Bombay University, she came to New York to get her doctorate in theoretical physics at Columbia University. Currently, she is a physics professor at George Mason University in Fairfax, Virginia. Her recent areas of research include topological insulators, Bose–Einstein condensates, and solitons. She has published numerous scientific articles; this, however, is her first book.

Physics is Indu's first love, and the outdoors is her second. She lives in Potomac, a suburb of Washington, DC, with her husband Sushil, a physicist at the National Institute of Standards and Technology. Both Indu and Sushil are marathon runners, and they enjoy hiking and biking as well. They have two children: Rahul, who is a biologist, and Neena, who is an investigative reporter.

Preface

A bird doesn't sing because it has an answer;
it sings because it has a song.

—Joan Walsh Anglund

In physics as in life, most fashions come and go. "Classics"—problems that continue to fascinate for more than a generation—are rare. Superconductivity, for instance, remains at the frontiers of physics thanks to the perennial hope that it will revolutionize the world. String theory feeds the craving for a deeper unification, for mathematical rules and laws that encode nature in some complex but beautiful way. The Dirac equation for relativistic electrons, which led to the discovery of antiparticles, and Einstein's equations of general relativity, which have now penetrated every household via the Trojan horse of GPS, are eternal poetic gospels of physics, and are testimony to the unimaginable power and richness of theoretical physics.

The Hofstadter butterfly, discovered some forty years ago, is destined to be immortal. In addition to its great visual appeal, it encodes one of the most exotic phenomena in physics, the quantum Hall effect. The Hofstadter butterfly combines the most fascinating mathematical aspects of fractals with the equally fascinating physics associated with the quantization of conductivity. These two aspects are intricately merged in a self-similar fractal energy spectrum. As Hofstadter stated in his PhD thesis, he coined the term "Gplot" after his friend David Jennings, struck by the infinitely many infinities of the surreal-looking spectrum, dubbed it *a picture of God*.

The physical system represented by Gplot is deceptively simple: an electron is moving in the "flatland" of a two-dimensional crystal lattice that is immersed in a magnetic field. The strange-looking graph shows the allowed and the forbidden energies of the electron, as a function of the strength of the magnetic field. The butterfly is formed exclusively of smaller copies of itself, nested infinitely many times, and thus it forms a fractal, which is a very rare phenomenon in quantum physics. Gplot's intricacy is the outcome of a "frustrated" physical system, which results when nature is confronted with two distinct problems, each characterized by its own natural period: on the one hand, an electron constrained by the square tiles forming a perfect crystal lattice, and on the other hand, an electron moving in perfect circles in a homogeneous magnetic field. Nature's elegant reconciliation of these two opposing situations was a beautiful surprise.

At the time of its discovery, Gplot was appreciated by many for its visual charm and mathematical intricacy, but it was generally considered by physicists to be an object of mere theoretical interest. However, the recent experimental confirmation of some of its properties has turned this once-exotic spectrum into one of the hottest topics in condensed-matter and cold-atom laboratories around the world. Certain aspects of the spectrum's complexity were recently demonstrated in experiments involving measurements of samples of matter that were cooled down to very low temperatures and subjected to very high magnetic fields. The key to observing the butterfly pattern of energy bands and gaps was the fabrication of a special material composed of graphene and boron nitride, two substrates that have similar lattice structures, and which, when overlapped at an angle, form a "moiré superlattice". This novel and ingenious experimental technique has pumped new energy into the field, as is evident from many recent papers in some of the most prestigious technical journals as well as from excited press releases for popular media.

The tale of the butterfly features stories inside stories. In the Prologue, readers will get a first taste of this in the inspiring story of the discovery of Gplot by its discoverer, Douglas Hofstadter. However, let me fast-forward almost forty years and tell the tale of my own suddenly awakened interest in this famous object.

My love affair with Hofstadter's 1976 creation (or discovery, as he would put it) came out of my recent studies, when it dawned on me that self-similar fractals that encode some topological features exhibit a new type of order as topology gets encoded at all length scales. I found this "reincarnation" of topology as a new kind of length scale to be absolutely fascinating. I pictured it as a vivid drama in which each of two powerful but opposing effects—namely, the global self-similarity and the topology—tried to assert their authority, each one struggling to prevail, but ultimately having to make some compromises. In short, the tale of fractals dressed with topology is a tale of two competing forces, a tale of reconciliation and accommodation, in which both parties not only contribute but cooperate in creating something new, unexpected, and astonishing.

My revisiting of the butterfly fractal started with my attempt to understand the interplay between topology and self-similarity, which turns out to be a chicken-and-egg problem. The wild idea of devoting an entire book to the butterfly was conceived when I found a relation between the butterfly and the beautiful ancient mathematical object called an *Apollonian gasket*. This structure starts out as four mutually tangent circles, and then it grows stage by stage, in the end becoming an infinite set of tangent circles on all scales—a mathematical kaleidoscope in which the image of four touching circles is reflected again and again through an infinite collection of curved mirrors. In 1938, chemistry Nobel laureate Frederick Soddy fell under the spell of Apollonius's four-circle problem, and he glorified its charm in a small gem of a poem entitled "The Kiss Precise".

As I explored the butterfly, I had the exquisite pleasure of seeing that the precise tangency of infinitely many circles (the "kiss precise" taken to its limit) and the precise quantization of Hall conductivity were connected in a subtle way, and this first insight opened a pathway for me that I subsequently followed and explored, and that led to this book. My presentation of the butterfly points out certain of its features that are reminiscent of other well-known fractals, such as the Mandelbrot set. One of the highlights of the butterfly landscape, decorated with integers, is the lovely way that it is related to the rich family of Apollonian gaskets. Although quite a few elusive mysteries still remain about the butterfly, I am excited to share my "$\hbar$-butterfly" story with others—science students, young researchers, and even lay readers attracted to fractals and intrigued by quantum physics.

As we approach the fortieth anniversary of the publication of Hofstadter's paper (September 1976), it is timely to share its magic with a broader audience. On the theoretical side, many ideas of solid-state physics and of fractal geometry are packed into Gplot, and on the experimental side, it is related to a new class of materials. All of this breathes new life into this object of stunning beauty. The butterfly fractal is thus a potential medium for bringing some of the joy of frontier physics to science enthusiasts and for exposing them to the hidden beauty of the quantum world.

The task of writing each chapter in the book began with finding a quotation[1] that I hoped would convey the spirit of what I wanted to say in that chapter. I may not have succeeded entirely in this effort, but the search for piquant quotes certainly stimulated me and catalyzed my writing process. I also devoted considerable time, effort, and thought to the creation of all sorts of figures and illustrations, believing strongly that a picture is worth more than a thousand words, even at the risk of my book's being labeled a "picture book". Some of my life's most challenging moments have been when I attempted to explain my love of physics to people having no background in science, including my father, who often quizzed me about what kind of science I do. The book will show whether I have made any headway in this quest.

As I began writing this book, my childish instincts resurfaced, and I found that I loved "dressing" the butterfly with Ford circles, trying out various color combinations, some of which readers will encounter in the book. My favorite happens to be the green–blue combination on the book's cover, which is tied to a nostalgic anecdote. In the good old days when I was a graduate student at Columbia university, a loving American couple hosted me for my first American Christmas. When I arrived dressed in a blue and green silk sari, my host spontaneously said to her husband, "Didn't I tell you blue and green form a perfect color combination? They're the colors of nature—the trees and the sky!"

Following Douglas Hofstadter's Prologue (in which he recounts the strangely meandering and lucky pathway that eventually led him to Gplot) and my Prelude (in which I give a brief overview of what is to come), the book begins with the above-described problem of mutually tangent circles, remembering the great mathematician Apollonius, who not only explored this problem around 300 BC, but who also coined the terms "ellipse" and "hyperbola". The story continues with the rediscovery of this ancient problem by French philosopher René Descartes in 1643 and with the poem written by Frederick Soddy.

Capitalizing on the geometrical visualization of rational numbers in terms of Ford circles, I reveal the hidden nesting-structure of the butterfly graph, accompanying my readers through many refreshing physical and mathematical wonderlands. These include three Nobel-Prize-winning discoveries—namely, the quantum Hall effect (1983), quasicrystals (2010), and graphene (2011)—as well as the topological spaces of the Platonic solids, Foucault's pendulum, and the Berry phase. With peeks into the quantum world, the book mostly follows a geometrical

---

[1] My love for little quotes originated during my high-school exam days when my uncle Arjun told me a few quotes, suggesting that I use them in my exam essays. I distinctly remember one of those quotes:

Beauty is to see but not to touch;
A flower is to smell but not to pluck.

path, suggesting a relationship between the "kiss precise" and "precise quantization", explaining many subtleties of the butterfly plot using mathematics that is as simple and elementary as possible, although of course at times it is neither simple nor elementary.

This book—my first book ever—is my sincere attempt to share my personal joy of discovery and understanding in relatively accessible language. It is hard to pin down the exact audience for the book. I have tried to remain at the level of *Physics Today* or lower, in the hopes that the book would attract a broad group of curiosity-driven readers. It will of course be helpful if readers have some background in physics and mathematics, but what is more important is simply that they be interested in science, and fascinated by the beauty and power of mathematics to predict the way that nature behaves.

Number theory—the mathematics of positive integers—is a universally appealing field, and the book exploits this fact. There are, inevitably, numerous technical discussions, which I have included for the sake of completeness, but which can be skimmed or skipped by general readers. I hope that students, teachers, readers of lay-level scientific articles, and even some professional physicists will find the book intriguing. May my small book help this exotic butterfly, today familiar to just a tiny community of physicists, spread its colorful wings and fly on to unknown and distant lands!

Of course, even a book devoted entirely to the Hofstadter butterfly cannot exhaust all its aspects. I extend my apologies to those whose favorite facet of the butterfly has been left out, acknowledging that my discussion of the butterfly primarily reflects my own personal understanding and taste. The presentation of the butterfly in this book is extremely visual. The lucky fact that such an intuitive approach exists is what allows a book on the subject to be aimed at nonspecialists. It is my hope that this book will help the butterfly fractal to awaken as much interest as have other fractals, such as the Mandelbrot set, and that this will in turn help quantum science to reach a broader audience.

Although some important aspects of the butterfly graph have not been mentioned in this book, I have tried in the Selected Bibliography to include all of the most important references, and interested readers who access these articles will get a sense of the vast sea of ideas from many facets of physics and mathematics that are hidden in the subject. Needless to say, these additional ideas related to the Hofstadter butterfly further "speak" and reveal the beautiful mathematics that runs through the literature on this subject. It is my hope that readers will be able to appreciate this beauty even if they do not fully comprehend it. As T S Eliot wrote, "Genuine poetry can communicate before being understood...".

Despite the remarkable progress that has been made since 1976, many aspects of the butterfly graph are still not understood. There is no gainsaying the fact that a number of profound new mathematical ideas have been unearthed and put to use in the quest to understand the butterfly's fractal magic. Nonetheless, these mathematical formulations have not yet fully characterized the very complex nature of the graph. Attaining a complete understanding of the Hofstadter butterfly still remains an open challenge. In my personal view, although much beautiful mathematics has been done, it does not yet glow with the purest type of mathematical beauty. One is almost reminded of what Paul Dirac once said about quantum electrodynamics: "I might have thought that the new ideas were correct, if they had not been so ugly."

Many years ago, my ten-year-old son Rahul, who in his childish innocence believed that "book-writing makes you famous", asked me why I wrote articles instead of books. "Perhaps I will do that when you go to college," I replied. So the book-writing task is long overdue. Attempting to write a popular book about a fractal in solid-state physics might be a crazy idea, but might it also be a path to sanity? Let me quote the Chilean poet Vicente Huidobro: *Si yo no hiciera al menos una locura por año, me volvería loco*—"If I didn't do at least one crazy thing each year, I would go mad."

Finally, I want to say that I am writing this book because I feel I have a story to tell. However, I am by no means an expert in all the topics touched on in this book, and it is undoubtedly imperfect in all sorts of ways. I therefore welcome all suggestions, comments, and critiques, and I will be extremely grateful to anyone who brings any errors that they may find in this book to my immediate attention. Fortunately, the "e-book" version will allow me to make changes at any time, even after the book has been published, and I hope that this freedom will be useful in improving the presentation of various scientific ideas and results, and in correcting errors in the future. I also invite readers to send me their poetic verses about the Hofstadter butterfly, if they happen to compose any. Their verses will find a home in some cozy corner of my web page, and may even appear in revised versions of the book.

Writing this book has been an incredible experience, exposing me to parts of myself that I had never dreamed of. This book is not only about the science that I love dearly, but also about everything else that I admire and adore deeply. I have been stunned by the intensity with which it has engrossed and consumed me, constantly posing ever deeper challenges and revealing new heights to transcend. It has been a rare joy that can only be experienced and cannot be expressed in words.

No journey is truly fulfilling unless one dares to take unpredictable little detours, leading one to stumble across quaint spots whose existence one would never have suspected otherwise. In the spirit of such a search of the unknown, I came across a bigger picture, in which poetry, music, and the joys of nature added to my originally purely scientific approach, filling out the picture in a richer way.

I am truly blessed to have a family that enriched and shaped my life with many such treasures. This book is a tribute first of all to my father, who gave me the precious gift of the love of poetry; it is to him that I dedicate this book. The book is also a tribute to my two children, Rahul and Neena—my everlasting joys who, with their music, have brought such profound harmony to my life. Finally, the book is a tribute to my dear husband Sushil, who introduced me to the boundless love of the outdoors, which has allowed me to take delight in the endless beauty of real butterflies.

So here is my very first book—and I dream of an audience touring a historic site dating from 300 BC, relaxing now and then with bite-sized items picked from a savory smorgasbord, and with exotic cocktails of quotations, poetry, art, and music!

My mother used to say that life begins at forty. That was her age when she had her first baby. I say that life begins at fifty-five, the age at which I published my first book.
—Freeman Dyson.

IIS MMXVI

Prologue
The grace of Gplot
Douglas Hofstadter

When I was in my late teen-age years, as a young mathematics major at Stanford, I intoxicatedly explored the endlessly rich world of integer sequences. This several-year odyssey was the first, and probably the most deeply rewarding, period of scientific research in my entire life. It was all launched one day in February of 1961, shortly after I turned 16, when I decided to take a look at how the triangular numbers (positive integers of the form $1 + 2 + 3 + \cdots + n$) are distributed among the squares. On a sheet of paper, I wrote out the first few dozen triangular numbers—1, 3, 6, 10, 15, 21, 28, ... (calculating them by hand)—and also the sequence of squares—1, 4, 9, 16, 25, 36, 49, ... Then I proceeded to count how many triangles there were between successive squares.

The sequence I thereby got—21211212112121121212112...—seemed to be composed solely of 1s and 2s, and it hovered fascinatingly between regularity and irregularity. Its way of closely approaching but always evading periodicity tantalized me no end. Soon I was unable to resist going upstairs to my Dad's little study, where on his desk he had a Friden electromechanical calculator (kind of like a cash register, but a little more sophisticated, since it could multiply and divide large numbers). For a few hours, I punched buttons on the Friden machine (which, back in those days, was quite fancy technology), and in return I got back many more triangles and squares, which I dutifully copied down on a much larger piece of paper. Counting the former between the latter confirmed my earlier observations that my sequence was made of just 1s and 2s and that it continually skirted but ever eluded periodicity.

After playing around for quite a while with the 100 or so terms that I had generated of my sequence, I eventually discovered that if you break it into two types of chunks, as follows:

21 **211** 21 21 **211** 21 **211** 21 21 **211** 21 **211** 21 **211** 21 21 **211**...

then if you count the black 21s between the red 211s, you get the following sequence: 212112121211212112...—and this is *exactly the same sequence* all over again! This purely empirical discovery was absolutely electrifying to me. (Had I been a Pythagorean, I'm sure that 40 oxen would have been sacrificed in honor of the discovery of this astonishing unexpected pattern!) A year or two later, having gained considerably in mathematical sophistication, I was at last able to prove rigorously the lovely fact that I had discovered, but proving it wasn't nearly as exciting or as important to me as the experience of finding the beautiful, unforeseen pattern.

The addictive excitement of this first number-theoretical discovery of my life pushed me to try to make *analogous* empirical discoveries, and I thus embarked on a very long voyage, in which I created a series of leapfrogging analogies that led me from one empirical discovery to another to another. I was tremendously excited—not only by the exquisite patterns of *numbers* that I was uncovering, but also by the intricate patterns of *ideas* that I was creating, which formed a dense web of mathematical analogies that I had never dreamt existed. (I talk about this in more detail in the book *Fluid Concepts and Creative Analogies.*)

Those years (roughly 1961–65) were an amazingly exciting and magically fertile period of my life. Many of the new discoveries I made at that time were made by my writing computer programs and running them late at night on Stanford University's only computer at the time—a Burroughs 220, hidden in the basement of the old, decrepit, and in fact mostly abandoned Encina Hall. Hardly anyone on campus even knew of this computer's existence, let alone how to program it. In those days of the early 1960s (or actually, those nights, since day in, day out, all day long, the B220 was used by some bank down in San Jose that co-owned it), I was doing what later would come to be known as "experimental mathematics" (in this case, experimental number theory), and I made literally hundreds of small, interrelated discoveries, some of which, *ex post facto*, I was able to prove, but most of which I never bothered to prove or never was able to prove.

To my mind, I was doing mathematics (or if you prefer, exploring the very real, concrete, down-to-earth world of integers) very much as a physicist explores the real, concrete, physical world. My Dad, an experimental physicist at Stanford, was my prototype for this analogy. Using a powerful 400-foot-long linear accelerator (huge for those days!), he sped electrons up to very close to the speed of light and then made them "scatter" off of atomic nuclei; from the angular distribution of the scattered electrons, he and his graduate students and post-docs were able to deduce the hidden inner structure of nuclei and even, eventually, of the mysterious proton and neutron. This research was extremely fascinating to me, and in my analogy likening myself to my Dad, the Burroughs 220 computer ensconced deep in Encina Hall's entrails was my "linear accelerator", while my various computer programs (written in the elegant and then-new language called Algol) were carefully designed experimental setups that revealed to me unsuspected truths of nature. Using a powerful tool (and for those days, the B220 was indeed quite powerful!), I was doing my own kind of "scattering experiments" and uncovering deeply hidden truths about elemental "objects" in this world. I loved this analogy, and the more I did my number-theoretical explorations, the truer it rang for me. After all, integers, to me, were every bit as real and as tangible as nuclei and subnuclear particles were to my Dad.

I have to stress once again that for me, deduction, or theorem-proving, was only a very small part of the act of "doing math", and not nearly as exciting or important a part of it as computational exploration. The main parts of "doing math" were: (1) using my fervent analogy-driven imagination to invent new number-theoretical concepts galore to explore, and then (2) performing the computer experiments and seeing how they came out. I was thus a dyed-in-the-wool experimentalist in number theory, not a theorist, and as such, I stumbled upon many marvelous miniworlds of mathematical ideas to explore.

It turns out that a fair percentage of the phenomena I was investigating with my metaphorical "linear accelerator" had never been explored before, and so I was breaking brand-new territory, although I unfortunately didn't publish any of my findings. (At the time, I didn't have the foggiest idea about how to publish an article, nor about the importance of doing so—I was just intoxicated with the fervor of my explorations. To me, that sublime joy was all that really counted.) Eventually, though, and to my gratification, a very small handful of teen-aged Doug's ideas became somewhat well-known, since in 1979 I published a tiny sprinkling of them in my book *Gödel, Escher, Bach*, and in subsequent years, I had conversations about some of them with a few influential mathematicians.

One of my favorite discoveries of that magical, unforgettable period of my life was a function of a real variable x that turned out to have a very odd, almost paradoxical kind of behavior. I dubbed this function "INT(x)" because in order to calculate it, you had to interchange two infinite sequences of integers—**coun**(x) and **sep**(x)—that were derived from x. I won't explain here *how*, given a specific value of x, these sequences were calculated, but I'll give two examples. For $\sqrt{2}$, the **coun**-sequence was simply 1, 1, 1, 1,... and the **sep**-sequence was 2, 2, 2, 2,... Complementarily, for the golden ratio $\frac{1+\sqrt{5}}{2}$, the **coun**-sequence was 2, 2, 2, 2,... and the **sep**-sequence was 1, 1, 1, 1,.... This symmetrical "partnership" of two very important real numbers fascinated me. Indeed, it made me wonder about the "partners" of other famous real numbers, such as π and e, and this in turn inspired me to define a new function of an arbitrary real number x. Specifically, I defined INT(x) to be that real number y such that y's **coun**-sequence was x's **sep**-sequence, and vice versa. Of course, this meant that whenever $y = $ INT(x), then symmetrically, $x = $ INT(y), and thus, for any x, INT(INT(x)) = x. I explored INT's nature empirically, using the good old Burroughs computer once again, and my first blurry visions of INT came from my very crude hand-done plots of it.

Back then, there were no computer screens to see anything on, and not even any plotters of any sort; instead of displaying any shapes, the 220 merely printed out long tables of *real numbers* for me, which were the Cartesian coordinates of points making up the graph of INT. I was interested in the shape of the graph between any two successive integers (e.g., 1 and 2, or 11 and 12), since the way INT was defined, that shape was exactly the same for all such pairs. And so, when for one such length-1 interval on the x-axis, I plotted these points by hand, using a pencil on a piece of graph paper, they seemed to form something like a diagonal line broken up into perpendicular rib-like pieces of different sizes, but I didn't really understand what I was seeing, so I then naturally asked the 220 to calculate for me the coordinates of lots of points belonging to a single particular rib that I chose. When I plotted *those* points, I was very surprised to see perpendicular "sub-ribs" of that rib starting to come into focus—and so on. It was quite painstaking work, but it was truly exciting to my teen-aged mind when I started to catch onto what was happening.

Seen from very far away, the graph of INT looked like the infinite 45-degree line $y = x$ (see figure P.1). But if you zoomed into it a bit, you would see that this upwards-sloping line was more like a picket fence than a line, since it was made up of an infinite number of identical, non-touching "backslashes" (the downsloping diagonals of all the 1×1 squares climbing up the line $y = x$, much like the steps of a staircase). Figure P.2 shows just one of these infinitely many backslashes, nestled inside the square whose southwest and northeast corners are the points (0, 0)

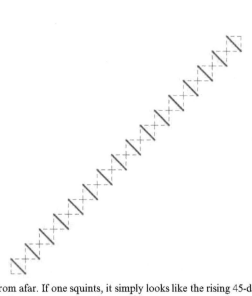

Figure P.1. INT as seen from afar. If one squints, it simply looks like the rising 45-degree line $y = x$, but if one looks more carefully, one sees that it is broken up into many short "backslashes" (these are shown in red).

and (1, 1). Each such backslash, when plotted in detail, turned out to be made of an infinite number of yet smaller, upsloping, and slightly bent "ribs", and the ribs were made of "sub-ribs"—and on and on this went, infinitely far down. In short, INT(x) had revealed itself to me to be an infinitely nested visual structure—and what a thrill it was to discover that!

Another big thrill for me was when I saw that at every rational value of x, INT(x) took a discontinuous jump, and that, roughly speaking, the "more rational" x was, the bigger the jump—thus, the jumps at $x = 1/2$ and $x = 1/3$ ("very rational" points) were very large, while the jump at $x = 5/17$ (far "less rational", in a certain sense) was very tiny. Of course what I meant by the phrases "more rational" and "less rational" had to be worked out, but once I had done that, I realized that my catch phrase about jump sizes implied that at *irrational* values of x, INT's jumps were all of size *zero*, which meant that at *those* points, INT was perfectly continuous! Coming to understand how this crazy-sounding kind of behavior was actually perfectly possible was one of my life's most exciting moments.

Today, wild shapes such as INT are known as "fractals", and because of their eye-catching nature, some of them, such as the Mandelbrot set, are familiar even to people who have never studied mathematics. Back then, however (I'm speaking of roughly 1962), this kind of visual structure, nested inside itself over and over again without end, was extremely unfamiliar and fascinatingly counterintuitive. Over the next year or two, thanks to my exploration of INT and its close relatives, such nested behavior, though initially terribly weird-seeming, became part of my most intimate mental makeup, and many years later, this deep knowledge would have a totally unexpected and marvelous payoff.

Well, I could tell much more about that exciting period of my life, but this brief sketch gets the main idea across: as a math-intoxicated teen-ager in the early 1960s, I discovered a marvelous bouquet of beautiful truths about the world of mathematics

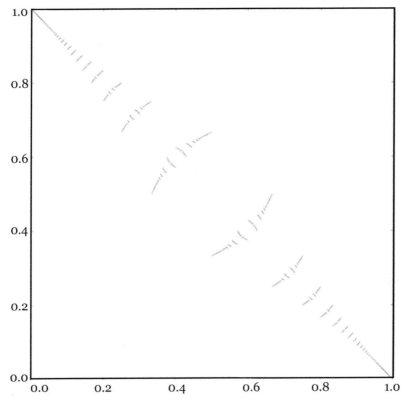

Figure P.2. One backslash of the graph of INT, between $x = 0$ and $x = 1$, made out of infinitely many "ribs", which are in turn made out of "sub-ribs", and so forth, *ad infinitum*. The largest rib on the left side is nestled between $x = 1/2$ and $x = 1/3$, the next-largest one between $1/3$ and $1/4$, and so on. The nearer you get to the upper left-hand corner, the shorter the rib is (and the straighter). Each of the ribs is in fact a perfect "copy" of the entire backslash, except, of course, for being smaller and gently bent a little bit. The right half of the backslash is identical to the left half, only rotated by 180 degrees around the exact center of the containing square.

by inventing new ideas via a long, joyous series of leapfrogging analogies, and then by doing computational experiments to explore these new ideas.

In 1966, I entered graduate school in mathematics at Berkeley, shooting for a PhD in number theory (what else?). Unfortunately, my brief time there turned out to be very traumatic. In my first year, I had to take several required courses, all of which were almost forbiddingly abstract. I managed to get decent grades in them all, but only by the skin of my teeth, and the *ideas* didn't sink in at all. I couldn't visualize anything, and most of the time my eyes just glazed over. In my second year, hoping for relief, I took a non-required course that announced itself as being about number theory, but I soon found out that actual *numbers* (that is, my old friends the integers) were essentially never mentioned by the professor. The integers came up only once in a blue moon, and even then, always as a mere *example*—usually a trivial example!—of far more general theorems about very abstract kinds of number systems that were totally nonvisualizable. I, who loved the visual and the concrete, was repelled by all of this. It was becoming apparent to me that mathematics—at least math as done at Berkeley—was hugely different from what I had thought it was during my intoxicating period of discovery in number theory at Stanford.

After roughly a year and a half of this very discouraging beating of my head against what I later came to call my "abstraction ceiling" (a phrase I coined just a few years ago), I finally realized that I was going to have to bail out of mathematics. I was caught wholly off guard by this turn of events, because my heady undergraduate years at Stanford had led me to the absolute certainty that mathematics was my "manifest destiny". I had been as sure as sure could be that I would just sail through math graduate school and become a top-notch number theorist, and yet now all these youthful dreams were being revealed to have been pipe dreams. I desperately asked myself, "What on Earth can I do in life, if not mathematics?"

As I mentioned earlier, my Dad was a physicist, and I had always found the ideas of physics (or at least those of particle physics) mesmerizing, but unfortunately I had had pretty bad experiences in high-school physics—and later, as an undergraduate math major at Stanford, in Halliday-and-Resnick physics courses. Math majors were required to take four quarters of physics, and those four quarters were really rough for me, their nadir being hit when I received an F in Physics 53 (Electricity and Magnetism)—and deservedly so! During the quarter, I had not applied myself whatsoever, and I went into the final exam as a know-nothing, and thus got as low a grade as could be gotten. The snag was, I could not graduate from Stanford without having a passing grade in E&M, and so, one year later, I had to retake the E&M course (and from the same professor, who, to make matters worse, was a good friend of our family's). That second time around, by determinedly working my tail off in a way that I had never before done in my life, I replaced that mark of shame with an A–, which remains in my memory as one of my life's proudest moments. Nonetheless, that hard-won A– didn't at all inspire me to consider pursuing physics as a possible profession. I had had far too many hard knocks to my self-esteem in physics, and from high school on, math had always seemed my forte, hands down. But now that the math option had been painfully ruled out by even harsher knocks to my ego in grad school in Berkeley, physics had to come back into the picture as a possibility.

Berkeley was quite wild in the 1960s, full of revolution and upheaval and chaos, and my personality was such that I badly needed a far calmer environment. It happened that I had a very close friend who had just started graduate school in biology way up north, at the University of Oregon in Eugene, and when, late in the fall of 1967, I visited him there, partly in order to test the waters of my idea of switching to graduate school in physics, I was impressed by the campus's beauty. Moreover, in contrast to Berkeley, both the town and the university felt friendly, even serene. Best of all, I was given a very warm welcome as a potential grad student by all the physics professors I met, and this was extremely encouraging. The upshot of it all was that in January of 1968 I jumped ship in two ways—first, in jumping from Berkeley to Eugene, and second, in jumping from number theory to physics.

When, in early January of 1968, I moved from California's intense, tumultuous Bay Area to Oregon's tranquil, rural Willamette Valley to open up this radically new phase in my life, my unquestioned goal was that of becoming a particle theorist, since particle physics struck me as the sole truly *fundamental* area of physics. Other areas of physics weren't even on my radar screen. More than anything, I yearned to be part of the noble quest to find out the deep, hidden nature of the mysterious entities that make up the finest fabric of our universe. For my first couple of years in Eugene, I was ecstatic about my choice to become a physicist, and I reveled in all my courses, learning a huge amount of physics from them, and from some excellent professors and some excellent books (most of all, volume 3 of the *Berkeley Physics Series*, called simply "Waves", by Frank S Crawford, Jr). I will always remember that period of my life, marked by my profound excitement about the beauties of physics, with great nostalgia. Not only was a dream coming true, but I was going to be following in my father's footsteps. This was a very rare and precious gift.

However, after about two years, things slowly started to change flavor, catching me off guard once again. The relatively recent ideas that I was supposed to be learning in my more advanced courses started to seem a bit shakier and more confusing, more and more arbitrary, and less and less beautiful. These ideas just didn't "take", in the way that ideas in my earlier physics courses had taken. Worse yet, the talks on particle physics given by distinguished visitors started to sound a bit like science fiction, or even pseudoscience, rather than like solid science. This was deeply shocking to me. My head started spinning, and as time went by, the spinning only got worse.

Particle physics, aside from quantum electrodynamics, just didn't make sense to me. The theories of the strong and weak interactions seemed filled to the brim with horrendous and arbitrary ugliness, rather than sparkling with sublime and pristine beauty, and I just couldn't swallow them, sad to say. I was so profoundly skeptical that I felt a kind of emotional nausea every time I picked up some random high-energy preprint in the library of the Physics Department's little Institute for Theoretical Science. Particle physics was unfortunately turning out not to be at all what I had thought it was, only a few years earlier.

It may sound as if I had once again hit my "abstraction ceiling", and perhaps that was part of what was going on, but I actually think this experience was quite different from what happened with math in Berkeley. Rather than an abstraction ceiling, I would say I was hitting up against an "absurdity ceiling", or an "implausibility ceiling", or even a "grotesqueness ceiling", if I may use such strong language. The sad truth is, the strange ideas I was surrounded by night and day made me reel with an almost visceral disgust. I won't go so far as to claim that my aesthetic revulsion was caused by the objective *wrongness* of the ideas; after all, I am not privy to the ultimate nature of the laws of physics. But I do know that what I was reading in papers and hearing in talks clashed violently with my personal sense of the beautiful—and from so much time spent with my Dad, I had absorbed an unshakable belief in the great beauty and simplicity of the laws of nature. To be sure, my professors often felt that the ideas I was railing at for being "as ugly as sin" overflowed with beauty, so what could I say? We just had to agree to disagree.

All this put me in the very weird position of essentially having to ask myself whether I trusted my own mind over the collective minds of all the particle physicists in the world—which included so many undeniable geniuses! It would seem the height of arrogance for me to insist that I was right and that they were all wet, and yet I was unable to knuckle under and accept the ideas that repelled me. After all, doing so would have amounted to sacrificing my own belief system at an altar that felt as if it had been created by aliens—and worse yet, if I gave up on my faith in my own deepest ideas and my most precious intuitions about the nature of the world, then what could I trust? What could I turn to? You can't just jump in bed with a set of ideas you hate! In short, the situation was very unstable. I churned and churned for a long time, desperately seeking some ideas that I found beautiful and could believe in, but no matter how hard I searched, none turned up. And since I was constitutionally unable to believe in the reigning doctrines of particle physics, there was no way I could do research in the field, because there was nothing to guide me, nothing solid to grasp onto, nothing to place my faith in. It was all treacherous quicksand, to me!

This period of dramatic change was a horrible, bitter time in my life. I jumped from one potential advisor to another to another, but although I intellectually admired them all and was personally very fond of each of them as well, nothing that I tried with any of them worked out, to my true dismay. After four years of extremely painful struggle, I realized that my dream of joining the ranks of particle theorists, noble though it had been, was going up in smoke. I became very frightened that I would never get a PhD at all. All of a sudden, my entire life's future felt completely up in the air.

Things really hit the boiling point on one fateful day—December 14th, 1973, to be precise. That day, as I was approaching my 29th birthday, I took a huge risk and bailed out once again. The occasion was a "brown-bag lunch" at the Institute for Theoretical Science, in which I was presenting to the small, very friendly particle-physics group the ideas in a recent *Physical Review* article whose three authors, in order to forge some kind of more abstract unity in the "particle zoo" (and, I must admit, this was undeniably a type of aesthetic quest on their part), had the gall to propose a huge family of hypothetical new particles—over 100 of them!—at one fell swoop. Well, I couldn't help but compare this shocking audacity with the great Wolfgang Pauli's striking timidity, back in 1930, about proposing just *one* new particle (which Enrico Fermi later dubbed the "neutrino"). Pauli was extraordinarily reticent about making this suggestion because to him it seemed so extravagant, but he finally took the plunge and dared to go out on that shaky limb because he knew that this far-fetched idea might be able to save, simultaneously, all three of the most central conservation laws in all of physics (conservation of energy, conservation of momentum, and conservation of angular momentum). In other words, Pauli did what he did only out of supreme desperation, in an attempt to save the deepest laws of physics. The three authors, however, were just making a wild guess, a random shot in the dark, and trying to justify their chutzpah by couching it in pages of fashionable, virtuosic, grand-sounding group-theoretical language. What a contrast! I had studied this three-author paper for weeks but had found it so outrageous, so implausible, and so unbelievably ugly that at the end of my talk, I threw the paper down and bitterly cried out, "These people have *no sense of shame*! I'm getting out right now. I'm quitting particle physics! I'm done with it!" And then I bolted out of the room, trembling at what I had just done. Of course I wanted to jump ship once again, but I saw no other ship to jump to. Particle physics had been my Holy Grail, and giving up on it was deeply traumatic to me. Echoing my anguished thoughts six years earlier upon dropping out of math, I desperately asked myself, "What on Earth can I do in life, if I can't study what is truly *fundamental* about nature?"

Once again, I felt nearly lost. Here I had invested six years in physics graduate school, yet I had absolutely nothing to show for it. I didn't even have an advisor or a research area any more. In deep confusion, I went around to several professors in other branches of physics, asking them what kinds of problems they might be able to give me if I were to become their student. I had brief chats with a couple of solid-state theorists who were very nice to me, but nonetheless, talking with them about possibly working with them felt like a horrible blow to my ego. After all, I had long been convinced that solid-state physics was essentially just glorified engineering, and in such practical matters I had zero interest. I wanted to think only about the deepest, most basic things in nature! I feared that I was wandering into the slums of physics, and I felt deeply ashamed of myself for doing so. So prejudiced was I that I almost had to hold my nose. This may seem very funny to you, but it is absolutely true.

Very luckily, right at that time, my close Chilean friend Francisco Claro (figure P.3 shows us forty years later), who had gotten his PhD at Oregon a couple of years earlier under the distinguished solid-state theorist Gregory Wannier, came back to Eugene from Santiago to work again with Wannier for a few months. At this point I really should pause for a moment to say a few words about Professor Wannier, since

Figure P.3. Douglas Hofstadter and Francisco Claro savoring Bach in Santiago, Chile, 2014.

he played an absolutely central role in what was to follow, although of that I of course had no inkling whatsoever, at the time.

Gregory Wannier was born in Basel, Switzerland, in 1911 (four years before my Dad was born in New York City), and he grew up in the German-speaking part of Switzerland. He came to the United States as a post-doc in the mid-1930s, and stayed here for good. Although he was in his mid-twenties when he arrived, his English nonetheless sounded pretty much native, which was always very impressive to me. (I do, however, remember—and with great amusement—that one time, when he was probably very tired, he slipped and referred to hydrogen as "waterstuff"—a compound word precisely mimicking the German word "Wasserstoff"!) Wannier spent a year or two at Princeton as a post-doc when my Dad was there as a grad student, and they knew each other from those old days. When I first went to Oregon in 1968, my Dad made sure that I got in touch with his old friend Gregory, who he described to me as "a bit of an odd duck, but an excellent physicist and a friendly, gentle person". Indeed, Professor Wannier was immediately friendly to the son of his old friend Bob, and over the next few years I enjoyed many a dinner at his home. He loved history, and I remember him boasting one time that no matter what year I named after 1000 AD, he knew some historical fact about that year. So I duly named some random pre-Renaissance year, and he instantly came out with some fact about it, which was confirmed by the encyclopedia that he kept ready at hand, right next to the dining-room table. Yes, Gregory Wannier was an odd duck, all right, but a very pleasant one. And many people considered him to be the most distinguished physicist who was ever at the University of Oregon.

Now where was I? Ah, yes—Francisco Claro's return visit to Eugene in early 1974. Well, for those few months, Francisco and I spent almost all our free time together. During this crucial period in my life, he told me that my prejudice against solid-state physics was wrong, and that doing solid-state actually meant participating in the building of a deep and subtle bridge linking the alien microworld of quantum mechanics to the everyday macroworld of tangible phenomena. He said that solid-state physics explained how the familiar properties of matter emerged from deeply hidden properties of particles and atoms. And thus Francisco managed to make solid-state physics sound fundamental and perhaps even beautiful, after all.

What finally turned the trick were the casual words of a Stanford physics grad student whom I met one day around that time. This person, whose name I never even learned, offhandedly said to me, "Particle physics is physics that's done in a *continuous* vacuum, whereas a solid—that is, a crystal—is a *discrete* vacuum, like a periodic lattice as contrasted with a perfectly smooth, homogeneous space. Particle physicists have only *one* vacuum to explore, whereas solid-state physicists have a vast number of *different kinds of vacuum* to explore—as many kinds as there are different crystals." To me, this characterization of crystals was amazing and hugely provocative, and it quickly brought to my mind the image of the integers as a periodic one-dimensional crystal lattice along the real line, and that in turn made me ask myself, "So solid-state physics is to particle physics as number theory is to analysis? Really!? Wow!!!" This off-the-wall analogical insight suddenly and radically changed my perspective—and eventually, my life.

Encouraged by Francisco Claro, I went and had a conversation with Gregory Wannier, who welcomed me very warmly as a potential doctoral student, suggesting that if I were to work with him, I might tackle, for my thesis, one of his all-time favorite problems—namely, the long-standing enigma of the allowed energy values of crystal electrons—or *Bloch electrons*, as they are often called—in a uniform magnetic field.

As I type this famous physics name, I feel the need to make yet another brief digression. Although it has nothing to do with physics *per se*, I cannot fail to mention with pride and pleasure that I grew up knowing the great Swiss-American physicist Felix Bloch (see figure P.4) and his entire family. Like Gregory Wannier, Felix Bloch was born in German-speaking Switzerland (Zürich, in Felix's case), and he grew up there. He was Werner Heisenberg's first PhD student, and the two of them used to ski in the Alps together. Also like Wannier, Felix Bloch came to the United States in the mid-1930s, but unlike Gregory, he spoke English with a strong Swiss-German accent. Among many other achievements, Felix was the founder of solid-state physics (and yet, astoundingly, that is not what he won the Nobel Prize for!). He also was the first Director-General of CERN, the famous European Center for Nuclear Research, located in Geneva. From 1950 onwards until his death in 1983, Felix was my father's closest colleague in the Stanford Physics Department,

Figure P.4. Felix Bloch and Robert Hofstadter sitting atop a peak in California's rugged Sierra Nevada mountain range, summer 1953. (Photo taken by Leonard Schiff, and reproduced here by courtesy of Laura Hofstadter.)

and the Bloch family and our family were intimate friends. In fact, the Blochs were so close to us that they were practically family. We spent oodles of time at each other's houses, and each family had a huge influence on the other one. Just as one tiny example, the Blochs introduced us to skiing in 1956 at the Sugar Bowl resort in the Sierras, and after that, we often skied together. That was a lifelong gift. I gained a taste for what the earliest days of quantum mechanics had been like by being around Felix Bloch for over three decades, and it affected me profoundly. When he died, a part of me died. I deeply missed Felix then, and even today I still do. Aside from being a genius in physics, he was artistic, musical, athletic, funny, Jewish (though not observant), from the Old World, cosmopolitan, political, opinionated, sophisticated, stubborn, principled, wise, and always exceedingly generous to me. From as early as I can remember, Felix clearly respected me and liked my ideas. Being held in esteem by such a great mind was a wonderful thing for me. Thinking of Felix as dead was very hard for me. But if the loss of Felix Bloch hit me hard, the blow to my parents was even stronger, alas. But let me get back to my story.

Actually, before I do that, let me insert right here a very short poetic interlude (generously assuming that what follows merits the label "poetry"). For a brief period while I was a grad student in Eugene, I got sucked into a personal limerick-penning binge, and I penned (with my pen) quite a few limericks about fellow grad students as well as a few about various physics professors. Some of these alluded to technical ideas, which made them fun for the in-crowd but a bit opaque for the out-crowd. Among these latter were one that I wrote about Felix Bloch and a related one about Gregory Wannier. With out-crowd apologies and -out further ado, I present those two poems here (and I advise readers in advance that "Wannier" is pronounced "wan-*yay*"):

> *A physics-freak who was called Felix*
> *Thought crystals were swell psychedēlics.*
> *It boggled his mind*
> *When he happened to find*
> *That a Bloch-state repeats (mod a helix)!*

> *A physi-Swisst known as Wannier*
> *Left old Basel for new USA.*
> *In solids, with lots*
> *Of—not waves—but dots,*
> *He transformed Bloch-functions away!*

So much for my student-day limericks. And now, at last, back to the main story...

The nature of the energy values of Bloch electrons in a magnetic field was a very fundamental quantum-mechanical question first posed in the late 1920s or early 1930s, and yet, even 40 or 50 years later, no one understood the spectra of such electrons at all well, even in the most idealized and simplified of crystal lattices. There was a conflict between the *continuum* of possible energies represented by a Bloch band (that is, the set of allowed energy levels of an electron in a crystal with no magnetic field), and the *discrete ladder* of evenly spaced Landau levels (that is, the set of allowed energy levels of an electron in a magnetic field in empty space). How did nature resolve this sharp conflict between the continuous and the discrete? No one was very clear on this, although it was known that when a magnetic field was applied, a Bloch band would somehow split up into vaguely Landau-level-like subbands.

But by far the strangest thing that Wannier said to me on that fateful day, when he was describing this important and profound problem, was that there was apparently a deep and fundamental difference between two kinds of situations: (1) when the magnetic field (as measured in the natural, dimensionless units of *how many flux quanta* passed through a unit cell of the crystal) was *rational*, and (2) when the magnetic field was *irrational*. People at that time believed that these two kinds of situations gave rise to entirely different kinds of wave functions and energy spectra. This odd belief, which was held by virtually all the people who were doing research on this problem, sounded like absolute nonsense to me, since nothing in this world can possibly depend on whether the value of a physical quantity is *rational* or *irrational*. (Do you prefer rooms with *rational* temperatures, or *irrational* ones? Traveling at *rational* speeds, or *irrational* ones? I can hear you smiling! After all, no physical quantity possesses a well-defined infinite decimal expansion!)

The very idea of such a rational/irrational distinction in physics made no sense at all to me, and yet it was the reigning dogma of experts all around the world at that time (with one notable exception—the Russian physicist Mark Azbel', who had written a paper proposing a different view, which I eventually read and which was deeply inspirational to me). So when I heard these strange words of Gregory's, my ears perked up, and I said to myself, "Maybe, just maybe, though who knows how, my old number-theoretical interest in rationals and irrationals, which I left behind long years ago, could somehow turn out to be relevant to this enticing mystery in the field of solid-state physics." Thus was I lured to this venerable paradox-grazing problem in the theory of the solid state, and I decided to take the risky plunge of becoming a graduate student of Gregory's.

When, in the fall of 1974, he and I traveled to the University of Regensburg, an hour's train ride north of Munich, to spend six months there exploring this problem, I found myself in a cozy little research group consisting of three professors—Gregory Wannier, Gustav Obermair (our host), and Alexander Rauh—and myself (a humble grad student). A nostalgic photo of us all is shown in figure P.5. Our little research group would meet a few times a week in the office of one of the professors, and the three of them would eagerly toss back and forth all sorts of fancy mathematical techniques for trying to wring subtle secrets out of Harper's equation (the equation governing this mysterious spectrum). They were all past masters at such things as real analysis, differential equations, group theory, linear algebra, and even Bessel functions and hypergeometric functions, and they knew dozens of mathematical tricks that were way, way above my head. I could wave no such magic wand, and so in those high-falutin' discussion sessions, I would just sit there, silent and confused. I felt like a tiny child amidst brilliant adults, and was very disheartened.

One fine day, however, I noticed a cute little Hewlett-Packard desktop computer (very lowly compared to the mainframes of those days) sitting on a rollable cart in

Figure P.5. The "Regensburg group" (left to right): Douglas Hofstadter, Alexander Rauh, Gustav Obermair, and Gregory Wannier, in the Lehrstuhl Obermair, Fachbereich Physik, Universität Regensburg, 1974.

the hallway just outside my office (figure P.6), and I timidly asked Gustav if anyone was using it. Of course no one was, since back then, few self-respecting theoretical physicists ever looked at concrete *numbers*. Theorists looked at *equations*, manipulating them formally and proving abstract theorems about them. I was far less sophisticated and far more down-to-earth than the trio of professors towering over me—but at least I knew how to program this little machine (an H-P 9820A, just for the record).

This actually was a most fortuitous and crucial coincidence—an amazingly lucky break in my life. Five years earlier, when my Dad had been invited to the Aspen Center for Physics for a couple of weeks and I had tagged along just for the fun of it (in large part to be surrounded by the beautiful Rocky Mountain scenery), I ran across an H-P machine nearly identical to the one I would encounter years later in Regensburg, and I found it wonderfully easy to program. For a few days that summer, I took great delight in playing around with that machine in the Aspen Center, using its plotter to graph some simple Fourier series and other functions that I was curious about. That joyful period of computer-aided math-play in Aspen, 1969, was the sole reason that the little machine sitting idle in the Regensburg hallway, 1974, tempted me. Had I not played around with a cousin computer five years earlier in Aspen, I would never have been tempted by the one sitting idle in the hallway in Regensburg. And had it not been sitting there, or had it not been available to me, my life's entire subsequent course would have been radically different (and I cringe whenever I think about how it might have gone).

And so, having no other way to attack this problem, I decided to try doing the only thing I knew for sure how to do: to numerically *calculate* the specific spectrum

Figure P.6. Rumpelzstilzchen, the fifth member of the Regensburg group. Copyright David G Hicks http://www/hpmuseum.org/hp9820.htm.

of quantum-mechanically allowed electron energy-values for several rational values of the magnetic field. I rolled the little machine into my office and took it over completely. From my seniors in the Regensburg group, I knew that calculating the spectrum for a rational value p/q amounted to locating the zones in which a wildly oscillating q^{th}-degree polynomial passed through a thin horizontal strip centered on the x-axis, x being energy. I called these zones of the x-axis the "fat roots" of the polynomial, since each one of them was centered on a root, but was a small x-axis *interval* rather than a single point. These "fat roots" were none other than the Landau-like subbands belonging to the magnetic-field value p/q. Eventually I figured out how to coax the little machine to give me, for any specific (rational) magnetic-field value, the endpoints of the various allowed energy subbands for the crystal electrons.

For each rational magnetic-field value that the machine tackled, I plotted its energy subbands in my little notebook, using colored felt-tip pens. (This machine, unlike the one in Aspen 1969, had no attached plotter.) After a couple of weeks of calculations, I (or rather, my mechanical friend, which Gustav Obermair jovially dubbed "Rumpelstilzchen", since, like the odd little goblin in the Grimm Brothers' famous fairy tale, it wove "gold" overnight) had calculated the spectrum for perhaps 15 or 20 rational magnetic-field values, and I had hand-plotted perhaps 50 or 100 colored horizontal line segments (energy subbands) in my little notebook, which together filled out a vaguely butterfly-like shape (see figure P.7).

One day in November of 1974, quite out of the blue, I suddenly had an eerie *déjà vu* experience. I recognized a familiar and magical pattern starting to emerge: a delicate, lacelike filigree (or "faint fantastic tracery", as American music critic James Huneker once poetically described Frédéric Chopin's ethereally wispy étude Opus 25, No. 2), and to my eye it seemed that this "faint fantastic tracery", if it could be fully realized, would be made of nothing but many—*infinitely* many—smaller,

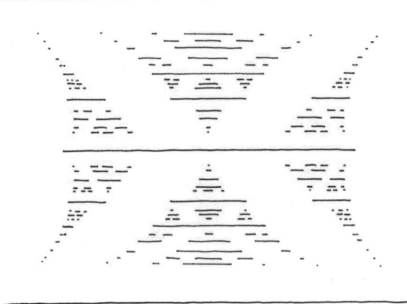

Figure P.7. The diagram above shows roughly how Gplot, as calculated by Rumpelstilzchen and hand-plotted by me in Regensburg, Germany, looked in its earliest incarnation, in November of 1974. This is unfortunately not the original plot itself, since the notebook in which I drew it has been lost, to my great chagrin. This is just my attempt to reconstruct how it looked. Most likely, the earliest version of Gplot was considerably sparser than this and yet I still somehow "sniffed" its recursivity, since I knew my deeply recursive function INT, from twelve or so years earlier, like the back of my hand.

distorted copies of itself. At first I was amazed by this idea, but soon I realized that for such an endlessly nested pattern to crop up in this context made perfect sense.

After all, in my long-gone days of intoxicated number-theoretical exploration, back in the early 1960s at Stanford, I had similarly hand-plotted a different, and much simpler, infinitely nested "faint fantastic tracery", and had come to understand it intimately. Passionate exploration of that graph over months had given me a deep understanding of the conflict between rationality and irrationality as well as of its resolution, and my new thesis problem in physics was all about that exact same conflict and resolution, only in a wildly different setting—a solid-state physics setting, of all things! My old graph of INT(x) enjoyed the subtle property of having smaller curved copies of itself located between the "very rational" points $\frac{1}{2}, \frac{1}{3}, \frac{1}{4}, \frac{1}{5}$, etc, on the x-axis—and in those early years I had explored and figured out just how the littler copies, and their varying amounts of distortion, came straight out of the rationality/irrationality fight. Likewise, in my new and still very coarse-grained plot— the mysterious eigenvalue spectrum of Harper's equation as seen through a very blurry lens—I now saw shrunken, distorted copies of the whole graph starting to shyly poke their noses out between the "very rational" magnetic-field values of $\frac{1}{2}, \frac{1}{3}, \frac{1}{4}, \frac{1}{5}$, etc—and as you approached the limiting x-value of 0, these copies grew smaller and smaller, and also less and less distorted, all exactly as had been the case with my old friend INT. This *déjà vu* feeling was mind-blowing to me.

Should I have been surprised, or not? Well, I had originally been drawn to the Bloch-electron-in-magnetic-field problem precisely when Gregory Wannier had first spoken of a mysterious kind of rationality/irrationality fight that nobody understood well, and now I was finding that essentially the same phenomena were cropping up in this new graph as had cropped up some twelve years earlier with my INT function, whose nested filigree resulted from a very similar rationality/irrationality fight. So maybe I shouldn't have been all that surprised. But surprised I certainly was, and my overwhelming feeling was that, through an amazing stroke of luck, I happened to be exactly the right person in the right place at the right time! In any case, I knew for sure that I had stumbled upon a deep analogical connection between my old INT graph and the new graph I was just beginning to unveil. I lightheartedly dubbed this emerging graph "Gplot", the "G" standing for either "God" or "gold", whichever way one preferred to think of it.

Where Gregory Wannier, Gustav Obermair, and Alexander Rauh had been systematically attacking the huge granitic boulder of Harper's equation with a powerful kit of refined mathematical tools but just breaking off smallish chips from it, I, a newcomer with only a blunt computational "axe", had somehow managed to split the boulder wide open and had found, to my amazement, that it was an exquisite geode, with a magically beautiful recursive structure hidden inside its very hard shell.

And yet… when one day, later in that month, I told Gregory Wannier about my findings, he didn't believe a word of what I was saying. Even when he was looking straight at the hand-done plot in my notebook, he had nothing but words of disdain for my empirical claim (which I was basing solely on my eyeballing of this butterfly-like shape) that the graph consisted of infinitely many tiny copies of itself, nested down infinitely many levels. He sadly shook his head and bewilderedly said to me, "This isn't physics! You're merely doing *numerology*!" Now that, coming completely out of the blue, was a real slap in the face, since to any self-respecting scientist, numerology is synonymous with pseudoscience, and that, in turn, is synonymous with sheer nonsense. To put it a bit more bluntly, *numerology* is to *number theory* as *astrology* is to *astronomy*. And so, in Gregory's eyes, I was a practitioner of astrology's closest mathematical cousin? Whew!

I couldn't believe Gregory had said what he had said, but he wasn't yet done. "I'm sorry to say that you won't ever be able to get a PhD thesis for your work on this problem, Doug. I guess if you want a PhD, you'll have to settle for writing a library thesis." I didn't know what he meant by this worrisome term, so I asked him, and he said, "It means summarizing other people's research on the problem in a thorough and scholarly manner—and then, at the end, if you really insist, you can put your idiosyncratic numerological speculations in an appendix."

As you might expect, I was pretty shocked by Gregory's harsh words, but luckily, I was neither dissuaded nor deeply wounded by them, because I knew in my bones that what Rumpelstilzchen and I had empirically discovered together was correct, and so I simply persevered, though now in private. The point is, many years earlier, I had had long personal experience with an analogous but much simpler structure (my graph INT), whereas Gregory had never had any experience with such structures, and so, even when peering straight at such a visual pattern, he was unable to recognize its hidden recursive essence, whereas to me, its essence was blindingly clear. (Incidentally, I still feel enormously grateful to Gustav Obermair and Alexander Rauh, for they at least remained agnostic about my unorthodox "numerological" claims, rather than opposing them or heaping scorn on them.)

To be quite honest, it makes me feel a bit sad and even somewhat guilty to tell this story, as it paints my good old *Doktorvater* in an unfavorable light, but this is exactly what happened that day in Regensburg, and I think it's important to tell it just as it was, for it reveals a lot about how profoundly science depends on nonverbal intuitions, mostly aesthetically based ones. Sometimes such gut-level feelings guide you beautifully down just the right track, and that's great, but other times (as in Gregory's case this time) they throw you way off.

We are all riddled through and through with unconscious preconceptions that determine our reactions to things we have never seen before; indeed, progress in science depends intimately on one's being powerfully guided, in the face of unfamiliar phenomena, by such unspoken biases. Whether inside or outside of science, it is important for all of us to believe in and trust our inner voices, but it's always a risk, and sometimes those voices, much though we trust them, will wind up misleading us.

In any case, let me make it clear that both before and after this troubling conversation, I greatly admired and was very fond of Gregory. He was a marvelously insightful physicist (after he died, I wrote an article praising him to the skies, called "A Nose for Depth"), and he was often very kind to me personally. During my seven years in Eugene, Gregory and his wife Carol invited me over for dinner at their chalet-like house many a time, and I always enjoyed those homey evenings with them immensely. I wouldn't want anyone to conclude from my tale that Gregory was cruel; he wasn't at all. He could be quite insensitive (as he was that day), but then so can we all. What I think was going on, that day, was that Gregory simply was convinced that young Hofstadter, being but a greenhorn in solid-state physics, had very naïve ideas that were way off the mark, and so the old hand was just trying to be as frank as he could with the greenhorn, in order to spare the latter considerable pain further down the line. It just happened, though, that this time the old hand was grievously in error, as he later came to see.

But back to my story... In early 1975, as my stay in Regensburg was winding up, I was preparing to make a brief trip by train to far-off Warsaw, Poland, as I had been invited there a few months earlier by Marek Demiański, a very friendly young physicist whom I had met once or twice in America. Marek had invited me purely out of friendship, thinking I might like to see Warsaw, but in his letter of invitation (a *real* letter—no email back then!), he also asked if I might want to give a talk about my doctoral research (of which he knew nothing) to an audience at the University of Warsaw's Instytut Fiszyki Teoretycznej, where he worked. Well, I accepted his invitation but with considerable trepidation, for never before had I given a departmental colloquium anywhere on any subject at all.

My trip to Poland was absolutely unforgettable, for not only was Poland the birthplace of both of my father's parents, but it was also the homeland of Frédéric Chopin, who, ever since childhood, had been my greatest hero. I was deeply excited, for both these reasons, to be going there. My first two days were spent in Kraków, with elderly relatives whom I had never met before, and one evening, when they went out and left me alone in their apartment for a couple of hours, I sat down at their small piano and played my heart out, doing my best with many Chopin pieces I loved. My relatives also told me (in Yiddish) harrowing tales of how they had survived the raging antisemitism in Poland during World War II. That was incredible. Then I took the train to Warsaw, where Marek met me, and in the next few days he took me to a Chopin concert in Warsaw as well as to Chopin's birthplace out in the country. Altogether I had a wonderful time with Marek.

But looming up ahead, and frighteningly soon, was my talk at his Instytut. It was scheduled as part of the Instytut's "Konwersatorium"—a series of talks given exclusively in English, even when given by Poles solely to Poles, just to make them practice. In my case, of course, I was not worried about whether my English would be up to par; my worry was whether they were going to scoff at the ideas I would tell them, perhaps making me doubt my own beliefs. After all, I still had only a crude hand-drawn graph of my crazy spectrum, and no proof whatsoever of my "numerological speculations" about the graph's infinite nesting inside itself. I was very fearful that I might be heading straight into the lions' den and might be eaten alive. However, when the fateful day rolled around, the audience of my first-ever physics colloquium was not only interested but even enthusiastic, and in the long series of questions afterwards, I didn't detect the slightest trace of skepticism. What a relief! That Polish audience's warm reception was a fantastic mitzvah for me (or *piece of good luck*, as they say in English), and it boosted my self-confidence considerably.

A couple of weeks later, Gregory Wannier and I returned to Eugene, Oregon, and I had no more Rumpelstilzchen to help me out. This was a major setback. One day, however, as I was ambling through the corridors of the Physics Department, I chanced to espy a familiar "face"—namely, that of a Hewlett-Packard 9820A desktop computer, sitting on a counter in the laboratory of Russell Donnelly, a well-known experimental low-temperature physicist. What a godsend! Moreover, this Oregonian cousin of Regensburg's Rumpelstilzchen had a plotter attached to it, just like the one in Aspen had had, back in 1969. Wow! This was very promising! I excitedly asked Russ if I could use his lab's computer for a while, and he was very happy to let me do so. And so, only a few days later, Rumpelstiltskin (as I dubbed my new friend) started slaving away for me, day and night—and for three whole weeks it wove its mathematical gold, plotting out all by itself, in beautiful colors, the band-structures, according to Harper's equation, belonging to about 200 different rational values of the magnetic field. What it finally came out with is exhibited in figure P.8. I was sure that this time around, thanks to the far larger number of bands and the far more precise plotting, the recursive, self-similar pattern formed by these many line-segments would be unmistakable, even to the most naïve, most untrained, pair of eyes.

Figure P.8. Gplot, as calculated by Rumpelstiltskin in Eugene, Oregon during the late spring of 1975. Energy is plotted on the x-axis, and magnetic field (in flux quanta per lattice cell) on the y-axis.

Of course I was hoping that when I showed my new computer plot to Gregory, the fog would start to lift. However, Gregory was a pretty tough customer, or at least very set in his ways, and to my disappointment, he still didn't catch on. It took a few painful weeks for him to see the light. It turned out that in order to fully convince him, I had to ask the obedient 9820A to "undistort" a couple of very tiny regions of Gplot, and when it plotted these regions in an "undistorted" and greatly magnified way, it was glaringly obvious that they were copies of the entire graph (even though they were far less detailed than the full plot). At that point, the scales at last truly fell from Gregory's eyes, and he soon came to revere the idea of infinite nesting that I had described in Regensburg and that he, at that time, had mercilessly slammed as "mere numerology".

From being my highly acerbic critic, Gregory soon metamorphosed into my most stalwart champion! That was quite a change, although for several years, he never acknowledged his harsh words of criticism, nor the rough way he had treated me for nearly a year. I truly thought that he had totally forgotten that he had ever opposed my ideas at all. However, in the summer of 1979, quite out of the blue, I received a very gentle and humble letter from Gregory in which he all but apologized for his earlier behavior, and thanked me for having been so patient with him during that rough period. I was deeply touched by this note, and it remains my strongest memory of his character. Just as my Dad had originally told me, Gregory was "a bit of an odd duck, but an excellent physicist and a friendly, gentle person."

My experimental exploration of Harper's equation, modeled on my teen-age computer experimentations in number theory, had revealed what no amount of deductive theoretical mathematizing had done in that whole sub-area of solid-state theory, over the course of several decades, with the key exception of the work by Mark Ya. Azbel', which was far ahead of its time but which virtually no one understood. The Gordian knot had been cut by someone who, to his own great frustration, didn't have the talent to throw fancy high-powered transforms at the equation, but who instead merely instructed a computer to calculate its eigenvalues numerically, and who then looked at the visual pattern that they formed—looking with a trained eye, to be sure—an eye that many years earlier, in a totally different field of exploration (number theory, a branch of pure mathematics!), had become deeply sensitized to a certain type of self-similar recursive pattern, a pattern made up of infinitely many distorted copies of itself.

You can see why this might at first have agitated, even offended, a greatly talented Old World theoretical physicist who had always done things in the old-fashioned way, who had never used computers, and who probably didn't even trust them. It was almost as if I was an impudent Wild West upstart—a cocky computer cowboy!—with no respect for the older generation. This wasn't true at all (I had great reverence for all the figures who founded quantum mechanics, for instance, not to mention Felix Bloch and of course my Dad!), but it may have seemed that way to Gregory. Eventually, however, he realized that, though he could run circles around me in physics, I had simply seen some things that had eluded him.

One of the most beautiful discoveries coming out of the nested nature of Gplot was the revelation that for any irrational value of the magnetic flux, the spectrum consisted not of bands, but of infinitely many isolated points forming an intricate pattern called a "Cantor set" — a quite wild kind of topological beast invented by the great German mathematician Georg Cantor in the late nineteenth century. This idea was truly stunning, and yet it was practically handed to me on a silver platter, once I had clearly understood Gplot's recursive structure. Making this connection was a very special moment in my life.

In addition, my newly gained insights into Gplot soon enabled me to undercut the previously accepted wisdom (or lack of wisdom) that, in this physical problem, there was a fundamental distinction between supposedly "rational" and supposedly "irrational" magnetic fields. True, there was a natural way to *calculate* the spectrum for any rational value p/q, using a qth-degree polynomial, whereas calculating the spectrum for an irrational value was trickier, but the key point was that the actual spectra, as you slid smoothly along the real axis of magnetic-field values, changed perfectly continuously, whether you were at a rational or an irrational value, and indeed, in my thesis, I was able to prove this fact (one of the few facts about the Harper's-equation spectrum that I actually was able to establish in the "normal"— i.e. deductive—mathematical fashion).

Another way of putting this is that if you "smear" the graph, by jiggling it up and down a tiny bit (here I'm thinking of the magnetic-field strength as being on the y-axis), so that the energy bands belonging to extremely close magnetic fields are superimposed, then the "smeared graph" that results is an ordinary spectrum

Figure P.9. Two "smeared" versions of Gplot, resulting from *less* jiggling (top) and *more* jiggling (bottom).

without any trace of infinite nesting or "fractality". No matter how little you jiggle the graph, the distinction between "rational" and "irrational" fields goes entirely away, as indeed it must, if we're talking about the real physical world. In preparing my thesis, I actually constructed two smeared graphs with different amounts of jiggling, painstakingly tracing them out completely by hand while sitting at my desk for many hours. Although they are both a bit spooky-looking, especially the one involving less jiggling (see figure P.9), they no longer look like "faint fantastic traceries" or infinitely delicate lacelike filigrees.

When I finally handed in my thesis, in December of 1975, it included, of course, a very high-quality reproduction of Gplot, which folded out in a lovely way. This foldout was inserted right after the last page of text of the book, almost as if it were an appendix. And it happened, perhaps by coincidence, or perhaps not, to be page 137 of my thesis. (For those readers who don't know the meaning to physicists of the hallowed number 137, suffice it to say that it is the reciprocal of the fine-structure constant, a key number in quantum electrodynamics, and why that dimensionless number has that strange value has been a beckoning mystery to physicists for 80 years or more. It turns out that it is not actually precisely the integer 137, but more like 137.02, although for a long time many people wondered if this wasn't a mystical property of the integer 137, which also happens to be a prime number. If the reciprocal of the fine-structure constant had actually turned out to be exactly an integer, that would indeed have been a most astounding fact.) And so, in the end, by making Gplot be page 137 of my thesis, I *did* wind up making a mischievous numerological gesture—and as I look back on the whole crazy story, I can't help smiling.

My doctoral defense took place in December of 1975 in Eugene, and although I was somewhat nervous, since I did not consider myself by any means a true solid-state physicist, I did just fine, and it was a huge relief to know that I was finally completely over the long, tortuous roller-coaster ride of my graduate career. Shortly before my defense, I'd had my thesis professionally printed up and bound in the classic manner, making about 25 copies of it altogether, half of them with green bindings (for the University of Oregon), and half of them with red bindings (for Stanford). The first copy went, of course, to Gregory Wannier, and then I gave copies to my other committee members, and to Russ Donnelly (Rumpelstiltskin's guardian), and to a few very dear grad-school cronies (including, of course, Francisco Claro in far-away Chile). Copies were also sent to the two other members of the Regensburg group, needless to say, and I gave one to my Dad, whose deep belief in the beauty of physics and whose moral support during these grueling years had meant everything to me. Indeed, the closing sentence of the "Acknowledgments" section in my thesis reads as follows:

> *Finally, I would like to say that my eternal faith in the beauty and simplicity of nature comes straight from my father, Robert Hofstadter, and has here acted and will always act as the main guiding principle in my view of the universe.*

And of course, one very special copy was reserved for Felix Bloch. I'll never forget the day I went to Felix's office in the Stanford Physics Department and handed it to him, and then, at his request, signed it. This was a very meaningful, tangible bond between our two families, which had so long been linked in friendship, and which now were linked in a new way by this creation of mine that owed so much to Felix's own doctoral thesis, way back in 1928, about the application of quantum mechanics to solids, in which what soon came to be known as Bloch waves, Bloch electrons, and Bloch bands were first introduced to the world. What a symbolic event for me!

Gplot was an incredible gift to me. Not only did it graciously grant me a PhD, which I had so feared I would never get, and which became the precious visa allowing me, a few years later, to enter the privileged land of academia, but it also looped the loop for me, allowing me to come full circle and get a doctorate in number theory (although it was disguised as a doctorate in solid-state physics). Such an ironic turn of events was the furthest possible thing from my imagination when, eight years earlier, I had desperately bailed out of math grad school in Berkeley!

Once my little Gplot, born as a hand-drawn figure in Regensburg in the fall of 1974, had made an appearance in the pages of *Physical Review* in the fall of 1976, in the only physics article I ever wrote, it slowly became known, and after a few years it was known far and wide. Eventually, people started calling it "the Hofstadter butterfly" (though I myself still call it just "Gplot").

In 1980, a few years after I had left physics, Gregory wrote me a letter politely asking me if I would send him a bunch of copies of my *Physical Review* article, explaining that I, having changed fields, presumably didn't need them any more. Then he added, "For me, on the other hand, the article is like one of my own." Wow—now *that* was going a bit far. Still, I couldn't help but feel touched by this unwitting revelation of how deeply and totally Gregory had turned around since the days of our intense clash. I wound up sending him about half the copies I had.

Today, hundreds of papers have been written about the butterfly, most of which I don't understand at all. Over these past 40 years, Gplot has been extensively generalized and connected to phenomena in many other areas of physics— connections that I could never have imagined, not even in my wildest dreams! Let me just mention the quantum Hall effect, the renormalization group, topological insulators, Apollonian gaskets, Berry phases, Chern numbers, cold-atom lattices, noncommutative geometry, Bose–Einstein condensates, Majorana fermions, and last but not least, "anyons" (of both abelian and nonabelian varieties, of course!), which are strange quasiparticles that live only in two-dimensional worlds and are somehow poised somewhere in between being bosons and being fermions. My head spins when I read about these kinds of surrealistically abstract notions linked to my little Gplot; to my great dismay and considerable embarrassment, many of them lie quite a long way above my abstraction ceiling (although, thank God, not above my absurdity ceiling!).

Aside from all this theoretical work connecting Gplot to other physical phenomena, just a couple of years ago some of the key properties of the energy spectrum of Bloch electrons in magnetic fields, which I'd speculated about in my doctoral thesis way back in 1975, were finally experimentally verified in the no-nonsense physical world. This remarkable experimental work relied on the creation of artificial lattices with extremely large unit cells, so that the amount of magnetic flux passing through them (as measured in flux quanta) was non-negligible. (Back when I wrote my thesis, the amounts of magnetic flux that could be made to pass through a crystal's unit cell, even with the most intense magnetic fields, was extremely tiny, so it was a major breakthrough when clever ways were devised of making "artificial" lattices that had much bigger unit cells.)

In January of 2015, at the Kavli Institute for Theoretical Physics in Santa Barbara, California, I had the honor of giving back-to-back talks about the butterfly with Philip Kim, one of the pioneers of the superlattice techniques that had allowed the butterfly to be glimpsed experimentally. My talk was mostly just reminiscences about how I'd discovered Gplot some 40 years earlier, and I felt a bit out of my depth during Kim's talk, but I nonetheless basked in the glory of having come across Gplot when I was young.

Two months later (on March 3rd, to be precise), I was very surprised to receive, out of the blue, the following remarkable and very friendly email from someone of whom I had never heard, in the Physics Department at George Mason University in Virginia:

Dear Professor Hofstadter,

I am writing a short book (Physics Today level), to be published in the "IOP Concise Physics" collection, titled "Butterfly in the Quantum World, Story of a Most Fascinating Quantum Fractal". I will be extremely grateful if you could comment/critique on the nearly complete draft (low resolution version) that I am attaching with the email. Of course any suggestions from your side will be extremely useful. It will be nice if you could write a few words reminiscing your thoughts when you first saw the butterfly, and how much the recursions of the Gplot influenced your future work or anything else about the graph that you may wish to say or share.

I thank you in advance for your consideration and look forward to hearing from you.

Sincerely
indu

P.S just a little note that the attached draft is a working draft and I am planning to finalize it within a week or so. However, it always takes longer than you expect, even when you take into account Hofstadter's Law.

I was quite amazed and thrilled to find out that someone was writing—indeed, had already written!—a whole book about my little Gplot, once just a hand-drawn diagram in a little notebook of mine. As an attachment, Professor Satija had included the latest draft of her book, and when I looked through it, I saw that she had spelled out many stunning links between Gplot and other phenomena, both in math and in physics, and had done so in an infectiously lively manner. I was flattered that she had asked me to contribute some personal reminiscences to this charming book, and so I replied to her with great enthusiasm. By coincidence, I was scheduled to give a talk in Maryland the next month, and I suggested that I could kill two birds with one stone by visiting George Mason as well. Indu Satija was gung-ho about this idea, and so it wasn't long before we met in person at George Mason University, and I had the distinct pleasure of giving her one of the very few copies I still had of my PhD thesis—a green-bound one, as I recall. A couple of hours later, I gave more or less the same talk of reminiscences to her department as I had given in Santa Barbara a few months earlier.

Indu realized that her book, like any book, was going to need some copy-editing, and during my visit to George Mason, I volunteered to do that for her, partly because I am a very perfectionistic writer and have lots of experience in writing for the wide public, and partly because I have a great deal of background in the subject matter itself (although of course I had not been involved in the physics for forty years). In the following months, then, I helped polish the prose that Indu had so painstakingly crafted for her readers. In suggesting changes to her delicate, elaborate construction, I often felt guilty, so I nicknamed myself "the bull in the china shop", a label that Indu delighted in. Indeed, she gave the warmest welcome to this "bull in the china shop", because, as she told me, she didn't really trust people who praised her book. She trusted only words of criticism, and the more critical they were, the merrier! I thought that was pretty funny.

For a few months, then, I had the amazing experience of reading in great detail about Gplot's many connections to all sorts of phenomena, and of thinking about these connections very carefully. I had never expected I would come into such close contact with all these modern developments in condensed-matter physics, and it was really enlightening to learn so many things from this book, and a great pleasure to be able to contribute here and there to its local flow and to its clarity—and also to contribute one invited "guest chapter", in which I recount some of the technical details in my thesis. To have a chapter of my own in this book is a great privilege. But the marvelously rich overall vision of this book and the myriad crisscrossing connections that it spells out all come from Indu herself, of course. I was just a humble copy-editor! Her enthusiasm for the magic of these connections is truly unbounded, and I am so grateful to her for allowing me to come on board and be, in my own small bull-in-china-shop way, a part of this glorious celebration of the butterfly. (By the way, Indu's suspicion that Hofstadter's law might apply to her book was confirmed in spades, since it was finally sent off to the publisher not a *week* later, but a *year* later…)

There are two other Gplot events that I feel a bit reticent about describing here, since to relate them might seem self-serving, but since they were both touching to me when they happened, and since I think they add human interest to my story, I'll take the risk of including them.

The first involves the review process of the article based on my thesis that I submitted to *Physical Review* in early 1976. A few months later, I received two anonymous reviews in the mail, both of which recommended publishing it, to my relief—and to my astonishment, one of them also said, "This paper reads like a Mozart *divertimento.*" This was, shall I say, music to my ears. Even if Mozart was not among my favorite composers, I knew exactly the sentiment that the unknown reviewer meant to express with those words, and I was extremely moved by them and grateful for them. I had never expected any such reception, and was rather bowled over by it.

Well, some sixteen years later, in the summer of 1992, I gave a colloquium about Gplot in the Physics Department at Berkeley. It was attended by only about a dozen people, if even that many, which was a little disappointing, but afterwards, one tall, thin, gray-haired gentleman came up to me and said, with a refined-sounding accent, "My name is Leo Falicov, and I'm a member of the department here." I recognized his name instantly, because he was a famous condensed-matter theorist. (In fact, in 1969, one of Falicov's research associates, Dieter Langbein, had published an article about the energy spectrum of Bloch electrons in a magnetic field, which included a graph, but its level of detail wasn't fine enough for Langbein or anyone else to pick up on its recursive nature. Whew! Was I ever lucky!) Falicov then said to me, "Do you by any chance remember an anonymous review of your *Physical Review* article that likened it to a piece by Mozart?" I replied, "How could I ever forget such a great compliment?" He smiled and then said, "Well, I was the one who wrote that review, and I still feel that way about this work." That was an unforgettable, albeit very brief, encounter. I never saw Leo Falicov again, but this meeting will forever stay with me.

The other anecdote is rather amusing. In the fall of 1998, I went to Saint Petersburg, Russia for about ten days, thinking of possibly spending a sabbatical year there. I wanted to explore various departments at the university, and so I arranged to give a couple of lectures, one of which was in a mathematics institute, on a topic in Euclidean triangle geometry, which at the time was a hot passion of mine. Talk about small audiences—this time I had only about five people who came to hear me! And a little more disturbing was the fact that they all seemed distinctly underwhelmed by what I was telling them. (I later discovered that all Russian mathematicians know Euclidean geometry like the backs of their hands, so no wonder my ideas hardly seemed Earth-shaking to them.) Despite the cool reception, the organizer of my talk kindly invited me out to lunch and tea afterwards, and as we talked, he said, with a very strong Russian accent, "You should know, my specialty is mathematics of physics, and I am great admirer of work in physics of your father." I was of course very pleased to hear this, but since my Dad had done many things in physics, I asked him which work he was referring to, and he said, "Butterfly graph, of course!" At that point, I suddenly felt deeply embarrassed, and I had to disabuse him of his illusion by saying, "Well, actually, that's something that I myself did..." At that, he looked most astonished, and spontaneously blurted out, "Oh! I see! In such case, you are *very important person*!" In roughly one second, I had leapt upward in his estimation by a factor of a thousand! Very funny.

If being the discoverer of Gplot made me a "VIP" in physics, then wouldn't it have been wise for me to stay in physics? I don't think so. These days, as I look back on my painful graduate-school struggle with physics and my subsequent non-physics career, I don't in the least regret having given up, right after finding Gplot, on my once-fervent dream of making physics be my profession. For me to have gone on in physics would have been a big mistake, even though the fact that I had discovered Gplot, with its great visual appeal and its counterintuitive properties, was a nearly surefire guarantee that I could have gotten a job in a fine physics department somewhere.

Why would going on in physics have been a mistake? Well, during my graduate-school days, it only slowly dawned on me, as I was struggling like mad to understand particle physics and later on solid-state physics, that I was not cut out to be a physicist. Of course this realization at first scared me and disappointed me, and I tried to fight it with all my might—but fighting it just didn't work. I simply was who I was. Although I had grown up in a physics family and had always loved the deep ideas of physics—as an eight-year-old I'd even yearned to "be a neutrino" (what eight-year-old *wouldn't* want to be a massless, spin-one-half particle that could go sailing unscathed right through planet Earth at the speed of light?)—I came to realize, in my nearly eight years of graduate school in physics, that being a creative physicist took a certain combination of talents that I just did not have, alas. Gregory Wannier and Alexander Rauh and Gustav Obermair all had those talents, but for some reason I didn't. Very luckily, though, I came to see that I had talents in other domains, and so, having done something in physics of which I knew I could be proud, but knowing that such a success story was very unlikely to repeat itself, I had the sense to get out while the getting-out was good. I thus set sail for other lands to which I was more suited, going out in a small blaze of fractal glory. That was a lovely, lucky feeling!

Speaking of luck, I think it would be fitting for me to close my prologue by pointing out that it was only thanks to a stunning series of strokes of luck that I found the recursive secret of this elusive butterfly. Here, then, are eleven pieces of luck that wound up revolutionizing my life:

- the excellent luck that, after traumatically quitting particle physics with a sense of revulsion, I ran into two people whose eloquence convinced me that to do solid-state physics was the furthest thing in the world from "going slumming";
- the excellent luck that Gregory Wannier, who for many years had studied the crystal-in-magnetic-field problem, was a professor at the University of Oregon, and that in my hour of need I had the good sense to approach him;
- the excellent luck that many years earlier, I had dreamt up the INT function, thus plunging myself into the world of number-theoretical problems featuring, at their core, rationals versus irrationals;
- the excellent luck that I trusted my instinct to "go for it", when I heard Wannier mention the irrational-versus-rational fight inherent in the problem, even though at the time I had no idea how I might contribute;
- the excellent luck that I came across and did my best to understand Mark Azbel"s difficult 1964 paper, which reinforced my inchoate hope that ideas of number theory might somehow play a key role in unraveling the problem's secrets;
- the excellent luck that I was incapable of keeping up with the three other members of the Regensburg group, and thus was forced to try out a radically different, far more concrete, far more humble, pathway;
- the excellent luck that I had learned to program computers when I was a teenager and had become an "experimental mathematician", unashamed of exploring mathematics as a physicist explores nature;
- the excellent luck that there was a Hewlett-Packard 9820A computer sitting on a cart in the hallway just outside my office in Regensburg, and that no one else in the department had the slightest interest in using it;
- the excellent luck that I had played around with an identical Hewlett-Packard computer in Aspen, Colorado five years earlier, and had quickly come to love the elegant simplicity of its user interface;
- the excellent luck that when I returned to Eugene from Regensburg, I found yet another identical H-P computer, this one, however, being luxuriously equipped with a mechanical pen-on-paper plotter;
- the excellent luck that nobody in the entire worldwide solid-state physics community had ever computed and plotted this graph in sufficient detail

before, for otherwise it might easily have been known as the "Langbein butterfly"...

It is astonishing to me to see how many truly tenuous links of luck in the chain there were, and yet, somehow, it's only thanks to all those links of luck taken together that I passed from near-failure to the dizzying discovery of that glowing gem called "Gplot". It was a long time, though, before I fully grasped the extent of the role of luck in my story. Only after twenty years or so did it finally hit me how terribly thin a thread my entire future had hung from back then, and how terribly close I had come to never getting a PhD at all, and thus never being able to go into academia, never becoming a professor, and everything that that extremely different pathway would have entailed. This sobering realization, once it finally came, gave me a lasting sense of humility and a profound respect for the vastly underestimated role of luck in human life.

To be sure, the happy ending of my story was not due in its entirety to sheer random luck. After all, there is no denying that, several times over the course of several years, I sensed that something about the pathway down which I was heading was deeply wrong, and that something had to give, in the sense of changing radically. Thus my bounce out of math-in-Berkeley and into physics-at-Oregon was one key turning point in my life, and the scary decision to take that fork in the road was not imposed on me at random, but came from within. Likewise, my leaps from one advisor to another in particle physics came from major shifts of tectonic plates inside my brain, as did my risky leap out of particle physics and into the realm of solid state—a move that would have struck me as crazy, absurd, and inconceivable when I first came to Oregon.

All these bounces—triggered by severe blows I received from the outside world—came about as results of the clash of many psychological forces inside me, not from outside. I was not just pushed around like a helpless ball in a pinball machine. No—my inner self was determined to search for a pathway that I felt truly comfortable with, and I kept on searching until—very luckily—I finally found one. All these abrupt veerings in my trajectory, all these desperate leaps of faith—they wound up saving me by guiding me away from certain disaster, from chasms and ravines in which I would without any doubt have perished. But by themselves, the leaps I took were no guarantee of success. They were just moves that let the game go on a bit longer. Each one gained me some time, restarting the clock in a new game situation. And luckily, after several such desperate leaps, I finally wound up in a situation where things were far more favorable. But the key word here is that adverb "luckily". Though I was the captain of my life's ship and did my best to steer it in the right direction in stormy weather, the random forces of nature were utterly out of my control, and I just had to deal with the harsh gusts and huge waves as they slammed into me. As it happens, though, I was also the beneficiary of a number of pieces of great luck, as spelled out in the list above. Somehow, as a combination of all these forces, inner and outer, I wound up sighting a beautiful desert island called "Gplot", calm and serene in the midst of the raging storm. And so, all I can do is thank my lucky stars that things turned out as well as they did.

When today I look at some of my friends who are every bit as talented as I am but to whom this fickle world has doled out less good fortune than to me, I think to myself, "Without that inexplicable string of great pieces of luck, that could oh-so-easily have been my fate, too. There but for the grace of God go I." Or sometimes, since I'm not a religious person, I say to myself, "There but for the grace of Gplot go I."

Envoi

Indu Satija very much hoped that I would include a poem about Gplot, and although I resisted her pleas for a long time, eventually she succeeded in convincing me to give it a try. It occurred to me that since the bulk of my doctoral research was carried out in the effervescent town of Eugene, Oregon, and since I have long been an aficionado of the effervescent novel in verse *Eugene Onegin* by Alexander Pushkin, it might be fitting to invoke the spirit of the latter as a way of celebrating a set of events many of which happened in the former. And so I decided to use the remarkable poetic medium that Pushkin devised for his novel—that is, to write what today is called an "Onegin stanza" (14 syllables of iambic tetrameter following an ABABCCDDEFFEGG rhyme scheme)—and in my Onegin stanza to give one last brief glance back at the tale told in this Prologue. So with the poem below I shall take my final bow and exit, stage left.

What happens if a crystal's laced with
The lines of a magnetic field?
What spectrum will the world be graced with?
What energies will nature yield?

It turns out that the matter's crux is
Determined by how great the flux is—
p-over-q q bands begets;
Non-ratios, though, give Cantor sets!

On hearing this, a physicist'll
Declare it numerology;
But once shown Gplot, all agree
Deep magic's lurking in a crystal!

This gem I found by luck. That's why
There but for Gplot's grace go I.

Prelude

The artist has to transcend a subject, or he loses the battle. The subject wins.
—Fritz Scholder

He looked at his own Soul with a Telescope. What seemed all irregular, he saw and shewed to be beautiful Constellations; and he added to the Consciousness hidden worlds within worlds....
—Samuel Coleridge Taylor, *Notebooks*.

"Let the dance begin" [1].[2]

The Hofstadter butterfly: home of the quantum Hall effect

In the fall of 1974, when Douglas Hofstadter discovered the butterfly graph, which revealed the allowed energies for a quantum system of electrons in a two-dimensional crystal, few people would have guessed that anything like this strange, eerie shape could have anything at all to do with the physical world. The quantum Hall effect had not yet been discovered, and the word "fractal" was unknown at the time.

---

[2] Copyright: The Library Company of Philadelphia (www.librarycompany.org).

A few years later, in his Pulitzer-Prize-winning book *Gödel, Escher, Bach: an Eternal Golden Braid* [2], Hofstadter discussed recursion in solid-state physics, and he connected it with recursion in other domains, including fugues in music, grammatical structures in human language, and the branching patterns of trees in mathematics and in nature. He also compared the intricate recursive beauty and complexity of the butterfly he had recently discovered with the recursions found in chess programming, with the abstract structures constituting human consciousness, with Dutch graphic artist M C Escher's prints, and with contrapuntal pieces of music by J S Bach—pieces that he acrostically described as *Beautiful Aperiodic Crystals of Harmony*.

Johann Sebastian Bach's Fifth French Suite, full of wonderful twists and turns, is a musical jewel. Its concluding Gigue is an example of the multifarious "recursive structures and processes" described by Hofstadter in chapter 5 of *Gödel, Escher, Bach*. In his words,

> *Like many Bach pieces, the Gigue modulates away from its original key of G and imitates a finish, but in the wrong key (namely, D). It then jumps right back to the start and retraces exactly the same pathway in harmonic space, winding up once again in the key of D. Thereafter, starting right up again with a new theme, which not coincidentally is the original theme upside down, it winds its way back home, modulating back to the original key of G and seeming to finish up in it. But just as before, Bach is not actually done, for at this point he jumps back to the midway point in D and reiterates the journey back home to G with the upside-down theme, only coming in for a true landing the second time around. These subtle and intricate maneuvers in musical space are like stories told within stories, where one goes "down" into new worlds (musical keys, in this case) and then pops back "up" out of them.*

Hofstadter used such examples in *Gödel, Escher, Bach* to illustrate the rich mathematical concept of recursion. At one point, he wrote: "A recursive definition never defines something in terms of itself, but always in terms of simpler versions of itself". If we allow "smaller" as a possible interpretation of "simpler", then this is certainly an appropriate way to describe the recursive structure of Hofstadter's butterfly graph, which is composed of infinitely many smaller copies of itself (with smaller copies yet inside those, and so forth, *ad infinitum*). The recursions in the butterfly, intertwined with its underlying geometry and topology, reveal a hidden beauty and simplicity, and that is what this book is all about.

To exploit once again the metaphor of music, this book can be thought of as a "butterfly symphony" consisting of four movements—a symphony that recounts the story of the butterfly in the quantum world. Short narratives of each movement, provided below, along with the schematics in figures P.10–P.13, are intended to provide readers with glimpses of the entire symphony—a bird's-eye panorama that will hopefully pique readers' curiosity.

- The first movement, entitled "The butterfly fractal", gives a feel for the fractal aspects of the butterfly. It uses chords and melodies familiar from

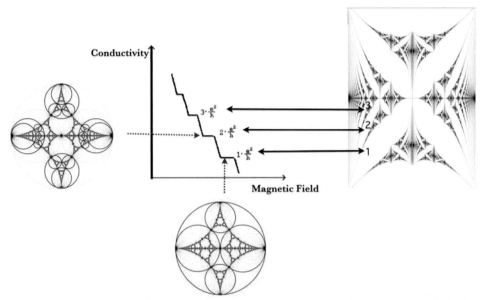

Figure P.10. This graph of electrical conductivity versus magnetic field reveals the integer quantum Hall effect. The horizontal plateaus in the graph show that electrical conductivity is *quantized*, which means that it assumes values that are integral multiples of a fundamental physical constant, $\frac{e^2}{h}$: here e is the charge on the electron and h is Planck's constant, a tiny quantity at the heart of all quantum effects. The panel on the right side is the Hofstadter butterfly, whose gaps, or empty white regions, are characterized by integers that are quantum numbers of the Hall conductivity; these same integers appear in the staircase on the left-hand side. Finally, the plots made up of nested circles show *integral Apollonian gaskets* — infinitely nested patterns of mutually tangent circles all having integer curvatures. The quantum Hall states filling the gaps of the butterfly fractal can be viewed as integral Apollonian gaskets in a reincarnated form.

some well-known geometrical fractals and conveys a sense of the profound and timeless beauty of this magical visual structure.

Fractals are fascinating objects. There is always something more to see as you look at them at smaller and smaller scales. In their never-ending nestedness, you will find similar structure no matter at what scale, large or small, you choose to look.

The Hofstadter butterfly is a very special fractal because, in a certain sense, it is "made out of integers". These integers are in fact the quantum numbers associated with the quantization of electrical conductivity—an exotic phenomenon known as the *quantum Hall effect*, the discovery of which was honored with a Nobel Prize in 1985. Interestingly, the roots of the integer recursions can be traced all the way back to some beautiful ancient Greek mathematics dating from roughly 300 BC. This adds a new dimension to the story, as it reveals some hidden symmetries of the butterfly.

The butterfly composition is a metaphorical fugue that has fascinated mathematicians and physicists for almost forty years [3]. Figure P.11 offers a pictorial introduction to the amazing intricacies of this object.

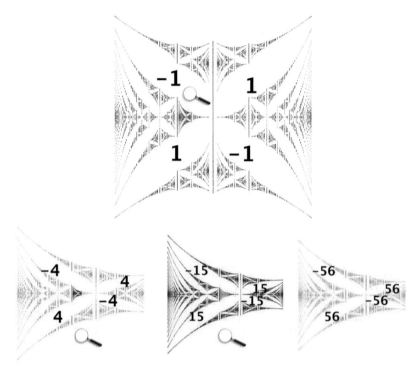

Figure P.11. Zooming into the butterfly fractal reveals identical patterns at all scales. The red butterfly is a blowup of the red region in the upper black graph. The blue butterfly is a blowup of the blue region in the red graph, and the green butterfly is, in turn, a blowup of the green region in the blue graph. The integers labeling the white gaps in these differently colored butterflies are the quantum numbers of the Hall conductivity, which form a quasiperiodic pattern with dodecagonal symmetry.

- The symphony's second movement, entitled "Butterfly in the quantum world", provides a peek into the quantum world, and also includes an introduction to the quantum Hall effect. This chapter, a preamble to the remaining parts of the symphony, emphasizes physical concepts such as quantization, and exposes readers to strange and highly counterintuitive physical phenomena.
- The symphony's third movement, "Topology and the butterfly", reveals the geometric aspects of the butterfly and relates them to the quantum Hall effect. This movement tells the most technically complex part of the story, escorting the audience to a very curious abstract curved space in which following a cyclic pathway does not necessarily return travelers to their original starting point.

  For quantum-science enthusiasts who love geometry and topology, the relationship between the butterfly fractal (made up of integers) and the quantum Hall effect is fascinating and also somewhat baffling. If the first two movements of the butterfly symphony aroused a sense of excitement, the third movement is bound to arouse a sense of profound mystery.

Figure P.12. The quantum Hall effect and topology. The unexpected emergence of trefoil symmetry at infinitesimal scales of nature (as described in chapters 2, 3, and 10) resonates beautifully with trefoil symmetry in macroscopic nature as can be seen, for instance, in redwood sorrels found in Montgomery Woods in Mendocino County, California (photos courtesy of Douglas Hofstadter).

- The abstract tone that pervades the symphony's first three movements is largely left behind in its fourth movement, as the quantum butterfly tantalizingly begins to show its footprints in the real world. As readers follow the decades-long journey toward the actual capture of the elusive butterfly, they will appreciate the art of scientific exploration in laboratories. This final movement of the symphony may be soothing or agitating, reflecting the as-yet unfinished task of definitively capturing this celebrated "beast" in the wild.

Finally, as an encore, we leave scientific discourse behind and let readers savor the exotic world described in this book with "Butterfly Gallery", "Divertimento" (butterfly tales), and some poetic verses in "Poetic Math and Science" inspired by the Hofstadter butterfly. We conclude with a Coda from "Lilavati", which was composed by Bhaskara 900 years ago.

The stunning form of the butterfly graph, when it first appeared in print in 1976, fascinated physicists and mathematicians, as it revealed new aspects of the beauty, simplicity, and harmony of the solid state of matter in the universe. The graph's infinitely nested structure was a great surprise at the time, but despite its visual appeal and its abstract fascination, it could easily have been forgotten, had it not

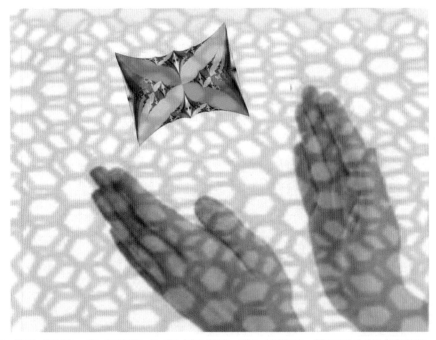

Figure P.13. Catching the butterfly in the laboratory using graphene. (Reprinted by permission from Macmillan Publishers Ltd, copyright 2013.) [4].

been for the laboratory realization, several years later, of two-dimensional electron gases manifesting the quantum Hall effect.

The discovery in 1980 of the quantization of resistance at low temperatures and high magnetic fields was wholly unexpected. The flat or plateau regions in figures P.10 and P.12 show where the conductivity is an integer multiple of the earlier-mentioned constant $\frac{e^2}{h}$. A schematic illustration of some of the links between the quantum Hall effect and the butterfly is shown in figure P.10, while figure P.12 is meant to convey a sense for how topology's abstractions play a key role in this drama.

Astonishingly enough, Hofstadter's butterfly plot turns out to be the home of the quantum Hall effect. Each empty region or gap in the butterfly represents a unique quantum Hall state, characterized by an integer that corresponds to the quantum number associated with the Hall conductivity. The butterfly is thus a dazzling structure built out of these integers.

The quantum Hall effect—or at least the *integer* quantum Hall effect—is now so well understood that it is often covered in undergraduate textbooks in quantum and solid-state physics. Even so, the subject remains at the forefront of physics, since two-dimensional electron gases exhibiting the quantum Hall effect are the simplest examples of *topological insulators*—new exotic states of matter that are revolutionizing fundamental physics and that offer the potential of groundbreaking technological applications.

The butterfly graph is testimony to the fact that breathtaking complexity can arise in very simple systems. It is a one-of-a-kind example of how nature, when subjected to competing scales, responds in a magical-seeming way, creating patterns rife with hidden mysteries. It is particularly fascinating when this happens in the real quantum world, a world that can be created in a laboratory. For this reason, the quest for experimental proof of the butterfly's reality is one of the hottest activities in many condensed-matter laboratories today, and indeed telltale signatures of some of the complex patterns making up the butterfly have recently been observed.

References

[1] Tomlinson K 1735 *The Art of Dancing Explained by Reading and Figures* London: Printed for the author engraving from http://www.librarycompany.org/visualculture/pop01.htm
[2] Hofstadter D R 1979 *Gödel, Escher, Bach: an Eternal Golden Braid* New York: Basic Books
[3] Hofstadter D R 1976 Energy levels and wave functions of Bloch electrons in rational and irrational magnetic fields *Phys. Rev.* B **14** 2239
[4] Dean C R *et al* 2013 Hofstadter's butterfly and the fractal quantum Hall effect in moiré superlattices *Nature* **497** 598–602

Part I

The butterfly fractal

Reprinted with permission of Jin Denevan.

IOP Concise Physics

Butterfly in the Quantum World
The story of the most fascinating quantum fractal
Indubala I Satija

Chapter 0

Kiss precise

"A mathematician," said old Weierstrass,
"who is not at the same time a bit of a poet
will never be a full mathematician."

Apollonius of Perga, René Descartes, and Frederick Soddy

Given three mutually tangent circles, how does one draw a fourth circle that is exactly tangent to all three?

Three great thinkers who independently discovered the solution to this mathematical puzzle were Apollonius of Perga (3rd century BC), René Descartes (17th century), and Frederick Soddy (20th century)[1].

Apollonius of Perga, a Greek geometer who was born around 262 BC in the town of Perga, now a part of Turkey, was the first mathematician to seriously consider this question. Among his few surviving works is a book on conic sections, which was the first systematic study of many geometrical shapes and curves that have remained central to mathematics ever since. In that great work, Apollonius coined the terms "ellipse", "parabola", and "hyperbola". He also made important contributions to astronomy and optics. The lunar crater Apollonius was so named to honor his work on the history of the Moon.

One of Apollonius' lost works is a book called *Tangencies*, purported to provide methods of constructing circles tangent to various combinations of lines and circles—more specifically, the problem of drawing a circle that is simultaneously tangent to three given geometrical entities, where those entities are either (1) three straight lines, (2) two straight lines and a circle, (3) two circles and a straight line, or (4) three circles. Tragically, all copies of this important work were destroyed in a great fire at the Library of Alexandria, and as a result, no one knows what solutions Apollonius gave to these problems, or whether his solutions were correct. After many of the writings of the ancient Greeks became available again to European scholars of the Renaissance, thereby evoking much renewed interest in geometry, the problems that Apollonius had posed and had tried to answer became, once again, a great challenge.

Geometrical problems involving tangent circles have intrigued people the world around for millennia. In the Edo period, the Japanese had a tradition of posing geometrical puzzles by writing them on public surfaces. One of the more popular of these riddles was how to calculate the radius of a circle tangent to three given circles that were tangent to each other. The solution to this was rediscovered by René Descartes, a French philosopher, mathematician, and writer who was born in 1596, between Poitiers and Tours. He has been dubbed the founder of modern philosophy, and much of today's Western philosophy can even be thought of as a collective response to his writings. Descartes started out with a strong interest in physics, but decided to concentrate on mathematics after hearing that Galileo had been arrested for claiming that the Earth orbits the Sun.

One of the appendices to Descartes' famous book *Discours de la Méthode* was an essay on mathematics called "La Géométrie". Although its title means "geometry", it actually focused on the connections between geometry and algebra. In it,

---

[1] A number of other people rediscovered Descartes' theorem and also generalized it. See [1].

Descartes' circle theorem, 1643

Theorem (in a letter to Princess Elisabeth of Bohemia)

$$(\kappa_0^2 + \kappa_1^2 + \kappa_2^2 + \kappa_3^2) = \frac{1}{2}(\kappa_0 + \kappa_1 + \kappa_2 + \kappa_3)^2.$$

Descartes introduced what is now known as the Cartesian coordinate system, and many modern algebraic conventions were invented by him and first appeared in this essay. For example, Descartes used letters from the beginning of the alphabet for constants and known quantities, and letters from the end of the alphabet for variables. So Descartes is the reason we solve for x, and not some other symbol!

The solution to the problem of calculating the radius of a circle tangent to three given circles that are tangent to each other is today known as "Descartes' theorem". Descartes discussed the problem briefly in 1643 in a letter to Princess Elisabeth of the Palatinate (1618–80) [2]. An intelligent and accomplished woman, she influenced many key figures and philosophers, most notably Descartes, and she corresponded with him extensively. Their epistolary conversation touched on many topics, including Descartes' idea of dualism, or the mind being separate from the body. In the above-mentioned letter to Elisabeth in November of 1643, Descartes posed the Apollonian challenge of finding a circle that touches each of three given circles in a plane. Elisabeth's solution no longer exists, but in a later letter, Descartes remarks that hers possessed a symmetry and transparency that his lacked. Unfortunately, her writings were never published, but she clearly was a woman with many intellectual interests, ahead of her time, and deserving of recognition for her contribution to this problem.

Descartes' own formula has never become widely known, even among mathematicians. Interestingly, it was rediscovered around 1936 by Frederick Soddy, who was awarded the 1921 Nobel Prize in Chemistry for the discovery of isotopes. As a chemist, Soddy had a great interest in how to pack spherical atoms of differing sizes, and this led him to the study of mutually tangent (or "kissing") circles. One particularly elegant configuration starts with a trio of three kissing circles, with a larger circle *outside* the trio and kissing them all, as well as a smaller circle *inside* the trio and kissing them all. So smitten was Soddy with the formula he found relating the radii of all these circles that he published his formula in the unusual form of a poem, in the journal *Nature* in 1936. Below we quote the first two (out of three) stanzas of the poem (reprinted from [6] by permission from Nature Publishing Group), and show an artistic rendition of the situation drawn by Sarah DeBauge.

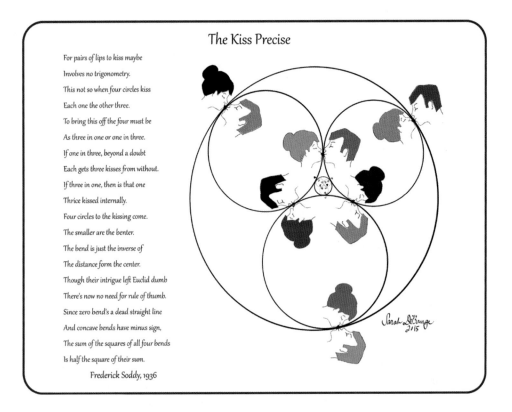

The poetic piece of mathematics that is described in the last two lines is none other than *Descartes' theorem*[2]:

$$\left(\kappa_0^2 + \kappa_1^2 + \kappa_2^2 + \kappa_3^2\right) = \frac{1}{2}(\kappa_0 + \kappa_1 + \kappa_2 + \kappa_3)^2. \tag{0.1}$$

Here the four κ_i are the *curvatures*—that is to say, the reciprocals of the radii—of the four circles.

Descartes' version of the same formula expresses the curvature of the fourth circle in terms of those of the initial three mutually tangent circles. We can solve the above equation in terms of κ_1, κ_2, and κ_3 as follows:

$$\kappa_0(\pm) = \kappa_1 + \kappa_2 + \kappa_3 \pm 2\sqrt{\kappa_1\kappa_2 + \kappa_2\kappa_3 + \kappa_3\kappa_1}. \tag{0.2}$$

---

[2] For a simple proof of Descartes' theorem see [4].

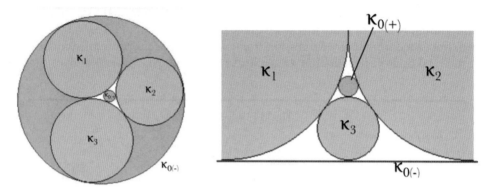

Figure 0.1. A five-circle configuration whose outermost circle has curvature $\kappa_0(-)$ and whose innermost circle has curvature $\kappa_0(+)$. Both the outermost and the innermost circles are shown in red, and both are tangent to all three of the gray circles, whose curvatures are κ_i, with $i = 1, 2, 3$. The outermost circle can also be a straight line (as can be seen in the panel on the right), since a straight line can be thought of as a circle with infinite radius, which means that its curvature equals zero.

The "$\pm$" in Descartes' formula reflects the fact that there are two solutions (figure 0.1):

(1) The "$-$" solution corresponds to an *outer circle* with smaller curvature (i.e., larger radius), and we denote its curvature as follows:

$$\kappa_0(-) = \kappa_1 + \kappa_2 + \kappa_3 - 2\sqrt{\kappa_1\kappa_2 + \kappa_2\kappa_3 + \kappa_3\kappa_1}. \tag{0.3}$$

(2) The "$+$" solution corresponds to an *inner circle* with larger curvature (i.e. smaller radius), and we denote its curvature as follows:

$$\kappa_0(+) = \kappa_1 + \kappa_2 + \kappa_3 + 2\sqrt{\kappa_1\kappa_2 + \kappa_2\kappa_3 + \kappa_3\kappa_1}. \tag{0.4}$$

For the solutions to these two equations to be consistent, the outermost, or bounding, circle, must be considered to have *negative* curvature. It's also important to point out that given any three mutually tangent circles, there are always two other circles (analogous to the two solutions above) that are tangent to the three circles.

0.1 Apollonian gaskets and integer wonderlands

A beautiful manifestation of Descartes' theorem appears in patterns describing the close-packing of circles, called *Apollonian gaskets*, in honor of Apollonius of Perga. Apollonian gaskets are fascinating patterns obtained by starting with three mutually tangent circles and then recursively inscribing new circles in the curvilinear triangular regions between the circles.

We start with three circles of any size, with each one touching the other two, and then we draw a larger circle snugly enclosing them, as well as a smaller circle that fits snugly in the space between them. This creates four roughly triangular spaces between the circles. In each of those four spaces, we draw a new circle that kisses all three sides. We keep on going forever, or at least until the circles become too small to see. The resulting foam-like structure is an Apollonian gasket.

Figure 0.2. Example of an integral Apollonian gasket, where the integer inside any circle indicates the curvature of that circle. All curvatures are positive except for that of the outermost bounding circle, which has curvature −1. Every trio of mutually tangent circles has two other circles to which the three are tangent, and their curvatures satisfy equation (0.5). An example is the Descartes configuration with curvatures 2, 3, 2, −1, 15, easily visible in the figure.

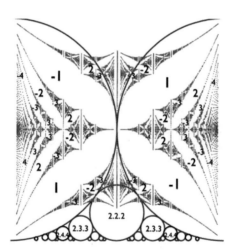

Figure 0.3. The butterfly with gaps labeled by integers that are quantum numbers associated with the quantum Hall effect. The lower part of the figure shows an underlying Apollonian gasket of the type shown on the right side of figure 0.1 (explained in chapter 2), where the product of the three integers inside the circle equals the curvature of the circle. For hierarchical aspects of this type of Apollonian packing, we refer readers to a short paper by Greg Huber [8].

One of the most remarkable aspects of an Apollonian gasket is that if the first four circles have *integer* curvatures, then every other circle in the packing will, as well. This fascinating result can be seen by adding the two roots $\kappa_0(-)$ and $\kappa_0(+)$ of equation (0.2):

$$\kappa_0(-) + \kappa_0(+) = 2(\kappa_1 + \kappa_2 + \kappa_3). \tag{0.5}$$

This linear equation implies that if κ_1, κ_2, κ_3 and, say, $\kappa_0(-)$ are integers, then $\kappa_0(+)$ will also be an integer. Figure 0.2 illustrates the construction of one of these integer wonderlands, which are called *integral Apollonian gaskets*. It turns out that there are infinitely many distinct integral Apollonian gaskets, where the curvature of every single circle at every single level of nesting is an integer. These are fractals that are "made of integers", so to speak, and their fascinating properties are discussed in considerable detail in chapters to come.

The butterfly landscape, first found by Douglas Hofstadter in 1976 [7], and later discovered to be capable of being decorated with integers (see figure 0.3), is subtly related to an integral Apollonian gasket. The integers used in decorating the butterfly are, in fact, the quantum numbers associated with Hall conductivity, an exotic physical phenomenon where the quantization of electrical resistance occurs with astonishing precision. This book tells the tale of the marriage of the Apollonian gaskets' *kiss precise* to the *precise quantization* of the Hall effect.

Appendix: An Apollonian sand painting—the world's largest artwork

Black Rock Desert in Nevada hosts a sand painting of circumference over nine miles—the world's largest single artwork. It is visible from 40 000 feet in the sky, as is shown in the photo that opens part 1 of this book. Created in 2009 by Jim Denevan and a team of three colleagues—Caleb, Nick, and Zach—the painting consists of one thousand circles that form an Apollonian gasket. It was constructed with a roll of chain-link fence six feet across, which was dragged repeatedly around by a truck, thus digging trenches into the desert. The result was a set of circular lines etched into the sand that are 28 feet wide and almost three feet deep in places. Using GPS technology to give them points on the circumferences of perfect circles, the artists took fifteen days to complete their "painting".

Jim Denevan is planning his next project in Antarctica, and he also hopes to push the boundaries yet further, by collaborating with NASA to produce artworks on the plains of Mars.

References

[1] Mackenzie D 2010 A tisket, a tasket, an Apollonian gasket *Am. Sci.* **98** 10 [this is the link http://www.americanscientist.org/libraries/documents/20091241321 07602-2010-01CompSci_MacKenzie.pdf]
[2] Descartes R 1901 *Oeuvres de Descartes Correspondence IV* ed C Adam and P Tannery (Paris: Léopold Cerf)
[3] http://plato.stanford.edu/entries/elisabeth-bohemia/
[4] http://euler.genepeer.com/from-herons-formula-to-descartes-circle-theorem
[5] http://www.shaba.co/wa?s=Apollonian_Gasket#D3_symmetry
[6] Soddy F 1936 *Nature* **137** 1021
[7] Hofstadter D R 1976 Energy levels and wave functions of Bloch electrons in rational and irrational magnetic fields *Phys. Rev.* B **14** 2239
[8] Huber G 1990 in *Correlations and Connectivity* ed H E Stanley and N Ostrowsky (Dordrecht: Kluwer) p 322

IOP Concise Physics

Butterfly in the Quantum World
The story of the most fascinating quantum fractal
Indubala I Satija

Chapter 1

The fractal family

> *Philosophy is written in this grand book (the Universe) which stands continuously open to our gaze, but it cannot be read unless one first learns to understand the language in which it is written. It is written in the language of mathematics.*
> —Galileo Galilei, 1623

This chapter opens with the Mandelbrot set, one of the most famous and most aesthetically appealing fractal objects, and it also discusses simpler fractals, such as Cantor sets, the Sierpinski gasket, and Apollonian gaskets. The butterfly plot inherits its "genes" from these fractals. The Mandelbrot set as shown herein may appear to be merely a distant cousin of the butterfly fractal, but in fact it shares the butterfly's *soul*. However, that this is the case is subtly encoded in notions of number theory. Furthermore, one of the most important aspects of the butterfly—namely, its labeling with integer quantum numbers—is rooted in integer Apollonian gaskets, which are beautiful fractal shapes consisting of circles nested infinitely deeply inside each other. In other words, the butterfly fractal—a quantum fractal and the star player in this book—has its DNA encoded in a family of classic fractals reflecting the magic, the mystique, and the simplicity of laws of nature that underlie our endlessly strange quantum/classical world.

What is a fractal?

Many readers will have already been exposed to the awe-inspiring Mandelbrot set. Shown in figure 1.1, this astonishingly complex and intricate structure has visual characteristics with universal appeal, and it shares certain features with the butterfly fractal.

Some of us are reminded by the Mandelbrot set of the excitement we felt when we realized how greatly our world had expanded when we began to understand Euclidean geometry, calculus, or some other powerful mathematical concept. Fractals are indeed the end product of some deep and exciting mathematical ideas, and yet, like many great mathematical discoveries, they also have a feeling of child's play.

The very first explorations in this area of mathematics were made in the latter part of the 19th century, in an attempt to resolve a conceptual crisis faced by mathematicians when they encountered functions that were non-differentiable. Some of the pioneers in those early days were Paul du Bois-Reymond, Karl Weierstrass, Georg Cantor, Giuseppe Peano, Henri Lebesgue, and Felix Hausdorff. The important discoveries made by these and other mathematicians helped the Polish-French mathematician Benoît Mandelbrot, several decades later, to invent a new branch of mathematics that describes shapes and patterns in nature, shapes that fall outside the framework of Euclidean geometry as previously understood. As Mandelbrot put it in his book *The Fractal Geometry of Nature* [1]: "Clouds are not spheres, mountains are not cones, coastlines are not circles... Nature has played a joke on the mathematicians. The 19th-century mathematicians may have been lacking in imagination, but Nature was not...".

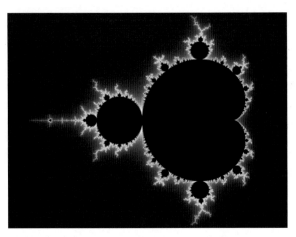

Figure 1.1. The Mandelbrot set. In this figure, the horizontal and vertical axes correspond, respectively, to the real and imaginary parts of the parameter c in equation (1.1). (Wolfgang Beyer/CC-BY-SA-3.0. https://en.wikipedia.org/wiki/Wikipedia:Featured_picture_candidates/Mandelbrot_set#/media/File:Mandel_zoom_00_mandelbrot_set.jpg)

In the 1970s, Mandelbrot coined the term *fractal*, which he derived from the Latin verb *frangere*'s past participle *fractus*, meaning "fragmented" or "irregular". The term was intended to describe objects and shapes that are crinkled and fragmented at all scales, no matter how tiny. In fractals, there is always something more to be seen as you look at them at smaller and smaller scales. In other words, they exhibit never-ending structure at all scales from very large to very small. In general, fractals are shapes whose dimensionality is not a whole number. But what could that possibly mean?

If you measure the length of a perfectly straight fence first in inches and then in feet, the first answer will obviously be 12 times greater than the second. This is because a fence is a *one*-dimensional entity. On the other hand, if you measure the area of a field first in square inches and then in square feet, the first answer will obviously be 144 (that is, 12^2) times greater than the second. This is because a field is a *two*-dimensional entity. Is it conceivable that there is some kind of shape that lies in between a fence and a field? Could the ratio of two different measurements of such a shape possibly yield a clue to its strange dimensionality?

Mandelbrot opens his book with the provocative question, "How long is the coast of Britain?" So let us imagine two surveyors, one with a ruler an inch long and another with a ruler a foot long, who both set out to measure the length of the coast of Britain. After they have completed their respective tasks and are comparing notes at the office, they expect, quite reasonably, that one answer will be 12 times greater than the other. They are enormously surprised, however, when they find that the first answer is in fact very close to $12^{1.3}$ times as big as the second answer. Mandelbrot showed that it was also true that the same coastline, when measured first in feet and then in yards, would give answers whose ratio was very close to $3^{1.3}$ (where the exponent, 1.3, is the same as before). He concluded that this meant that the British coast is not like a fence, whose dimension is 1, nor like a field, whose dimension is 2, but that instead it has an *intermediate* dimension of roughly 1.3. Mandelbrot thus coined the word "fractal" to denote any object having non-integral dimension, in this sense.

1.1 The Mandelbrot set

Benoît Mandelbrot formulated the concept of fractals in the 1960s and 1970s, although his first published work on the subject didn't appear until the mid-1970s. Ironically enough, however, it wasn't until 1980 that he discovered the fractal that bears his name, and which today is probably the most famous fractal in the world. We now will take a look at where it comes from.

Consider a sequence of complex numbers $z_0, z_1, z_2, z_3, z_4, \ldots$ (where $z = x + iy$, and where x and y are real numbers), defined as follows:

$$z_{n+1} = z_n^2 + c = f(z_n), \tag{1.1}$$

Here, c is a fixed complex number, and $z_0 = 0$. Each number z_n is fed into the very simple quadratic function f, and this gives a new number z_{n+1}, which in turn is fed back into f, and around and around it goes. Where will these numbers meander in the complex plane as the "self-feeding" process of iteration is repeated over and over again, *ad infinitum*? What kind of asymptotic behavior will they exhibit as n goes to infinity?

The way such an iteration behaves is of course totally determined by the choice of the parameter c. There are some values of c such that the sequence $\{z_n\}$ will diverge to infinity, and there are others for which it will remain bounded—in fact, it will never exceed 2 in absolute value—no matter how large n grows. This is a very fundamental distinction between two types of starting point c, and the Mandelbrot set is all about this difference. (We refer to c as a "starting point" since when the iteration is initiated, the very first spot it lands on, after 0, is the number c.)

Indeed, the Mandelbrot set is defined as the set of all values of c for which the sequence $z_0, z_1, z_2, z_3, z_4, \ldots, z_n, \ldots$ *remains bounded* for all n. In other words, it is the set of those complex numbers c for which the images of 0 under the iteration of the quadratic mapping $z \to z^2 + c$ remain forever bounded.

The sequence $z_0, z_1, z_2, z_3, z_4, \ldots$ is sometimes called "the orbit of 0", since the initial value is 0 and the subsequent pathway in the complex plane is something like the orbit of a planet—or rather (since the sequence is a set of discrete points rather than a continuous curve), it is like a set of periodic snapshots of a planet that is following a meandering orbit in the sky.

With the aid of a computer, one can check that $c = 0, -1, -1.1, -1.3, -1.38$, and i all lie in the Mandelbrot set, whereas $c = 1$ and $c = 2i$ do not. It turns out that for $|c| \leq 2$, if the orbit of 0 ever lands outside the circle of radius 2 centered at $c = 0$, then the orbit will inevitably tend to infinity. Therefore, the Mandelbrot set is a compact subset of the disc $|c| \leq 2$.

Some orbits, even if they are bounded, never settle down into a regular (periodic) pattern; they jump around rather randomly forever. Other orbits, however, eventually do settle down, drawing closer and closer to a single point, which is called the *fixed point* of the iteration. Yet other orbits gradually settle down into an alternation between just two points, bouncing back and forth between them. In other words, some orbits tend to become periodic, in the limit, while others do not.

A perfectly periodic orbit that comes precisely back to itself after n steps is called a *period-n cycle*. The simplest case is of course if $n = 1$, so that the orbit consists of just one point—a fixed point of the mapping. The next-simplest case is that of a period-2 cycle, where the orbit jumps back and forth between two different points in the plane. And for each positive integer n, there are period-n cycles in the iterative process that defines the Mandelbrot set.

Most orbits, however, are not exactly periodic, but many orbits *approach* periodic behavior as n grows large. In other words, there are many sequences $z_0, z_1, z_2, z_3, z_4, \ldots$ that grow closer and closer to a periodic orbit whose period is some positive integer n. Such limiting orbits are called *attractors*, or more technically, *period-n attractor cycles*.

The set of c-values for which the iteration $z_{n+1} = z_n^2 + c$, when launched from initial value 0, does not go to infinity is the Mandelbrot set. Given a specific value of c, the set of points that are hit along the way is known as a *Julia set*, after the French

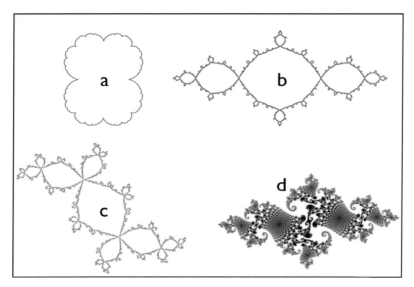

Figure 1.2. Parts (a), (b), and (c) of this figure show the Julia sets for $c = \frac{1}{4}$, -1, and $-0.123 + 0.745i$, with corresponding fractal dimensions 1.0812, 1.2683, and 1.3934, respectively. Part (d) shows a set of dimension 2, where the c-values include the boundary of the Mandelbrot set.

mathematician Gaston Julia. For a stunning view of the Mandelbrot set and some Julia sets, we encourage readers to watch the delightful animation [2], where each frame is a magnification of the previous frame, so more and more detail is visible, revealing new structures more clearly than words could.

Figure 1.2 shows some examples of Julia sets, along with their fractal dimensions. The fractal dimension (also known as the "Hausdorff dimension") of the boundary of the Mandelbrot set is equal to 2. In general, one needs to use a more abstract notion of dimension (beyond the scope of this book) to quantify the fractal aspect of the boundary of the Mandelbrot set and the corresponding Julia sets. Although Julia sets are fascinating and beautiful, we will not devote further attention to them here, since our main goal is to learn about the butterfly fractal, and that fractal has numerous provocative parallels with the Mandelbrot set.

The Mandelbrot set consists of a large cardioid-shaped region (see figure 1.3), off of which sprout numerous (in fact, infinitely many) "bulbs". The cardioid is the region of values of c such that, as n goes to infinity, the sequence z_n converges to a single point—the *fixed point* of the map—which, as was mentioned earlier, is called a "period-1 attracting cycle". What this means is that as n grows larger and larger, the points z_n in the orbit draw closer and closer together, eventually converging to a single value, denoted by z_∞. Symbolically, $z_n \to z_\infty$ as $n \to \infty$.

The boundary of the Mandelbrot cardioid is obtained by solving the simple quadratic equation defining such periodic behavior—namely $z = z^2 + c$. For the iteration to be stable at any such point, it is necessary and sufficient that the derivative of $f(z)$ should be less than or equal to 1 in absolute value. Therefore, to

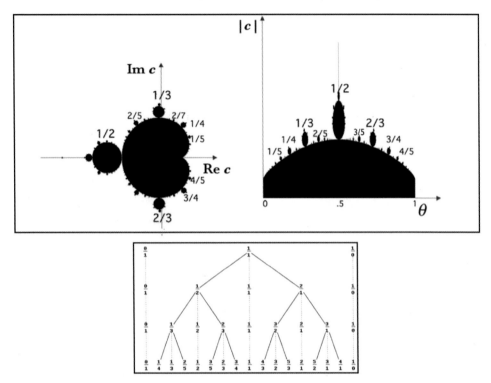

Figure 1.3. The upper two graphs show the Mandelbrot set in two different ways. In each plot one sees bulbs tangent to the cardioid, each one labeled with a rational number p/q. There is a unique bulb for each rational number p/q between 0 and 1. These fractions are ordered in a very interesting and very natural way as one rotates around the cardioid. The cardioid's boundary, although curved, can be thought of as a real line, where the complex numbers $c_{p/q}$ along it are arranged in the same order as the fractions p/q forming a *Farey sequence*. This number-theoretical notion is illustrated in the lower part of the figure, which shows the so-called *Farey tree*, each of whose rows is defined by first copying the row above it, then taking the *Farey sum* of each pair of neighboring fractions. The Farey sum of two fractions $\frac{p}{q}$ and $\frac{r}{s}$ is defined to be $\frac{p+r}{q+s}$. (Of course it is not a sum in the normal sense of the term; this is an extension of the word.)

find the values of c that form the cardioid's boundary, all we need to do is solve the following two simultaneous equations:

$$c = z - z^2$$
$$|f'| = 2|z| = 1.$$

The lower equation, $2|z| = 1$, implies that the complex number z can be written in polar form as $z = \frac{1}{2}e^{2\pi i\theta}$, where $2\pi\theta$ is a real number equal to the angle that z makes with the real axis (as measured in radians). Substitution of $z = \frac{1}{2}e^{2\pi i\theta}$ into the upper equation lets us solve for z. We will denote the solutions, which are parametrized by the real number θ, as c_θ, where $c_\theta = \frac{e^{2\pi i\theta}}{2}(1 - \frac{e^{2\pi i\theta}}{2})$.

The real number θ parametrizes the boundary of the cardioid. As θ runs from 0 to 1, the complex number c_θ swings in a clockwise direction around the boundary of the cardioid. For each rational value $\frac{p}{q}$ of θ along the way, there is a bulb that is tangent to the main cardioid. These bulbs consist of attracting cycles having period q, where $\frac{p}{q}$ describes the rotation number of the bulb as $z_{q+t} = z_t + p$. Here, t is the number of iterates needed to converge to the attracting cycle. Let $c_{\frac{p}{q}}$ be the parameter value that hosts a bulb of rotation number $\frac{p}{q}$, with p and q relatively prime.

$$c_{\frac{p}{q}} = \frac{e^{2\pi i \frac{p}{q}}}{2}\left(1 - \frac{e^{2\pi i \frac{p}{q}}}{2}\right). \tag{1.2}$$

As the parameter c moves out of the central cardioid into a bulb, the behavior of the orbit $z_0, z_1, z_2, z_3, z_4, \ldots$ changes in character. Whereas inside the cardioid, the orbit always approaches a 1-cycle (a fixed point), inside the bulb labeled by $\frac{p}{q}$, the orbit always approaches a q-cycle (a set of q distinct points that are traced out over and over again). This kind of mutation in the behavior of an iteration, where, when some parameter crosses a critical boundary, a stable period-1 cycle loses its stability and gives rise to a *longer* stable attracting cycle of period q, is an extremely important phenomenon in the study of iterative equations such as $z \to z^2 + c$, and is given the name of *bifurcation*.

Figure 1.3 shows some of the bulbs surrounding the Mandelbrot cardioid, revealing the high degree of complexity and order underlying the Mandelbrot set. We now summarize some of the key aspects of the bulbs of the Mandelbrot set, each one having a close correspondence with the butterfly fractal, as readers will come to see later in this chapter.

- Sandwiched between any two bulbs associated with the fractions $\frac{p_l}{q_l}$ and $\frac{p_r}{q_r}$, the next largest bulb is associated with the fraction $\frac{p_c}{q_c} = \frac{p_l + p_r}{q_l + q_r}$. This way of "adding" fractions is known as "Farey addition" and is illustrated in the lower part of figure 1.3.
- A period-q bulb has $q - 1$ "antennae" at the top of its limb.
- Numerical experiments have shown that the radii of these bulbs tend to zero like $\frac{1}{2q^2}$.

1.2 The Feigenbaum set

The Feigenbaum set, discovered in 1978 [3] by physicist Mitchell Feigenbaum (two years before the Mandelbrot set was found), is closely related to the Mandelbrot set. This set is defined by iterating the quadratic polynomial $f^F(x) = x^2 + c$, where both x and c are real. Note that this is the very same equation by whose iteration we defined the Mandelbrot set, except that in the latter case, the variable z and the constant c were complex. Here, however, we limit ourselves to real numbers. The phenomena will nonetheless be very rich and astonishing.

What Feigenbaum discovered, using a computer to do experimental mathematics, was that there are certain orbits of x-values that have period 1, and there are other orbits of period 2, and so forth. When the parameter c is slowly changed, there will be critical moments when some orbits of period 1 will *bifurcate* and become orbits of period 2. Likewise, there are some critical values of *c* where certain orbits of period 2 will bifurcate, becoming orbits of period 4—and so forth. This *period-doubling* behavior is the most important aspect of the iteration that Feigenbaum studied.

Beyond a critical value (where the fractal dimension of the Feigenbaum set is approximately 0.538), the iteration exhibits *chaotic* behavior in addition to having small windows where the orbits are periodic (corresponding to bulbs on the real axis of the Mandelbrot set, where the cycle length is not a power of 2). So-called "chaotic behavior" of the iteration is characterized by a *Lyapunov exponent*, which we will denote as γ, with $\gamma > 0$. This number quantifies the rate of exponential divergence of two nearby points at a distance Δx as we iterate the map n times.

$$\Delta x_n \approx \Delta x_0 e^{\gamma n} \qquad (1.3)$$

1.2.1 Scaling and universality

Table 1.1 exhibits the sequence of *bifurcation points* $c_0, c_1, c_2, \ldots$, which are those values of *c* where orbits double their periods—more specifically, for each n, c_n is that special value where the period changes from 2^n to 2^{n+1}. In 1978, Feigenbaum observed that as the period of the orbits N approaches infinity, the corresponding values c_N approach a finite limiting value. Moreover, to his great surprise, he discovered that this convergence was characterized by what is called a *power law*:

$$c_N - c_\infty \approx \delta^{-N}, \qquad \delta = 4.6692016091029909\ldots \qquad (1.4)$$

This constant δ is known as "Feigenbaum's number".

Feigenbaum later showed, even more surprisingly, that *all* one-dimensional iterations that have a single quadratic maximum (i.e. that are shaped like a parabola at their peak) are characterized by exactly the same value of δ. In other words, the details of the equation constituting the iteration don't matter; all that matters is the local behavior just at the function's very peak and nowhere else. Furthermore, this same number δ was also found to describe the period-doubling behavior of actual physical systems, such as dripping faucets and other phenomena involving fluids.

Table 1.1. Convergence of period-doubling bifurcation points and the universal ratio δ.

N	Period $= 2^N$	Bifurcation parameter c_N	$\delta = \dfrac{c_{N-1} - c_{N-2}}{c_N - c_{N-1}}$
1	2	−0.75	N/A
2	4	−1.25	N/A
3	8	−1.3680989	4.2337
4	16	−1.3940462	4.5515
5	32	−1.3996312	4.6458
6	64	−1.4008287	4.6639
7	128	−1.4010853	4.6682
8	256	−1.4011402	4.6689
∞	∞	−1.4011551890...	

Feigenbaum's discoveries thus provide one of the simplest examples of *universality*, a dream scenario for theoretical physicists, where very simple models, often called "toy models", can predict the behavior of far more complex systems. The scaling ratio δ is a *critical exponent*—a number that usually characterizes a transition point. In the example above, it is a number that describes the so-called "onset of chaos"—the transition from regular to chaotic dynamics.

The concepts of *scaling* and *critical exponents* originated in physics—specifically, in the theory of phase transitions (such as the transition from a magnetically disordered state to an ordered state). Later, these concepts were found to apply to many other areas of physics, ranging from turbulence all the way to particle physics. Originally introduced in 1960 and 1970 respectively, scaling and critical exponents are beautiful and revolutionary ideas, and among the principal early pioneers in exploring them were Michael Widom, Michael Fisher, and Leo Kadanoff. These nascent ideas were eventually extended into a comprehensive and elegant theory by Kenneth Wilson, who discovered a calculational method to describe the phenomena, for which work he was honored with a Nobel Prize in 1982. As readers will see later in this book, these ideas also can be applied to describe the behavior of fractal objects, including topological features of the butterfly.

> *Theorists should study simplified models.*
> *They are close to the problems we wish to understand.*
> —Leo Kadanoff (1937–2015)

1.2.2 Self-similarity

Another important feature of the Feigenbaum set is that the sequence of values c_N, shown in table 1.1, is a *self-similar set*. What does "self-similar" mean? A pattern or set is said to be self-similar if it is exactly similar (in the sense of Euclidean geometry) to some *part* of itself (i.e. the whole has the same shape as one or more of its parts). (The property of self-similarity is also frequently referred to as *scale invariance*.)

A trivial case of self-similarity is provided by a straight-line segment, since it looks just like any shorter segment that it contains. However, no further detail is revealed by blowing up such a segment. By contrast, a true fractal shape exhibits non-trivial self-similarity, because any time any part of it is blown up, new detailed structure is revealed. This process of revelation of finer and finer structure continues forever, on arbitrarily small scales.

The self-similarity of the Feigenbaum set is due to the geometric scaling of the parameter intervals, as is shown in equation (1.4). All of the infinitely many intervals $(c_{n+1} - c_n)$ are identical, once they are scaled by the factor δ between two consecutive iterations.

Many entities in the real world, such as coastlines, possess a more abstract type of self-similarity—that is, where certain parts of the whole shape are not precise copies of it, but have much in common with it. This is sometimes called *statistical self-similarity*, which means that although such a shape is not a precise larger copy of

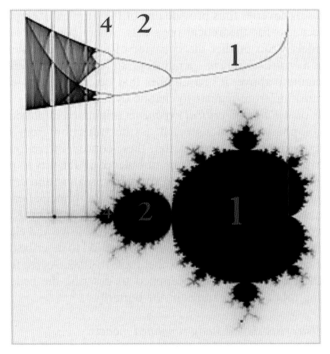

Figure 1.4. An illustration of the subtle relationship between the Mandelbrot and Feigenbaum sets, which are defined with the help of the equations $f^M(z) = z^2 + (c_R + ic_I)$ and $f^F(x) = x^2 + c_R$, respectively. Note that the first equation involves complex numbers, while the second involves only real numbers. The Mandelbrot set is the graph of c_R versus c_I when those numbers are such that the sequence $\{z_n\}$ remains bounded, while the Feigenbaum set is the graph of c_R versus x-values, once the x-iterates have settled on an attracting set. The two sets share the c_R values—the horizontal axis—and hence the parametric window of those c_R that correspond to the period-doubling of the Feigenbaum set also describes the horizontal projections of the two-dimensional parameter space (c_R, c_I), where the Mandelbrot set exhibits period-doubling. (https://commons.wikimedia.org/wiki/File:Verhulst-Mandelbrot-Bifurcation.jpg)

smaller parts of itself, its parts at many scales nonetheless are characterized by *statistical properties* that are identical to those of the full object.

Strictly speaking, the Mandelbrot set is not precisely self-similar, except at some points. A very enlightening comparison of the Mandelbrot and Feigenbaum sets can be made by superimposing them, as is shown in figure 1.4. The two distinct graphs share only the horizontal axis (real values of the parameter c), since the vertical axis in the Feigenbaum graph shows the values that the iterations eventually settle into, once they have gone into an attracting cycle. The red numerals show how all the x-values where the Feigenbaum set exhibits period-doubling correspond to special regions of the Mandelbrot set.

1.3 Classic fractals

We will now describe some simpler fractals that will prove to be essential in understanding the butterfly fractal. Some of these fractals exhibit self-similarity as defined above.

1.3.1 The Cantor set

We start with one of the first fractals ever discovered—the *Cantor set*, which is sometimes more picturesquely called a "Cantor dust". This notion is named after the German mathematician Georg Cantor, who introduced the idea in 1883.

In mathematics the art of asking questions is more valuable than solving problems.
—Georg Ferdinand Ludwig Philipp Cantor (1845-1918).

The simplest, most canonical case of a Cantor set is the 1/3-Cantor set (figure 1.5). Consider the interval [0,1]. (The square brackets mean that the interval includes its left and right endpoints: the same interval without its endpoints would be written "]0,1[".) Now erase the interval's middle third,]1/3,2/3[, leaving two smaller intervals, [0,1/3] and [2/3,1]. Now erase the middle third of each of these smaller intervals—and keep on carrying out this "elimination of the middle third" over and over again. If you do so with a physical pencil and a physical eraser, there is obviously a practical limit to how far you can carry such a process, but in the mind of a mathematician, it can be repeated indefinitely. What is left at the end (after an infinite number of eliminations of middle thirds) is merely a "dust"—an infinite (in fact, uncountably infinite) subset of the interval [0,1], but which does not contain any interval of non-zero length. In other words, all the uncountably many points in the Cantor set are disconnected—and yet it turns out this "pathological" set (as such sets were once called by skeptical mathematicians in the late 19th century) has non-zero dimension.

Each time we erase a middle third, the total remaining length (called the *measure* of the set) decreases by a factor of 2/3. The measure thus starts out at 1, then

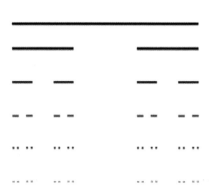

Figure 1.5. The construction process that produces the simplest fractal: the "1/3"-Cantor set. We begin with the line segment shown at the top, and then we chop out its middle third (the result of this is shown just below the segment). Then we eliminate the middle thirds of the two remaining segments. As we move downwards, we keep on eliminating more and more middle thirds. The end result of doing this infinitely many times would be the Cantor set. Of course the Cantor set cannot be drawn, since, although it consists of infinitely many points, it contains no intervals at all. So you just have to imagine the "slim pickings" that would be left at the very end.

becomes 2/3, then 4/9, then 8/27 etc. The measure clearly tends to zero, so the dimension of the dust is less than 1. The fractal dimension d_c of this Cantor set turns out to be

$$d_c = \ln 3 / \ln 2 \approx 0.63. \tag{1.5}$$

As is quite obvious from the way it is constructed, the 1/3-Cantor set is exactly self-similar.

1.3.2 The Sierpinski gasket

We now move from the Cantor set, which is a subset of a line, to the Sierpinski gasket, which is a subset of a plane. (The dictionary definition of "gasket" is "tight seal". Mathematicians use the term to describe a shape that is defined by its "holes", sealed tightly by the set.)

To define this strange object, we start with an equilateral triangle. We mark the midpoint of each side, and draw a new triangle whose vertices are those three midpoints. In so doing, we will cut the original triangle into four congruent pieces, the middle one of which is upside-down. Now we throw away the middle triangle (this is of course reminiscent of eliminating the middle third). Now we focus on the three remaining triangles, and carry out the same elimination process inside each one of them. As with the Cantor set, we will keep on repeating this process infinitely many times. In the end, we will have constructed a fractal in the Euclidean plane.

Figure 1.6 shows the construction of the Sierpinski triangle or gasket. The fractal dimension of the Sierpinski gasket is, once again, $\ln 3 / \ln 2$, and the Sierpinski gasket is likewise a fractal that enjoys the property of exact self-similarity.

Another fractal that bears a strong similarity to the Sierpinski gasket is sometimes called the *curvilinear Sierpinski gasket*; a few stages in its construction are shown on the right side of figure 1.6. It is made by an iterative process of removing circles from an initial area that is a concave curvilinear triangle. This type of fractal, also known

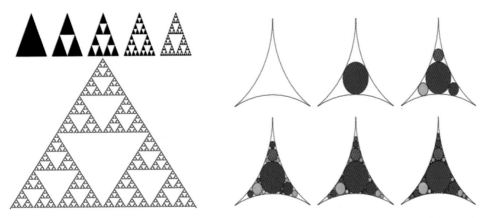

Figure 1.6. This figure shows the construction process that gives rise to both the Sierpinski gasket (left) and the curvilinear Sierpinski gasket (right). (The sole purpose of the colors is to give aesthetic pleasure.)

as an Apollonian gasket, was briefly touched on in the previous chapter, and it has a close relationship with the butterfly fractal, as we shall see later in this book.

We encourage readers to watch the following amusing video (especially with little kids around!) [4].

1.3.3 Integral Apollonian gaskets

Integral Apollonian gaskets, already discussed in chapter 0, are exquisite fractals made out of integers alone. To discuss such entities with precision, we need to define the *curvature* of a circle. If a circle has radius R, its curvature is defined to be $1/R$. With this definition, we begin our examination of Apollonian gaskets by imagining four mutually tangent circles whose curvatures κ_0, κ_1, κ_2, κ_3, satisfy the following equation:

$$\kappa_0 = \kappa_1 + \kappa_2 + \kappa_3 \pm 2\sqrt{\kappa_1\kappa_2 + \kappa_2\kappa_3 + \kappa_3\kappa_1}. \tag{1.6}$$

The two solutions, which we will denote by $\kappa_0(+)$ and $\kappa_0(-)$, satisfy the following linear equation:

$$\kappa_0(+) + \kappa_0(-) = 2(\kappa_1 + \kappa_2 + \kappa_3). \tag{1.7}$$

To understand how the above two equations give us an integer Apollonian fractal, we note that in figure 1.7,
- With $\kappa_1 = -1$, $\kappa_2 = 2$, $\kappa_3 = 3$, we find, using equation (1.6), that $\kappa_0(\pm) = 2, 6$.
- With $\kappa_1 = 2$, $\kappa_2 = 3$, $\kappa_3 = 2$, we find, using equation (1.6), that $\kappa_0(\pm) = -1, 15$.
- With $\kappa_1 = 2$, $\kappa_2 = 3$, $\kappa_3 = 6$, $\kappa_0(-) = -1$, we find, using equation (1.7), that $\kappa_0(+1) = 23$.
- With $\kappa_1 = 2$, $\kappa_2 = 3$, $\kappa_3 = 15$, $\kappa_0(-) = 2$, we find, using equation (1.7), that $\kappa_0(+) = 38$.

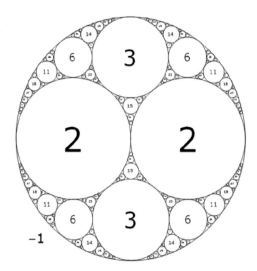

Figure 1.7. An example of an integer Apollonian gasket.

In other words, given four integral curvatures that correspond to four mutually tangent circles that satisfy equation (1.6), the whole gasket can be constructed using just the linear equation (1.7).

1.4 The Hofstadter set

The Mandelbrot and Feigenbaum fractals, along with other classic fractals discussed above, are abstract mathematical objects conceived to illustrate how structures having great complexity can emerge from very simple nonlinear equations or geometrical constructions. We now turn to a fractal set that was originally discovered in the context of quantum physics. It came from fundamental questions about crystals in magnetic fields—situations that in principle could be studied in a laboratory. Although this set came out of physics, it can nonetheless be described purely mathematically, with no reference to any concepts of physics. This butterfly fractal, which we will often call the "Hofstadter set", is the star character in the drama played out in this book.

Several years before fractals became famous, Douglas Hofstadter was studying how electrons in a crystal—so-called "Bloch electrons", named after Swiss-American physicist Felix Bloch—behaved in the presence of a magnetic field. At the time, this was a long-standing unresolved problem of quantum mechanics, and one reason it was so fascinating was that it concerned the behavior of an electron caught in the metaphorical crossfire of two highly contrasting types of physical situation—on the one hand, a crystal lattice, and on the other hand, a homogeneous magnetic field.

Below, we give a brief introduction to the Hofstadter set [5], bypassing all the quantum aspects of the problem that will be discussed in the coming chapters. Here we present this fractal as a purely mathematical object, along the lines of the

Mandelbrot set. The reason this is feasible is that Harper's equation—the quantum-mechanical equation that gives rise to the butterfly graph (equation (6.4))—can be recast in the form of two coupled equations [6] involving two real variables, r and θ, which together define a two-dimensional mapping:

$$r_{n+1} = -\frac{1}{r_n + E + 2\cos 2\pi\theta_n} \tag{1.8}$$

$$\theta_{n+1} = \theta_n + \phi \pmod{1}. \tag{1.9}$$

Here E and ϕ represent, respectively, the vertical and horizontal axes of the butterfly graph. The butterfly itself is the set of pairs (ϕ, E) that satisfy the following equation:

$$\gamma = -\lim_{N \to \infty} \frac{1}{2N} \log\left[\frac{r_1}{r_N}\right] = 0 \tag{1.10}$$

Equation (1.10) simply means that the Hofstadter set consists of values of ϕ and E for which the two-dimensional mapping given by equation (1.9) does not diverge, which is to say, it has a Lyapunov exponent of 0 (see equation (1.3)). (It turns out that negative Lyapunov exponents give the gaps—the empty regions in the butterfly—as is shown in figure 1.8, and those gaps are the complement of the Hofstadter set.) In the appendix to this chapter, we briefly outline the relationship between this map and the quantum map known as Harper's equation.

In spite of its complexity, there are some simple facts about this fractal. For any rational value $\frac{p}{q}$ of the flux variable ϕ, the graph consists of q band-like regions,

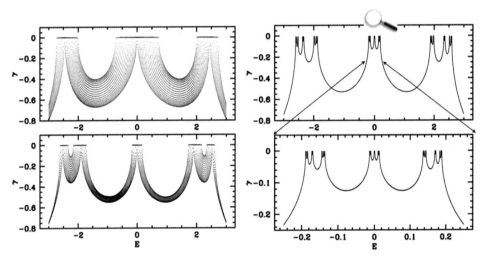

Figure 1.8. The left panel shows the energy bands for $\phi = 1/3$ (top) and $\phi = 2/5$ (bottom). These bands are values of energies E where the Lyapunov exponent $\gamma = 0$. The right panel is a graph of E versus γ illustrates the Cantor-set structure of the butterfly graph for the irrational flux-value $\phi = \frac{\sqrt{5}-1}{2}$. The lower curve is a blowup of the interval near $E = 0$, revealing the self-similar Cantor-set-like structure of the Hofstadter set. The Hofstadter set for flux-value ϕ consists of values of the energy E that correspond to $\gamma = 0$, obtained by iterating the entire set of θ_0 values in the interval $[0, 1]$.

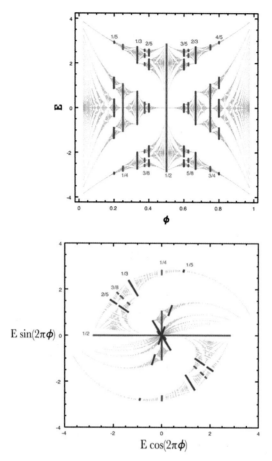

Figure 1.9. The upper graph, discovered by Douglas Hofstadter in 1976 (see [5]), shows the allowed energies E of Bloch electrons in an idealized crystal as a function of the magnetic flux ϕ in which the crystal is immersed. For a few specific rational values of ϕ, the band structure has been highlighted. Whenever the flux takes on a rational value ($\phi = p/q$), there are exactly q bands and $q - 1$ gaps (white regions) between them. The red bands and the blue bands, respectively, belong to even-q and odd-q cases. Whenever q is even, the two central bands "kiss" at $E = 0$. The lower graph shows the real versus the imaginary part of $Ee^{2\pi i\phi}$, with $0 \leqslant \phi \leqslant 0.5$.

shown as dark vertical line segments in the plot, in each of which the allowed energy varies continuously. These q subbands are separated by $q - 1$ *empty* regions— white gaps that represent forbidden energy values—as can be seen in figure 1.9.

To whet the reader's appetite for the exciting phenomena associated with the Hofstadter landscape, we note that the gaps of the butterfly fractal represent a highly sophisticated aspect rooted in destructive interference of matter waves. Such interference lies at the very heart of the quantum science describing microscopic particles. Added to this phenomenon is one of the most fascinating aspects of this graph—namely, the continuity of the gaps as ϕ varies. This is tied to some beautiful notions involving the topology of abstract quantum spaces.

1.4.1 Gaps in the butterfly

We now focus in on a key piece of terminology—namely, the word "gap", which we used intuitively above, but which needs a bit of spelling-out in order to avoid confusion in the future. In the remainder of this book, this term will be used with two closely related but subtly different meanings.

First, there are gaps in the butterfly belonging to any specific value of ϕ, such as $\phi = 1/3$. This means we are focusing our attention on the energy spectrum belonging to just one value of the magnetic flux. Such a spectrum is, by definition, always a subset of the original Bloch band (the interval $[-4, +4]$ of the energy axis, which is vertical). Thus for the flux-value $\phi = 1/3$, for example, there are exactly three smaller bands (three vertical line-segments, one above the other), which are separated by one-dimensional white zones that stretch from the top of one band to the bottom of the band just above it. Thus our first sense of the word "gap" refers to *one-dimensional gaps* in a one-dimensional spectrum.

The second sense of "gap" refers to *two-dimensional* zones in the butterfly as a whole (in other words, we are no longer limiting ourselves to just one value of ϕ). Basically, this second sort of gap is a diagonal white swath crisscrossing a large (or a small) portion of the graph. Such a gap is thus a *two-dimensional* region; in fact, it is the union of an infinite number of one-dimensional gaps belonging to different values of ϕ. Thus, for example, the four very salient white wings that meet at the butterfly's center (and that give the butterfly its name) are gaps of this second sort; indeed, *all* the gaps of this second sort are "wings" of smaller butterflies nested at various hierarchical levels inside the large butterfly. In future chapters, we will often use the word "swath" to refer to gaps of this second sort.

1.4.2 Hofstadter meets Mandelbrot

Figure 1.9 shows two different representations of the Hofstadter set, just as the right and left sides of figure 1.3 show two ways of displaying the Mandelbrot set. This sets the stage for an interesting comparison between the two sets. The two variables r and θ in the Hofstadter set play roles that are analogous to the roles played by the two real variables that define the complex variable z involved in defining the Mandelbrot set, and the two parameters E and ϕ can likewise be viewed as counterparts to the two real numbers that define the complex parameter c of the Mandelbrot set. The condition of a zero Lyapunov exponent in the Hofstadter set is analogous to the requirement on the variable c that the values of the z_n sequence should remain bounded.

An interesting common feature of the two sets is the emergence of regions (bulbs in the Mandelbrot set; bands in the Hofstadter set) associated with rational numbers. More technically put, a striking commonality between the Hofstadter and Mandelbrot sets is the Farey organization of periodic regimes. Therefore, notions from number theory turn out to play a key role in descriptions of both the Mandelbrot set and the Hofstadter set.

On the other hand, the Mandelbrot and the Hofstadter sets are mathematically quite different. The Mandelbrot set is a *dissipative* set, meaning that a blob consisting of an initial set of points shrinks down, in the iteration process, to a few disjoint sets of points that define attracting cycles, whereas the Hofstadter set is not dissipative. To obtain the entire butterfly graph, one needs to consider all possible initial values of r and θ. In contrast, recall that the Mandelbrot set is obtained by using $z = 0$ as the sole initial condition. We highlight the common aspects of these two fractals below.

- The *bulbs* in the Mandelbrot set are analogous to the *energy bands* in the butterfly fractal.
- A period-q bulb with $q - 1$ "antennae" at the top of its limb corresponds to the spectrum belonging to a rational value of ϕ with denominator q, which consists of q bands with $q - 1$ gaps between them.
- A quantitative parallel between the two sets is tied to the fact that the radii of the bulbs tend to zero like $\frac{1}{2q^2}$—that is, like the radii of the Ford circles that determine the horizontal size of the butterflies—as will be explained in chapter 2.
- The periods of the bulbs correspond to the topological quantum numbers—the Chern numbers of the butterfly.

The butterfly fractal exhibits many familiar characteristics explained in the context of the simpler fractals described above, and this fact will be highlighted through the book. As is illustrated in figure 1.8, the butterfly spectrum for any irrational value of ϕ is a Cantor set, where the total length (or more technically, the *Lebesgue measure*) of allowed energies is zero.

1.4.3 Concluding remarks: A mathematical, physical, and poetic magπ

The Hofstadter butterfly is the most central member of a rich family of fractals that is not yet very well known, despite the large worldwide community of fractal enthusiasts. As described in this book, the Hofstadter butterfly is a mathematical, physical, and perhaps also a poetic "Magπ". Among other things, it encodes the mathematics of nested tangent circles discovered around 300 BC by Apollonius, and also the quantum Hall effect, which was discovered in 1980. Furthermore, as readers of this book will discover, a handful of poems have been inspired by the butterfly theme. Various facets of this story will echo again and again throughout this book.

We will conclude this chapter with a note stressing the importance of the butterfly and its family, and focusing in on the fact that for irrational flux-values, the butterfly spectrum is a measure-zero Cantor set. Around 1980, this mind-boggling idea came to be known as the *Ten Martini Problem*.

The spell cast by this exotic notion has a complex history spanning several decades, and engaging a sizable community of mathematicians and physicists. The conjecture of a Cantor-set spectrum has its earliest roots in the important 1964 paper by Mark Ya Azbel' [6]. (Interested readers should check out Israeli mathematician

Yoram Last's 1995 paper "Almost Everything about the Almost Mathieu Operator" [7][1], which gives a historical summary.)

Mark Azbel lecturing to his physics class[2]

Perhaps it is important to point out that Azbel''s 1964 paper, which is often credited with conjecturing the Cantor-set hypothesis for the Harper spectrum, is not an easy paper to read, and many important aspects described there are not very transparent. However, some mathematicians and physicists who have carefully studied Azbel''s paper feel that they can clearly "smell" the Cantor set in his intricate analysis, and they thus credit him with that discovery. Azbel', without ever mentioning Cantor sets explicitly, pointed out the relationship of the problem's energy spectrum to the continued-fraction expansion of the magnetic flux-value, describing how energy bands split into sub-bands according to the denominators in the continued-fraction expansion of the magnetic flux.

According to Jean Bellissard, one of the pioneering investigators of the Harper spectrum and related phenomena, "Azbel''s paper does a serious job in describing the various levels of renormalization leading to an infinite number of gaps. It is not rigorous, it does not really prove things but he essentially understood what was going on." (private communication).

Michael Wilkinson, who has made important contributions to our understanding of the hierarchical nature of the Harper spectrum using the framework of the renormalization group, has the following things to say about Azbel''s work: "Azbel''s paper is quite remarkable. He understood the essentials of the structure

---

[1] In mathematics, the differential equation $\frac{d^2y}{dx^2} + (a - b\cos 2x)y = 0$ is known as Mathieu's equation (after French mathematician Émile Léonard Mathieu). Harper's equation can be viewed as a discrete version of Mathieu's equation, and is sometimes referred as the *almost Mathieu equation*.

[2] Freeman Dyson, in his foreword to Mark Azbel''s book *Refusenik: Trapped in the Soviet Union*, published in 1981 by Houghton Mifflin), wrote the following glowing tribute to Azbel': "Mark Azbel' is one of the genuine heroes of our time, worthy to stand on the stage of history with Andrei Sakharov and Alexander Solzhenitsyn. I met him first in Moscow in 1956 when he was shy and thin, a brilliant young physicist rising rapidly through the ranks of the Soviet scientific establishment. He and I had worked independently on the same problem in solid-state physics. His solution was more general and more powerful than mine. I knew then that he would become an important scientist. I had no inkling that he would become a famous dissident."

of the spectrum without apparently having access to numerical experiments. His method is, arguably, the first realization of a renormalisation group calculation (he produces a sequence of transformed equations of motion acting on successively longer length scales), appearing before the term was introduced in statistical mechanics. Azbel' probably didn't know the term "Cantor set", but his paper gives a nearly correct description of the form of the spectrum." (private communication)[3]

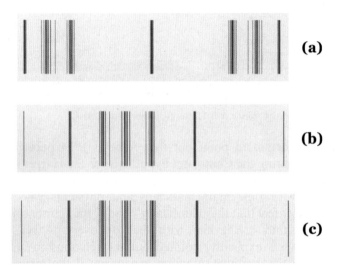

This figure shows the complex nature of the energy spectrum near an irrational magnetic flux where the energy bands form a Cantor set. Here, three panels illustrate results of a theoretical framework (renormalization group transformation). Panel (a) shows the full spectrum of Harper's equation near an irrational flux close to 3/7 where the energy spectrum (horizontal axis) forms seven clusters with complex fine structure. Panel (b) shows the second of the seven clusters after a linear transformation that is predicted to resemble Harper-like spectrum with renormalized magnetic flux. Panel (c) shows Harper-like model produced by renormalization. For further details, we refer readers to 1987 paper by Wilkinson in Selected Bibliography. (Graphs courtesy of Michael Wilkinson.)

---

[3] Below we quote Michael Wilkinson's reply when he was asked to pinpoint exactly where the "zero measure" aspect is pointed out in Azbel"s paper.
"I find Azbel"s paper extremely hard to read, but there is plenty of evidence that he understood Harper's equation very well. He does not mention 'zero measure' specifically, but rather describes a scheme to construct the spectrum that clearly leads to a zero measure set. The relevant discussion is in section 3, which discusses the semiclassical construction of the spectrum. The diagram, figure 2, implies that he considers the 4-fold symmetric case described by Harper's equation, for which the lack of open orbits implies that the leading semiclassical approximation is a discrete spectrum. At the top of page 642, column 2 he argues that the splitting, determined by tunnelling, has a universal form, parametrised by a tunnelling energy scale which is exponentially small in the transformed commensurability parameter. Because he argues that the equations describing the further splitting are universally equivalent to the ones describing the original model, his argument implies that each of the discrete levels splits into discrete levels with a smaller separation. This algorithm, involving the subdivision of a sequence of point sets, does construct a zero-measure Cantor set."

Eleven years later, Douglas Hofstadter, in his doctoral thesis [8], described in detail the nature of the spectra for rational flux-values, which he had found computationally. On the basis of those findings, he then showed that for irrationals, the spectrum must be a Cantor set of measure zero [5]. Although Hofstadter was inspired by Azbel"s paper, he did not learn about the Cantor-set aspect of the spectrum from that paper. His numerical proof of the Cantor-set conjecture for Harper's equation was the first time the notion of Cantor-set spectra was mentioned in print.

In 1981, in a talk at the annual meeting of the American Mathematical Society, mathematician Mark Kac humorously offered ten martinis to anyone who could rigorously show from first principles that the Harper equation "has all its gaps there". This led mathematician Barry Simon to name Kac's challenge "The Ten Martini Problem". In the years thereafter, many people strove to come up with a full and ironclad proof of the Cantor-set hypothesis, but the challenge was met only in 2009 by mathematicians Artur Ávila and Svetlana Jitomirskaya. All this shows the worldwide importance of the problem and the fascination and mania that have gripped a substantial community of mathematical physicists for many years.

Interestingly, the fascination with the subject continues. A recent paper by Ávila, Jitomirskaya, and C. A. Marx entitled "Spectral Theory of Extended Harper's Model and a Question by Erdös and Szekeres" [9] hints at many more mathematical treasures hidden in this problem. This study proves the Cantor-set nature of the spectrum for a wide range of lattice geometries, including setups that involve electrons hopping to both nearest-neighbor and next-nearest-neighbor sites. Their analysis requires some very delicate number-theoretic estimates, which ultimately depend on the solution of a problem posed way back in 1950 by the renowned Hungarian mathematicians Paul Erdös and George Szekeres.

We conclude this chapter with two paragraphs from a letter that Jean Bellissard, already quoted above, sent me recently. In them, Professor Bellissard's enthusiasm and passion for the subject come through vividly as he reminisces about the mathematical intricacies that underlie this problem:

The Ten Martini problem was more challenging, though. Not only did it ask about whether the spectrum was a Cantor set for ALL irrational flux-values, but also it was addressing implicitly the question of the nature of the spectral measure. The main new tool of study came during the last 15 years with the development of the theory of cocycles. The earliest hint in this direction came from Michel Herman before he died, followed by Raphael Krikorian and Hagan Eliasson. Another breakthrough came with the work of Yoccoz (another former student of Michel Herman) on Siegel disks; Yoccoz introduced a technique, due to Brjuno, for including all possible irrational numbers that were inaccessible before. Artur Ávila changed the game by developing the theory of cocycles. Svetlana Jitomirskaya, who worked for a long time with Yoram Last, then with Bourgain, jumped on this wagon and was able to finish the job with Ávila. Today, there are still tiny corners left over for which we do not know the nature of the spectral measure, but it is almost tight.

What is remarkable is that this problem has been worked on by a very large number of scientists, both in the physics community and in the mathematical community as well. *I once listed 200 seminal papers from physicists that could be counted as important, and I realized then that most of the leaders of solid-state physics had contributed to the problem. The mathematical community dealing with the problem used techniques coming from dynamical systems, from C*-algebras, and from PDE's, to fill up the multiple holes that remained over time. It is a remarkable topic. And the consequences will last for a very long time.*

To take wine into our mouths is to savor a droplet of the river of human history.
—Clifton Fadiman

Appendix: Harper's equation as an iterative mapping

As will be discussed in chapter 6, the difference equation known as "Harper's equation", discovered in the mid-1950s, describes the quantum-mechanical behavior of an electron in a crystal lattice in the presence of magnetic field. This equation—a version of the Schrödinger equation in a mathematically idealized situation—is as follows:

$$\psi_{n+1} + \psi_{n-1} + 2\cos(2\pi n\phi + k_y)\psi_n = E\psi_n \qquad (1.11)$$

Here, ψ_n is the so-called *probability amplitude* for finding the electron at the crystal lattice site n. (As will be explained in chapter 5, in contrast to the classical world, where a particle is described by its position and momentum, in the quantum world, a particle is described by a wave function whose square gives the "probability amplitude" for finding the particle at a given position.)

These days, the problem of a Bloch electron in a uniform magnetic field, as described by Harper's equation, is frequently referred to as "the Hofstadter problem", although occasionally, some articles also refer it as "Azbel–Hofstadter" problem. In fact, in recent literature, many people have taken to calling Harper's equation "the Hofstadter model". This vast oversimplification troubles Douglas Hofstadter, who rightly points out that he does not deserve any credit for coming up with the equation (after all, he was only ten years old when it was published!)—just for the discovery of the nature of its spectrum (plus some ideas about its wave functions).

Without further discussion, we simply note when the constant "2" in front of the cosine term in Harper's equation is replaced by a parameter λ, then the more general equation that results also describes the energy spectrum of electrons in a one-dimensional quasicrystal, and this has been used to study localization–delocalization transitions [10].

Harper's equation can be transformed into an iterative formula (somewhat analogous to the iterative formula that underlies the Mandelbrot set) by letting $r_n = \psi_n/\psi_{n-1}$. This operation transforms equation (A.1) into the following equation:

$$r_{n+1} = -\frac{1}{r_n + E - 2\cos(2\pi\phi n + k_y)}. \tag{1.12}$$

The relationship between Harper's equation and the above iteration is an example of the "Prüfer transformation" (for further details, we refer readers to the original paper [11]). The iteration defined in equation (A.2) can be written as a two-dimensional mapping by introducing a variable $\theta_n = 2\pi\phi n + k_y$:

$$r_{n+1} = -\frac{1}{r_n + E + 2\cos 2\pi\theta_n} \tag{1.13}$$

$$\theta_{n+1} = \theta_n + \phi \quad (mod\ 1). \tag{1.14}$$

We note that the Lyapunov exponent γ of this map can take on all possible values from $-\infty$ to 0. However, the Hofstadter set consists only of those pairs (E, ϕ) for which $\gamma = 0$. The energies that fall in the gaps of the Hofstadter set correspond to $\gamma < 0$. This is illustrated for a particular value of ϕ in figure 1.8.

To obtain the entire butterfly spectrum from this formula, we iterate it starting with various initial conditions for both r and θ in the interval [0, 1]; only those eigenvalues E that satisfy equation $\gamma = 0$ will belong to the set. The union of all the allowed values of E as a function of ϕ, for all values of ϕ, is referred to as the Hofstadter set. This explicit analogy linking the Mandelbrot iteration with Harper's equation (when conceived of as an iteration) brings the butterfly fractal a bit conceptually closer to other well-known fractals, such as the Mandelbrot set.

References

[1] Mandelbrot B B 1982 *The Fractal Geometry of Nature* (San Fransisco, CA: Freeman)
[2] http://upload.wikimedia.org/wikipedia/commons/a/a4/Mandelbrot_sequence_new.gif
[3] Feigenbaum M J 1980 Universal behavior in nonlinear systems *Los Alamos Sci.* **1** 4
[4] Hart's "Infinity Elephants" video https://www.youtube.com/watch?v=DK5Z709J2eoVi
[5] Hofstadter D R 1976 Energy levels and wave functions of Bloch electrons in rational and irrational magnetic fields *Phys. Rev.* B **14** 2239
[6] Last Y 1995 Almost everything about the almost Mathieu operator. I. *XIth International Congress of Mathematical Physics (Paris)* (Cambridge, MA: International Press) pp 366–72

[7] Azbel' M Ya 1964 Energy spectrum of a conduction electron in a magnetic field *JETP* **19** 634
[8] Hofstadter D R 1975 The energy levels of Bloch electrons in a magnetic field *PhD thesis (University of Oregon)*
[9] Avila A, Jitomirskaya S and Marx C A 2016 Spectral theory of extended Harper's model and a question by Erdös and Szekeres, arxiv. 1602.05111 (unpublished)
[10] Aubry S and André G 1980 Analyticity breaking and Anderson localization in incommensurate lattices *Ann. Israel Phys. Soc.* **3** 133
[11] Ketoja J and Satija I 1997 *Physica* D **109** 70

IOP Concise Physics

Butterfly in the Quantum World
The story of the most fascinating quantum fractal
Indubala I Satija

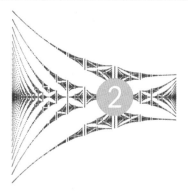

Chapter 2

Geometry, number theory, and the butterfly: Friendly numbers and kissing circles

I wish you ladies and gentleman out there knew some of this mathematics. It is not just the logic and accuracy of it all you're missing—it's the poetry too.
—Richard Feynman (BBC interview, "A Novel Force in Nature")

We begin narrating the butterfly tale by describing how the problem of four mutually tangent circles, first formulated in 300 BC, as described in chapter 0, plays an important role in the Hofstadter landscape. The fascinating geometry of four kissing circles not only underlies every butterfly in this fractal graph but also determines its recursive structure. This remarkable aspect of the story unfolds as we marinate the butterfly plot with *Ford circles* and the *Farey tree*—concepts that constitute important parts of the number theory. Therefore, the focus of this chapter is the number-theoretical aspects of the butterfly, and we leave this enchanting landscape's many connections with quantum physics for discussion in the chapters to come.

If one glimpses the butterfly for just an eyeblink, one cannot help but be struck by the convergence of four diagonal white swaths ("wings") at a central point, two of them arriving from the left and the other two from the right. If we study the graph more closely, we will see that essentially this same structure—namely, four white wings converging to a central point—appears at every scale. That is, there is a butterfly at every scale.

As can be seen in figure 2.1, showing the full butterfly, the four wings enjoy a fourfold symmetry. In other words, the Hofstadter butterfly possesses both an *up–down* symmetry (invariance under reflection in a horizontal mirror, which is the line $E = 0$) and a *left–right* symmetry (invariance under reflection in a vertical mirror, which is the line $\phi = 0$). This precise fourfold symmetry is partially lost as we zoom into the graph and look at smaller butterflies.

Consider, for example, the small butterfly whose left edge coincides with the left edge of the full butterfly (at $\phi = 0$), and whose right edge is the vertically centered member of the trio of bands at flux value $\phi = 1/3$. This butterfly's four wings all meet on the horizontal line $E = 0$, at the flux value of $\phi = 1/4$. Although the left half of this butterfly is obviously much bigger than its right half (so it lacks left–right symmetry), it still enjoys up–down symmetry (i.e. it is invariant when reflected in the

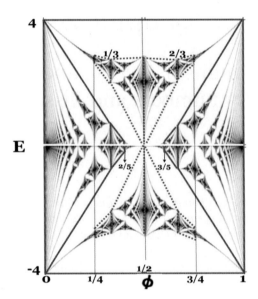

Figure 2.1. The Hofstadter butterfly, horizontally centered at $\phi = 1/2$, and enclosed in a red rectangle. The full graph possesses both a perfect *up–down* symmetry and a perfect *left–right* symmetry. This plot also highlights eight smaller butterflies inside the full butterfly. Four of them (those inside red trapezoids made of solid lines) are *central* butterflies, which possess only *up–down* symmetry, or equivalently, mirror symmetry about the axis $E = 0$ (that axis shows up as a horizontal white line exactly halfway up the plot). Also highlighted are four *off-center* butterflies (those inside red trapezoids made of dashed red lines), and these possess none of the exact symmetries possessed by the full butterfly. All central butterflies have four wings that meet at a point—a feature that is not shared by the off-center butterflies. Figure adapted with permission from [3] copyrighted by the American Physical Society.

horizontal line $E = 0$). More generally, all of the infinitely many butterflies whose centers are located on the horizontal line $E = 0$—we will call them "central butterflies"—have wings that meet in a point, and all these butterflies enjoy up–down symmetry but lack left–right symmetry.

A more extreme case of loss of symmetry is given by the small *non*-central butterfly whose left edge is the lowest of the three bands at flux-value $\phi = 1/3$, and whose right edge is the lower of the two (kissing) bands at flux-value $\phi = 1/2$. The four white wings of this sad butterfly do not meet in a single point. More generally, all of the infinitely many non-central butterflies have wings that fail to meet in a point, and thus these butterflies enjoy neither horizontal nor vertical mirror symmetry.

For the sake of simplicity, our discussion in this chapter will be mostly confined to *central* butterflies, which enjoy up–down but not left–right symmetry. In the butterfly landscape, off-center butterflies are continuations of the central butterflies, which control many of their characteristics, such as topology.

In what follows, we will be focusing on how such number-theoretical ideas as Farey fractions, Ford circles and Apollonian gaskets can be used to characterize the recursive patterns in the Hofstadter butterfly. In the first half of appendix A, Douglas Hofstadter briefly sketches the original ideas underlying the butterfly's recursive landscape in his own words.

2.1 Ford circles, the Farey tree, and the butterfly

2.1.1 Ford circles

In his 1938 paper entitled simply "Fractions" [1], Lester Ford, an American mathematician, begins with the following modest remark:

Perhaps the author owes an apology to the reader for asking him to lend his attention to so elementary a subject, for the fractions to be discussed are, for the most part, the halves, quarters and thirds of arithmetic. But the fact is that the writer has, for some years, been looking on these entities in a somewhat new way. Here will be found a geometrical visualization which will be novel to the reader and which will supply a visual representation of arithmetical results of diverse kinds.

Figure 2.2 illustrates Ford circles for two fractions p/q and P/Q associated with real numbers represented on the x-axis of the xy-plane. At each rational point $\frac{p}{q}$ is drawn a circle of radius $\frac{1}{2q^2}$ and whose center is the point $(x, y) = (\frac{p}{q}, \frac{1}{2q^2})$. This circle, known as a *Ford circle*, is tangent to the x-axis in the upper half of the xy-plane. This circle constitutes a geometrical representation of the fraction $\frac{p}{q}$.

Below, we summarize the key aspects of the mapping of Ford circles onto rational values of ϕ (and more details of the mapping are given in appendix B):
- Each rational number $\frac{p}{q}$ on the ϕ-axis can be pictorially represented by a Ford circle that kisses (i.e. is tangent to) the axis at the value $\frac{p}{q}$.
- The Ford circle located at the point $\frac{p}{q}$ has diameter $\frac{1}{q^2}$.

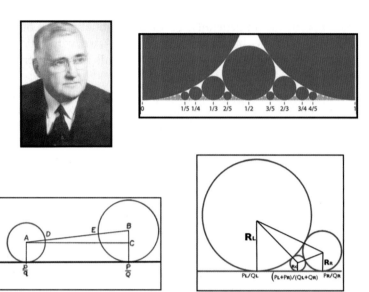

Figure 2.2. This figure shows how fractions can be represented by circles tangent to the x-axis, as was discovered by Lester Ford. The upper right panel shows the Ford-circle representation of all the fractions in a Farey tree with $q_{max} = 5$. The lower left panel shows Ford's graphical representation of two fractions p/q and P/Q, while the lower right panel shows the special case where the Ford circles touch.

- No two Ford circles ever intersect; the only way two Ford circles can meet is by *kissing* each other (being tangent at one point). These kisses, described in chapter 0's discussion of Apollonian gaskets, will emerge as a key ingredient in characterizing the nesting of the butterfly fractal and its topological properties.

Figure 2.3 shows the butterfly graph along with a few circles, some kissing each other. The significance of these "kissing" circles in characterizing the butterfly landscape will be discussed below, after a short introduction to the Farey tree.

2.1.2 Farey tree

Farey fractions were independently discovered in the early 1800s by Charles Haros and John Farey, and they form a beautiful part of number theory. One way of presenting them is in a Farey tree, as shown in figure 2.4. The tree is built up row by row, starting at the top, which contains only $\frac{0}{1}$ and $\frac{1}{1}$. Each successive row of the tree inherits all the Farey fractions from the level above it, and is enriched with some new fractions (all of which lie between 0 and 1) made by combining certain neighbors in the preceding row using an operation called "Farey addition". To combine two fractions $\frac{a}{c}$ and $\frac{b}{d}$, one simply adds their numerators, and also their denominators, so

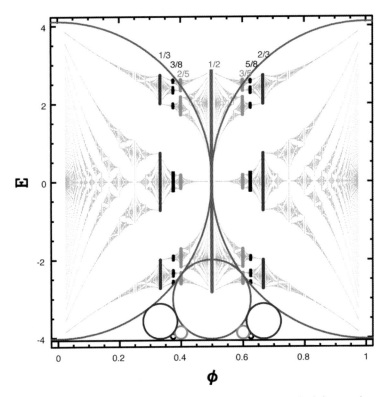

Figure 2.3. The butterfly graph shown with a handful of circles that are all sitting on the x-axis (which measures magnetic flux ϕ, so it is really the ϕ-axis). These circles belong to a special family, infinite in number, whose members are called *Ford circles*. A Ford circle always kisses the x-axis at a single rational point, and indeed there is one at *every* rational point, and yet no two Ford circles ever intersect each other! In this figure, each Ford circle is color-coded to match the set of energy bands directly above it. Above the flux-value of 1/3, for instance, are three highlighted dark-blue bands. More generally, at any rational flux-value $\frac{p}{q}$—for example, $\frac{1}{2}$, $\frac{1}{3}$, $\frac{2}{5}$, and $\frac{3}{8}$—there is a set of exactly q bands separated by white gaps. However, when q is even, the two central bands "kiss" on the horizontal line $E = 0$, so there is no gap between them, as can be seen for ϕ values $\frac{1}{2}$ and $\frac{3}{8}$. The butterfly's left and right edges, which are full Bloch bands corresponding to flux-values of 0/1 and 1/1, are associated with the large red semicircles. A key feature of Ford circles, hinted at in this figure and further discussed below, is how they are tangent to one another.

that the Farey sum is $\frac{a+b}{c+d}$. This Farey sum is often represented by the symbol "$\oplus$". Thus

$$\frac{a}{c} \oplus \frac{b}{d} \equiv \frac{a+b}{c+d}. \tag{2.1}$$

To make the nth row from the n–1st row, one takes neighboring Farey fractions in the n–1st row and computes their Farey sums; those whose denominator is equal to n become the new members of the nth row. Another way of characterizing the nth row of the Farey tree is as the set of all irreducible rational numbers $\frac{p}{q}$ with $0 \leq p \leq q \leq n$, arranged in increasing order.

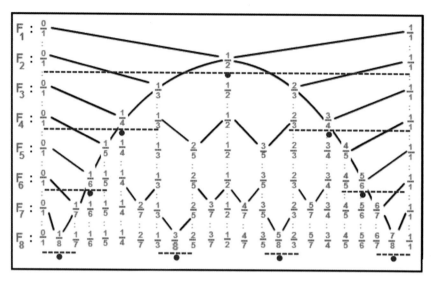

Figure 2.4. A row-by-row construction of the Farey tree, as described in the text. This figure highlights the special property of even-denominator Farey fractions (indicated by blue dots) that both of their Farey neighbors have odd denominators, and that there exists a unique pair of odd-denominator fractions (shown at the two ends of each dashed blue line) with denominators less than the even denominator.

In general, two neighboring fractions in a Farey tree—say, $\frac{p_1}{q_1}$ and $\frac{p_2}{q_2}$—obey the following identity:

$$p_1 q_2 - p_2 q_1 = \pm 1. \tag{2.2}$$

Equation (2.2) is known as the "friendship rule", and any two rational numbers connected in this fashion are said to be "friendly numbers". Thus the left and right neighbors of any rational number in a Farey tree are called its "friendly numbers". (However, two rational numbers that have this property need not belong to the same level of the Farey tree.)

2.1.3 The saga of even-denominator and odd-denominator fractions

For a given rational value p/q of the magnetic flux, it is quite surprising that the parity of the denominator—that is, whether q is even or odd—plays an important role in shaping the butterfly landscape.

First, the dynamics of the Hofstadter set (or of the quantum-mechanical equation that gives rise to it) responds quite differently to rational flux-values having even and odd denominators. This can be clearly seen in figure 2.5.

For any p/q where q is *even*, there are two central bands that touch each other, or "kiss", on the horizontal line $E = 0$. This is called "degeneracy" of the quantum state. The degeneracy is lifted as one moves away, either leftwards or rightwards, from this flux value. There is an energy gap above and below the line $E = 0$, resulting in the meeting of four wings at p/q, forming the characteristic butterfly pattern.

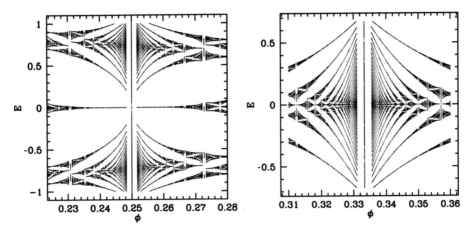

Figure 2.5. Zooming in on the butterfly near an even-denominator flux-value ($\phi = 1/4$) and an odd-denominator flux-value ($\phi = 1/3$).

By contrast, for any flux-value p/q where q is *odd*, the single central band breaks up into many tiny fragments as one moves away, either leftwards or rightwards, but the fragments all remain clustered very closely around the band at p/q.

It is almost magical how the quantum behavior associated with flux-values having even and odd denominators is perfectly "in synch" with the number theory of Farey fractions, where even-denominator and odd-denominator rational numbers behave differently, as we will now see.

We summarize below the differences between the even-denominator and the odd-denominator Farey fractions:

(1) The two friendly numbers of any even-denominator fraction (i.e., its left and right neighbors in its row) always have odd denominators. This is in contrast to odd-denominator fractions, whose two friendly numbers can have either even or odd denominators.

(2) For every fraction $\frac{p_1}{q_1}$ with even denominator q_1, there exists a unique pair of odd-denominator friendly numbers, $\frac{p_2}{q_2}$ and $\frac{p_3}{q_3}$, whose denominators are both less than q_1.

(3) A set of three fractions such that any pair of them satisfies the friendship rule will be referred to as a *Farey triplet*. Symbolically, a Farey triplet is a set of fractions $\frac{p_1}{q_1}, \frac{p_2}{q_2}$, and $\frac{p_3}{q_3}$ such that

$$\frac{p_1}{q_1} = \frac{p_2}{q_2} \oplus \frac{p_3}{q_3} = \frac{p_2 + p_3}{q_2 + q_3} \tag{2.3}$$

$$\left| p_i q_j - p_j q_i \right| = 1, \quad \text{for all } i = 1, 2, 3, \quad j = 1, 2, 3, \quad i \neq j. \tag{2.4}$$

As is illustrated in figure 2.2, the three Ford circles associated with any Farey triplet are all mutually tangent.

The above two conditions, along with Harper's equation (see chapters 1 and 7) define the butterfly landscape, in the manner described below.

Friendly numbers, kissing circles and the butterfly

All rational values $\frac{p}{q}$ of ϕ, where q is *even*, have two central bands that touch each other on the line $E = 0$. Such touching-points are the centers of all central butterflies. The left and right edges of any central butterfly are located at rational ϕ values with odd denominators, which are the friendly numbers of the center. This implies that, given the location (the *x*-coordinate, which is to say, the ϕ value) of the *center* of any central butterfly, the locations of its two *edges* are uniquely determined. If the center is at $\frac{p_c}{q_c}$, then the left and right edges are at $\frac{p_L}{q_L}$ and $\frac{p_R}{q_R}$. These edges' *x*-coordinates are the left and the right friendly numbers of the center's *x*-coordinate.

For any central butterfly, the locations of its center and its left and right edges are related to each other by Farey addition:

$$\frac{p_c}{q_c} = \frac{p_L}{q_L} \oplus \frac{p_R}{q_R} = \frac{p_L + p_R}{q_L + q_R}. \tag{2.5}$$

As was explained above, equation (2.5) is also the condition for three Ford circles to be mutually tangent as well as tangent to the horizontal axis. Therefore, the Ford circles representing the center and the edges of a central butterfly correspond to configuration (*c*) of Descartes' problem: a collection of four mutually tangent or kissing circles, commonly called a *Descartes configuration*, as is shown in figure 2.6.

Equation (2.5), which is the tangency condition for two Ford circles, applies both to central and to off-center butterflies. However, off-center butterflies do not necessarily have their centers located at flux-values with even denominators.

The Ford circles representing the two edges and the center of any miniature butterfly are all mutually tangent, as is shown on the right side of figure 2.7. These three circles, along with the horizontal *x*-axis to which the the circles are tangent, constitute a specific case of four mutually tangent circles, so they satisfy Descartes' theorem (see chapter 0), namely:

$$\kappa_\pm = \kappa_1 + \kappa_2 + \kappa_3 \pm 2\sqrt{\kappa_1\kappa_2 + \kappa_2\kappa_3 + \kappa_1\kappa_3}. \tag{2.6}$$

We can choose $\kappa_1 = 2q_L^2$ (left edge), $\kappa_2 = 2q_R^2$ (right edge), and $\kappa_3 = 0$ (the horizontal axis). Then equation (2.6) simplifies as follows:

$$\sqrt{\kappa(\pm)} = |\sqrt{\kappa_L} \pm \sqrt{\kappa_R}|. \tag{2.7}$$

Clearly, κ_+ corresponds to the Ford circle representing the center of the butterfly, and therefore we have:

$$\sqrt{\kappa_+} = \sqrt{\kappa_L} + \sqrt{\kappa_R} = \sqrt{\kappa_c}$$
$$\sqrt{\kappa_-} = |\sqrt{\kappa_L} - \sqrt{\kappa_R}| = ?\,.$$

The significance of the second solution—namely, $\sqrt{\kappa_-} = |\sqrt{\kappa_L} - \sqrt{\kappa_R}|$ (shown in red on the right side of figure 2.7)—will become apparent when we discuss the nesting of the butterflies, as it corresponds to a tangency condition between two different generations.

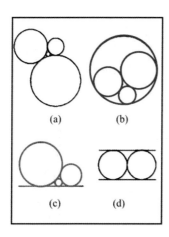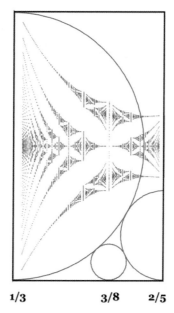

1/3   3/8   2/5

Figure 2.6. On the left are shown four different types of "Descartes configurations" of four pairwise tangent circles in a plane. This includes cases where one or more of the circles degenerates to a straight line (a circle with infinite radius). The butterfly nesting is related to configuration (c), as is shown in the right part of the figure, which shows a butterfly whose center is located at flux-value 3/8, and which is represented by the small Ford circle sitting on the axis at that same flux-value. The left and right edges of the butterfly are located at 1/3 and 2/5, and these values are represented by large Ford circles (in the figure, one sees only half of each large circle). These three circles, along with the horizontal axis, are all mutually tangent.

Echoing the terminology of chapter 0, where a tangency of two circles was poetically described as a "kiss precise", we will distinguish between two types of tangency—those between generations and those within a single generation. More specifically, an *inter-kiss precise* means an intergenerational kiss, which is to say, the tangency of two Ford circles belonging to *different* generations, while an *intra-kiss precise* means an intragenerational kiss, involving the tangency of two Ford circles belonging to the *same* generation.

2.1.4 The sizes of butterflies

The width (or the horizontal size) $\Delta\phi$ of a butterfly with center $\frac{p_c}{q_c}$ and left and right edges $\frac{p_L}{q_L}$ and $\frac{p_R}{q_R}$ is given by:

$$\Delta\phi = \left| \frac{p_R}{q_R} - \frac{p_L}{q_L} \right| = \frac{1}{q_L q_R}, \qquad |p_L q_R - p_R q_L| = 1. \tag{2.8}$$

2.2 A butterfly at every scale—butterfly recursions

As we examine the full butterfly at smaller and smaller scales, we note that there exists a central butterfly at every scale, and these miniature versions exhibit every

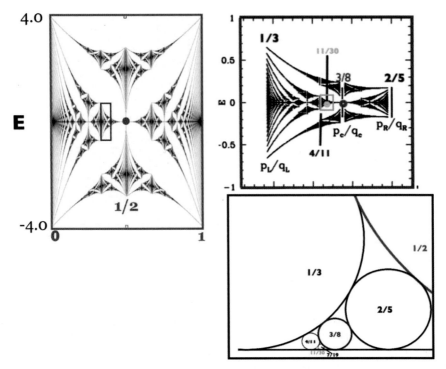

Figure 2.7. In the red box on the upper left, we see the central butterfly with center at 1/2 (the red "pin"). In the blue box on the upper right, we see the *first*-generation butterfly that stretches between 1/3 and 2/5, with its center (the blue pin) located at 3/8. This is a *second*-generation central butterfly. Inside it we see a *third*-generation butterfly centered on 11/30, and with left and right edges at 4/11 and 7/19. Of course we could continue going down, down, down, but three levels suffice to give the idea. These ideas involving generations of central butterflies map elegantly onto isomorphic ideas involving Ford circles, which are shown in the black box at the bottom of the figure. Ford circles representing the butterfly centers for three successive generations are shown in red, blue, and green (these colors match the colors of the pins representing the centers of the corresponding butterflies). The large black circles at values 1/3 and 2/5 correspond, of course, to the second-generation butterfly's edges (note their odd denominators), while the blue circle at 3/8 (with even denominator) corresponds to the blue pin at the center. Inside the second-generation central butterfly is found a third-generation butterfly centered on 11/30, and the corresponding Ford circle at 11/30 is a third-generation Ford circle. This figure illustrates both the "inter-kiss precise" and the "intra-kiss precise" conditions.

detail of the original graph. A nice illustration of this was shown earlier in the book (see figure 2.3 in the prelude).

Since the nesting of butterflies goes infinitely far down, it is useful to define a notion of *levels* or *generations*. The top level, or first generation, is the full butterfly, stretching between $\phi = 0$ and $\phi = 1$, with its fourfold symmetry. We will say that butterflies A and B belong to *successive generations* when B is *contained inside* A and when there is no intermediate butterfly between them (i.e., there is no butterfly both contained in A and also containing B).

Our discussion below includes only those cases where the larger and smaller butterflies share neither their left edge nor their right edge. In this manner, any miniature butterfly can be labeled with a positive integer telling which generation it belongs to. We will show that this class of butterflies is characterized by a nontrivial scaling exponent. By contrast, in cases where a nested set of butterflies from different generations all share a common boundary, the underlying asymptotic scaling exponent is trivial (see equation (3.12), and the discussion that follows it).

As we have already pointed out, there is a tight correspondence between central butterflies and Ford circles tangent to the *x*-axis. Just as each central butterfly can be assigned to a specific generation, so can Ford circles with even denominators. Recall that even-denominator ϕ values are the *centers* of butterflies, and recall also that the center "pins down" the butterfly completely. Figure 2.7 helps make clear this aspect of the mapping between butterfly generations and generations of Ford circles.

Butterfly recursions and more kissing circles

We now seek a rule for finding a sequence of nested butterflies as we zoom into a given flux interval $\Delta\phi$, a sequence that evolves towards an "invariant configuration" (i.e., a fixed point), where two successive butterfly zooms are exact copies of each other, except for a scaling factor. We start with a butterfly inside the interval, whose center is at ϕ value $f_c(l) = \frac{p_c(l)}{q_c(l)}$, and whose left and right edges are at $f_L(l) = \frac{p_L(l)}{q_L(l)}$ and $f_R(l) = \frac{p_R(l)}{q_R(l)}$. Let us assume that this butterfly belongs to generation *l*.

A systematic procedure to describe a nested set of butterflies that converge to the desired "fixed point" behavior involves three generations. To understand why three are involved, let us suppose that we begin with the entire butterfly landscape—the first-generation "mother" butterfly—and from the infinite zoo of smaller butterflies inside it, we pick one tiny butterfly, which we will refer to as the second-generation daughter butterfly. The next step is to zoom into this tiny butterfly and choose the third-generation butterfly—the granddaughter—and we will choose "her" in such a way that she has *the same location relative to the daughter butterfly as the daughter had relative to her mother*.

By repeating this process of zooming into ever-higher generations, we will converge to fixed-point behavior. The key trick assuring convergence to fixed-point behavior is always to stick to the *same* rule for the relative location of two successive generations of butterflies, as one carries out the successive zooms. Such a sequence of butterflies evolving towards a stable invariant structure will be called a "butterfly hierarchy". Figure 2.8 pictorially shows the process of finding a series of butterflies that converge to this kind of fixed-point behavior.

The three-step recursion is given by the following equations:

$$f_L(l+1) = f_L(l) \oplus f_c(l) \tag{2.9}$$

$$f_R(l+1) = f_L(l+1) \oplus f_c(l) \tag{2.10}$$

$$f_c(l+1) = f_L(l+1) \oplus f_R(l+1). \tag{2.11}$$

These equations relate fractions on the ϕ-axis. If we wish, however, we can instead focus on these fractions' numerators and denominators. Rewritten in terms

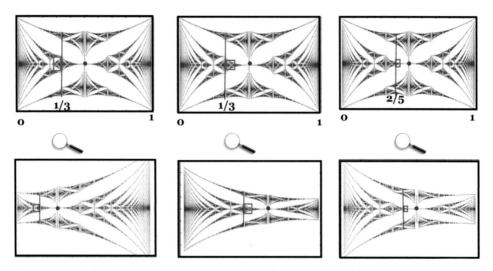

Figure 2.8. Illustrating the meaning of the notion of a "butterfly hierarchy", explained in the text. In the top row are seen first-generation "parent" butterflies, inside which three distinct second-generation butterflies have been selected, indicated by the small red boxes. The lower row shows these "daughter" butterflies—blowups of the regions indicated by the red boxes in the upper row. The red boxes in the lower panel in turn indicate the locations of third-generation "granddaughter" butterflies. The key point is that the boxed butterflies always have *the same relative locations* with respect to the centers and edges of their parent butterflies. These three distinct cases are examples of a single hierarchy characterized by a universal scaling ratio, as is described in the text.

of the integers $p(l)$ and $q(l)$, the above equations become the following recursion relation:

$$s_x(l+1) = 4s_x(l) - s_x(l-1) \tag{2.12}$$

where $s(l) = p_x(l), q_x(l)$ with $x = L, c, R$. In other words, the denominators $p(l)$ and the numerators $q(l)$ of the flux-values corresponding to the edges (L or R) or the centers (c) of a butterfly all obey the same recursion relation.

These recursion relations can also be written in matrix form as follows:

$$\begin{pmatrix} s(l+1) \\ s(l) \end{pmatrix} = \begin{pmatrix} 4 & -1 \\ 1 & 0 \end{pmatrix} \begin{pmatrix} s(l) \\ s(l-1) \end{pmatrix}.$$

Note that the 2×2 matrix in this equation has determinant 1.

Equations (2.10) and (2.11), linking two different generations of Ford circles, show the significance of the two solutions of Descartes' equation—that is, equation (2.7). We summarize these two types of tangencies as follows:

- An *intra-kiss precise* corresponds to $\sqrt{\kappa_+} = \sqrt{\kappa_c(l)} = \sqrt{\kappa_L(l)} + \sqrt{\kappa_R(l)}$ (using equation (2.11)).
- An *inter-kiss precise* corresponds to $\sqrt{\kappa_-} = \sqrt{\kappa_c(l-1)} = |\sqrt{\kappa_L(l)} - \sqrt{\kappa_R(l)}|$ (using equation (2.10)).

Both types of tangency, intra- and inter-, are visible in figure 2.7.

The butterfly nesting described by equation (2.12) represents the most natural and the simplest way to describe the self-similar character of the butterfly landscape. In appendix E, we introduce the notion of the *Farey path* and characterize this particular scheme by a sequence of letters—"L"s (for left) and "R"s (for right)—defining a Farey path in the Farey tree (see figure 2.13). This leads to many other ways to describe self-similarity, which we refer to as different possible hierarchies, each characterized by its own unique Farey path.

This idea is further discussed in chapter 4, where each hierarchy is associated with an irrational number. Just as the "LRLR" hierarchy is associated with the number $2 + \sqrt{3}$, for which we coined the term "diamond mean", there are other hierarchies, such as the golden and silver hierarchies (which are associated with the golden and silver means). The diamond hierarchy, apart from being the most important one, is also unique among all the infinitely many hierarchies, because it has a special hidden symmetry, which we will discuss below and also in chapter 3.

2.3 Scaling and universality

2.3.1 Flux scaling

Consider three butterflies in a hierarchy, belonging to three successive generations $l - 1$, l, and $l + 1$. Their centers are located at the rational numbers $f_c(l - 1)$, $f_c(l)$, and $f_c(l + 1)$. Recall that the radius of the Ford circle associated with any given rational number p/q is $1/(2q^2)$, and that the circle's curvature κ is defined to be the reciprocal of this radius. Let us denote the curvature of the Ford circle associated with the butterfly centered at $f_c(l)$ by $\kappa_c(l)$. The recursion relation (2.12) then tells us that the curvatures of the Ford circles associated with the centers of our three nested butterflies satisfy the following equation:

$$\sqrt{\kappa_c(l+1)} = 4\sqrt{\kappa_c(l)} - \sqrt{\kappa_c(l-1)}. \tag{2.13}$$

These Ford circles do not touch each other. How are their radii related? The ratio of their radii is (by definition) the reciprocal of the ratio of their curvatures. Let us define the scale factor $\zeta(l)$ as the square root of the ratio of the curvatures of two Ford circles belonging to generations $l + 1$ and l:

$$\zeta(l) = \sqrt{\frac{\kappa_c(l+1)}{\kappa_c(l)}}. \tag{2.14}$$

Plugging this definition into the previous equation, we obtain:

$$\zeta(l) = 4 - \frac{1}{\zeta(l-1)}. \tag{2.15}$$

For large l (that is, when we are many generations down), $\zeta(l) \to \zeta(l+1)$. We denote the limiting value of this sequence by ζ^*. This number is the fixed point of equation (2.15), and thus it satisfies the following quadratic equation:

$$(\zeta^*)^2 - 4\zeta^* + 1 = 0, \quad \zeta^* = \lim_{l \to \infty} \sqrt{\frac{\kappa_c(l+1)}{\kappa_c(l)}} = 2 + \sqrt{3}. \tag{2.16}$$

The other root of the quadratic equation, $2 - \sqrt{3}$, represents taking the $\zeta(l)$ in the reverse order.

It follows that Ford circles corresponding to even-denominator fractions form a self-similar fractal consisting of circles whose curvatures are asymptotically scaled by the factor ζ^*. No matter what even-denominator fraction we start with, we will get the same scaling factor. We further note that between two successive levels, the Ford circles of $f_L(l)$ and $f_L(l+1)$ are tangent, while those of the corresponding f_c and f_R are not.

We also note that

$$\sqrt{\frac{\kappa_c(l)}{\kappa_L(l)}} \to 1 + \sqrt{3}, \quad \sqrt{\frac{\kappa_c(l)}{\kappa_R(l)}} \to \frac{1+\sqrt{3}}{\sqrt{3}}, \quad \sqrt{\frac{\kappa_R(l)}{\kappa_L(l)}} \to \sqrt{3}. \tag{2.17}$$

We now calculate the scaling factor associated with the magnetic flux—that is, the ratio of the widths of butterflies at different generations. Let us denote the width of the generation-l butterfly by $\Delta\phi(l)$:

$$\Delta\phi(l) = f_R(l) - f_L(l) = \frac{1}{q_L(l) q_R(l)}. \tag{2.18}$$

The ratio of the widths at two successive generations is given by:

$$R_\phi = \lim_{l \to \infty} \frac{\Delta\phi(l)}{\Delta\phi(l+1)} = \lim_{l \to \infty} \frac{q_L(l+1) \, q_R(l+1)}{q_L(l) \, q_R(l)} = (\zeta^*)^2 = \left(2 + \sqrt{3}\right)^2. \tag{2.19}$$

Note that the scaling of the curvatures of the Ford circles representing the centers of the butterflies belonging to two consecutive generations is:

$$\frac{\kappa_c(l+1)}{\kappa_c(l)} = \frac{q_c(l+1)^2}{q_c^2(l)}, \quad \lim_{l \to \infty} \frac{\kappa_c(l+1)}{\kappa_c(l)} = \left(2 + \sqrt{3}\right)^2. \tag{2.20}$$

In other words, the Ford circles associated with the butterflies' centers scale exactly like the butterflies themselves.

Here we shall mention only briefly what may well be the most remarkable feature of the butterfly's scaling—namely "topological scaling" (which we will discuss in chapter 9). This is the scaling of the topological quantum numbers of the various gaps of the butterfly landscape. Interestingly, the topological scaling factor is found to be equal to $\zeta^* = \sqrt{R_\phi}$.

2.3.2 Energy scaling

So far, we have only discussed scaling properties along the ϕ-axis of the butterfly graph. However, the butterfly is a two-dimensional object, so now we turn to the question of scaling along the *energy* axis. Unlike the magnetic-flux window associated

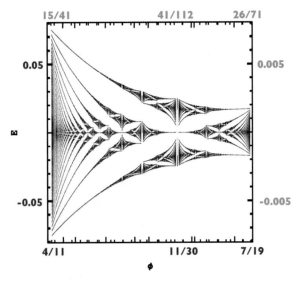

Figure 2.9. Two consecutive generations of butterflies, shown in blue and green, overlaid. The blue labeling of the axes tells us where the blue butterfly is found, and likewise for the green labeling of the axes. This nearly perfect alignment illustrates the asymptotically exact self-similarity of the butterfly graph. (However, only the flux scaling has been proven analytically.) The vertical or energy scale factor has so far been determined only by numerical calculation of the Hofstadter set, and it is approximately $9.87 \approx \pi^2$.

with a butterfly, which we can compute given the butterfly's center, the corresponding energy scale has to be determined numerically, as described in chapter 1.

Figure 2.9 illustrates the self-similarity property of the butterfly graph as we overlay two miniature butterflies—one belonging to the *l*th generation, and the other to the *l* + 1st generation—by magnifying the plot of the *l* + 1st generation by the scaling ratio R_E along the *vertical* direction, and by the scaling ratio R_ϕ along the *horizontal* direction. This figure shows the two numbers R_E and R_ϕ, which characterize the scaling of this two-dimensional landscape. The numerically computed value of R_E is approximately 10.

2.3.3 Universality

Characterizing the butterfly graph in terms of scaling ratios such as R_ϕ and R_E may constitute a first step in pinning down the graph's *universal* aspects. Appendix E discusses universality in a limited sense, commenting on the validity of these scalings for a class of magnetic flux-values characterized by the *tails* of their continued-fraction expansions. More generally, though, the concept of universality means that *other physical systems* that have a butterfly-like spectrum should possess exactly the same scaling ratios. For example, butterfly plots coming from non-square lattices might have the same scaling ratios as the Hofstadter butterfly does, or perhaps the Hofstadter butterfly's scaling ratios will turn out to be unchanged under other types of modification of the system. Recall the discussion of universality in chapter 1, where it was pointed out that the same scaling rule that characterizes period-doubling in quadratic maps (as studied by Mitchell Feigenbaum) was later found to characterize other complex systems, even including dripping water faucets. In that spirit, it is to be hoped that simple models, such as the system studied by Hofstadter, may have mathematical properties that will turn up in more complex systems as well. These observations are but the tip of the iceberg of this fascinating subject, and a detailed exploration of which features of the butterfly graph are universal remains to be carried out.

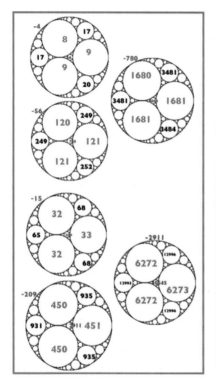

Figure 2.10. The six nested structures on the left are integral Apollonian gaskets that have *approximate* trefoil symmetry. As one descends ever more deeply into the gaskets, the smaller gaskets asymptotically approach *exact* trefoil symmetry. The curvatures of some of the circles are shown by the integers inside them. Note that in each gasket, two of the three biggest circles have identical curvature values, and the third one's curvature differs from the others by just 1. This is why all these gaskets come very close to having perfect trefoil symmetry. The outer curvature of each of the upper trio of gaskets is an even number, and for the lower trio, it is odd. For each of these two trios, the bottom gasket is a blowup of the circle in the middle of the top gasket, and the right gasket is a blowup of the one on the bottom. Thus the upper gasket's outermost circle has a curvature of 4, and at its center we see a very small circle with curvature 56. This circle (actually a gasket) is blown up underneath, and at its center we see a small circle with curvature 780. Once again, this circle is blown up, giving us the third member of the upper trio. A lovely fact is that as the magnification is increased, the ratio of the curvature of the innermost circle to that of the outermost circle of all these gaskets approaches a constant. Thus, in the upper trio, we see $56/4 \to 780/56 \to 10864/780 \to (2 + \sqrt{3})^2$. And similarly, for the lower trio, we have $209/15 \to 2911/209 \to 40545/2911 \to (2 + \sqrt{3})^2$. In fact, the ratios of the curvatures of any circles that lie in corresponding positions converge to exactly the same constant as one zooms inwards, an example on the upper side being $120/8 \to 1680/120 \to (2 + \sqrt{3})^2$. The photo on the right is of redwood sorrels found in Montgomery Woods in Mendocino County, California, showing trefoil symmetry in nature (photo courtesy of Douglas Hofstadter).

2.4 The butterfly and a hidden trefoil symmetry

We next show that the butterfly scaling ratio R_ϕ associated with the butterfly hierarchy characterized by the finite Farey paths "LRL" and "RLR", described in appendix E, is related to the nested set of circles in an Apollonian gasket that matches the Descartes configuration (b) in figure 2.6.

We consider a special case of Descartes' theorem (2.6), where $\kappa_1 = \kappa_2 = \kappa_3 = \kappa$. This corresponds to an Apollonian gasket that has perfect trefoil symmetry. The ratio of the curvatures of the inner and outer circles is determined by the equations:

$$\frac{\kappa_+}{\kappa} = \sqrt{3}\left(2 + \sqrt{3}\right); \qquad \frac{\kappa_-}{\kappa} = \sqrt{3}\left(2 - \sqrt{3}\right); \qquad (2.21)$$

$$\frac{\kappa_+}{\kappa_-} = \frac{2 + \sqrt{3}}{2 - \sqrt{3}} = \left(2 + \sqrt{3}\right)^2. \qquad (2.22)$$

The fact that the ratio of these two curvatures is irrational shows that there is no *integral* Apollonian gasket possessing exact trefoil symmetry. Interestingly, however, as can be seen in figure 2.10, in some integral Apollonian gaskets, perfect trefoil symmetry is asymptotically approached as one descends deeper and deeper into the gasket, thus getting larger and larger integral values of the curvature, which give closer and closer rational approximations to the irrational limit, $(2 + \sqrt{3})^2$.

A most surprising fact is that the ratio of the radii of innermost and outermost circles exhibits the same scaling as the ratio of the curvatures of the Ford circles corresponding to two successive generations of the butterfly, as is shown in equation (2.19). In other words, the scaling underlying *nearly* trefoil-symmetric Apollonian gaskets is identical to the magnetic-flux scaling associated with the butterfly graph.

The trefoil symmetry described above may appear rather mysterious, as it lacks any geometrical picture that may help in visualizing what this symmetry means for the butterfly landscape. Clearly, none of the butterflies nested inside the butterfly fractal exhibit this kind of symmetry. The question of where this hidden symmetry shows up in the Hofstadter landscape will be revisited in chapter 3, where we will describe the mapping between the butterfly fractal and integral Apollonian gaskets.

2.5 Closing words: Physics and number theory

That number theory plays a crucial role in describing certain physical phenomena is a quite new finding, and even today it remains relatively unknown among many theoretical physicists. Back in 1969, Richard Feynman, in replying to a letter sent to him by Robert Boeninger, wrote the following[1]: "I don't know why number theory does not find application in physics. We seem to need the mathematics of functions of continuous variables, complex numbers, and abstract algebra." In the same spirit, but much more recently, physicist and science writer Michio Kaku, in his 1995 book *Hyperspace*, declared: "[S]ome mathematical structures, such as number theory

---
[1] "The Quotable Feynman", edited by Michelle Feynman, Princeton University Press.

(which some mathematicians claim to be the purest branch of mathematics), have never been incorporated into any physical theory" [2] and Kaku then goes on to suggest that perhaps in the future string theory may do so[2]. Despite their usual insightfulness, both Feynman and Kaku turned out to be quite wrong, however, since, as we have just seen, number theory pervades the butterfly's structure. And thus that salient gap in the history of mathematical physics has been filled—without any need for string theory to come to the rescue!

Appendix A: Hofstadter recursions and butterfly generations

For the first few paragraphs of this appendix, Douglas Hofstadter tells, in his own words, how he characterized the recursive structure of the butterfly in his work in the 1970s.

Douglas Hofstadter's own story of his recursive breakdown of the butterfly

In my doctoral thesis and my subsequent 1976 *Physical Review* article, I partitioned the butterfly into smaller copies of itself. Each copy was located inside a region that I called a *subcell* (see figure 2.11). Having defined subcells, I then showed how the structure of the full graph could be mapped isomorphically (or more strictly speaking, *homeomorphically*) onto the structure found inside any such subcell.[3]

In the full butterfly, the flux-value ϕ runs between 0 and 1; what I did was to show that inside any subcell, there is an analogous "local variable"—here I will refer to it as "α"—that likewise runs between 0 and 1. The nature of the analogy between ϕ and α can be described as follows. When ϕ takes on, say, the value 2/7, there are seven bands in the full butterfly forming a "cluster-pattern" of the form "3–1–3" (meaning there are three bands close together on the left side, one isolated band in the middle, and three close together on the right). Precisely analogously, in any given subcell, when that cell's *local* variable α takes on the same value (2/7, in our example), we will find the same number of bands (seven, in this case) inside the given subcell, and moreover they will be arranged in exactly the same cluster-pattern ("3–1–3", in this case).

Of course this happy analogy between the full graph and a subcell doesn't hold merely in the specific case of 2/7; the same thing can be said for *any* rational number p/q—namely, when ϕ takes on that value, there is a particular cluster-pattern in the entire butterfly involving q bands, and inside any subcell, when the local variable α takes on that same value, there is an identical cluster-pattern. This elegant mapping between the full graph and each of its subcells can be further extended to irrational values of ϕ and α by taking limits using rational approximations. In this systematic fashion, I was able to show how the mapping between any subcell and the full graph is complete and perfect.

---

[2] I am grateful to Francisco Claro for bringing this to my attention.
[3] Note: In Hofstadter's thesis and article, the ϕ-axis was always oriented vertically, whereas in this book it is almost always horizontal. However, for the next few paragraphs, we will follow Hofstadter's convention, with the ϕ-axis vertical.

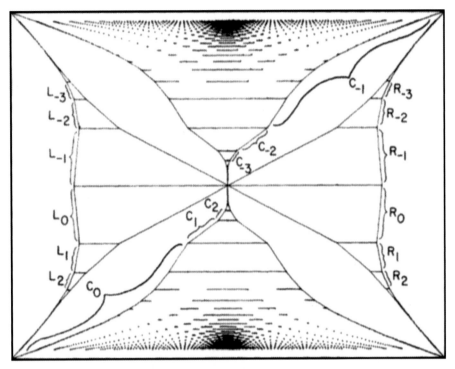

Figure 2.11. Partitioning of the butterfly landscape into three different infinite families of subcells—L-subcells, C-subcells, and R-subcells—as described by Hofstadter in his 1976 *Physical Review* article [3]. Figure adapted with permission from [3] copyrighted by the American Physical Society.

Incidentally, I was led to this insight by the behavior of my graph INT, which is displayed in the prologue (see figure P.2). In that much simpler graph, the idea of local variables is somewhat easier to grasp. Inside the full graph (where both x and y run between 0 and 1), there are very clear copies of it, such as (say) the small copy located between the x-values of 1/3 and 1/4. (Of course that specific copy of the full graph, like all the other copies, is tilted at 90 degrees relative to the full graph, and is also scaled down and slightly bent.) Now imagine that there is a local variable α inside the *little* graph that runs from 0 to 1 while x is running from 1/3 to 1/4 inside the *big* graph. In other words, we want the local variable α to play the same role inside the little bent copy of INT that x plays in the big INT graph. For example, in the full INT graph, the biggest jump-discontinuity of all takes place where $x = 1/2$, and so, in the little copy, we want the analogous biggest jump-discontinuity to happen where $\alpha = 1/2$. (Incidentally, this jump inside the little copy is located where $x = 2/7$.) Is there a precise formula that, given any value of x, tells you what the value of the local variable α is? Indeed there is, and in fact it's quite obvious:

$$x = \frac{1}{3 + \alpha}. \tag{A.1}$$

The "3" in the denominator reflects the fact that in this case, we are dealing with the little copy located between $x = 1/3$ and $x = 1/4$. Had we instead chosen the copy between 1/4 and 1/5, then the formula would have had a "4" instead of a "3" (and so forth and on).

This idea of a local variable playing the same role inside a subcell as the main variable plays inside the full graph constituted the essential idea in my recursive breakdown of the graph I had discovered. Many of the ideas that I had originally found in exploring my INT graph, many years earlier, carried through very straightforwardly to my new Gplot graph. Even the exact same algebraic formula relating the local variable α to the full-graph variable ϕ held for certain infinite families of subcells of Gplot (though not all of them).

My first task was to identify and label the subcells of the butterfly containing copies of the full butterfly (of course the copies were scaled down and distorted, just as were the copies in INT). I did this as is shown in figure 2.11. I gave the subcells along the graph's *left* side the names $L_{-2}, L_{-1}, L_0, L_1, L_2, \ldots$, those running down its *backbone* the names $C_{-2}, C_{-1}, C_0, C_1, C_2, \ldots$, and those on its *right* side the names $R_{-2}, R_{-1}, R_0, R_1, R_2, \ldots$.

Once I had precisely identified the small copies of the butterfly inside it, then the next step was to find formulas relating the local variables α inside the copies to the global variable ϕ. This actually was quite easy, and in the case of the L-subcells and R-subcells, I soon discovered that the formula was absolutely identical to the formula I'd found for the subgraphs of INT many years earlier. This gratifying discovery completely clinched, in my mind, the intuition I'd had that INT and Gplot were close cousins. In the case of the C-subcells, the formula connecting the local α variables to the global ϕ variable was a little more involved, but not greatly so. From my point of view, therefore, the L-subcells and the R-subcells were the "normal" cases, while the C-subcells were "exceptional". Curiously enough, this is essentially the flip side of how Indu Satija

l	(L, C, R)	ϕ_{l+1}	ϕ_l^C	$\phi_l^L = \phi_l^R$	Band clustering
1	L	1/3	0/1	1/1	(1-1-1)
1	C	3/8	1/2	2/3	(3-2-3)
1	R	2/5	1/1	1/2	(2-1-2)
2	L	4/11	1/3	3/4	(4-3-4)
2	C	11/30	3/8	8/11	(11-8-11)
2	R	7/19	2/5	5/7	(7-5-7)

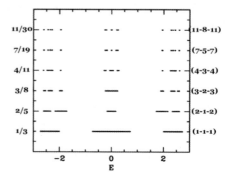

Figure 2.12. We will take the rational flux-values ϕ_{l+1} to be the edges and the center of the butterfly for two generations of the diamond hierarchy. The table above shows the corresponding ϕ_l for the C-cell, the L-cell, and the R-cell. (Incidentally, the boldface letters **L**, **C**, and **R** indicate the left edge, the center, and the right edge of the butterfly, and should not be confused with the letters L, C, and R that label the cells of the butterfly plot, as shown in figure 2.11.) In this example, the "inner" and "outer" variables (as they are called in Hofstadter's paper) coincide with two successive generations of the butterfly only for the C-cell. The table also shows the band-clustering pattern for ϕ_{l+1} as predicted by the recursive scheme. The figure on the right shows the complete agreement between the predicted clustering pattern, shown on the right side of the box, and the numerically calculated bands (plotted inside the box).

characterizes the recursions in Gplot, so our perspectives are perforce slightly different. But that's all right—variety is the spice of life! Let one hundred flowers bloom!

Connecting Hofstadter's breakdown with the butterfly-generations approach

As Hofstadter explained in detail in his 1976 article, the *C*-cell and *L*-cell recursions together describe the clustering patterns of bands and subbands (etc) for all values of ϕ. Figure 2.12 illustrates this for the set of ϕ values corresponding to the butterfly hierarchy characterized by the fixed point $\phi^* = \frac{\sqrt{3}-1}{2}$.

As *l* varies, the flux-values ϕ_l^C describe a sequence of butterflies from one generation to the next, as in the following case: $(\frac{0}{1}, \frac{1}{2}, \frac{1}{1}) \to (\frac{1}{3}, \frac{3}{8}, \frac{2}{5}) \to (\frac{4}{11}, \frac{11}{30}, \frac{7}{19})$. Here, each Farey triplet (explained in appendix B) contains the flux-values defining, respectively, the left edge, the center, and the right edge of the butterfly in question. Although the variables ϕ_l^L and ϕ_l^R play a crystal-clear role in Hofstadter's explanation of the nesting, how they would enter into a description of the hierarchies discussed above has not yet been elucidated.

As Hofstadter pointed out, the recursive decomposition of the spectrum tells us that the spectrum for any given rational flux-value ϕ_{l+1} consists of three parts, located inside *L*, *C*, and *R* subcells, separated by gaps from each other. The numbers of bands inside these cells are given, respectively, by the denominators of ϕ_l^L, ϕ_l^C, and ϕ_l^R, which are listed in the table in figure 2.12.

In chapter 3, we will use Hofstadter's idea of local variables inside subcells to obtain fixed points and scaling factors R_ϕ, and we will show that his recursions, which involve *two* generations, give the same scaling as the *p*–*q* recursions described above.

To conclude our discussion of Hofstadter's recursions, we note that the ϕ recursions connect two successive generations, while the recursions involving integers *p* and *q* are three-term recursions involving three generations. There appears to be no obvious way to obtain the ϕ recursions from the integer recursions for *p* and *q*. Furthermore, it is important to note a commonality between these two distinct type of recursions. Just like the integer recursions, the ϕ recursions are consistent with the organization of rational numbers involving friendly numbers. Without further discussion, we remark that this shared feature may be rooted in the fact that the ϕ recursions represent Möbius transformations. This is an open issue that needs to be investigated.

Appendix B: Some theorems of number theory [1]

(1) Each Farey fraction $\frac{p}{q}$ possesses two friendly numbers.
(2) Even-denominator fractions are never friendly numbers. (This implies that all friendly numbers of even-denominator fractions have odd denominators.) In contrast, odd-denominator fractions can be friendly numbers.
(3) Of all fractions that are friendly numbers of $\frac{p}{q}$ ($q > 1$), only two have denominators less than *q*. As was stated earlier in this chapter, this is the key property that pinpoints a butterfly at all scales in the butterfly diagram.

(4) The successive approximants of a simple continued-fraction expansion (see appendix D) form adjacent fractions.
(5) It is easy to prove that, given any two distinct irreducible fractions $\frac{p_1}{q_1}$ and $\frac{p_2}{q_2}$, the Ford circles associated with these fractions never intersect—that is, either they are tangent to each other or they touch each other nowhere at all.

From figure 2.2, the distance d between the centers of two Ford circles can be written in terms of the fractions as:

$$d^2 = (AC)^2 + (BC)^2; \quad AC = \left[\frac{P}{Q} - \frac{p}{q}\right]; \quad BC = \left[\frac{1}{2Q^2} - \frac{1}{2q^2}\right]$$

$$d^2 = \left[\frac{P}{Q} - \frac{p}{q}\right]^2 + \left[\frac{1}{2Q^2} - \frac{1}{2q^2}\right]^2 = \left[\frac{P}{Q} - \frac{p}{q}\right]^2 + \frac{1}{Q^2q^2} + \left[\frac{1}{2Q^2} + \frac{1}{2q^2}\right]^2$$

$$= \frac{(Pq - pQ)^2 - 1}{Q^2q^2} + (AD + EB)^2.$$

- If $|Pq - pQ| > 1$, the two circles are external to each other.
- If $|Pq - pQ| = 1$, the two circles are tangent.
- The only other possibility is $|Pq - pQ| = 0$, which is to say, $\frac{P}{Q} = \frac{p}{q}$.

This proves that the Ford circles belonging to two distinct fractions are either disjoint or tangent.

Appendix C: Continued-fraction expansions

Any irrational number α can be written as a *simple continued fraction* with infinitely many denominators n_k, all of which are integers:

$$\alpha = n_0 + \cfrac{1}{n_1 + \cfrac{1}{n_2 + \cfrac{1}{n_3 + \cfrac{1}{n_4 + \cfrac{1}{n_5 + \cfrac{1}{n_6 + \ldots}}}}}}.$$

The denominators $n_k > 0$ are generated by the following recursion relation:

$$\alpha_{k+1} = \frac{1}{\alpha_k - n_k}$$

where $n_k = [\alpha_k]$. (The notation "$[x]$" denotes the integer part of x; for example, $[\pi] = 3$.) Here $\alpha_0 = \alpha$. A concise notation for the continued-fraction expansion of an irrational number α is:

$$\alpha = [n_0; n_1, n_2, n_3, \ldots].$$

If one truncates the infinite simple continued fraction of an irrational number α after a finite number of terms, one obtains a *rational approximant* belonging to α. The kth rational approximant of α, denoted by $\frac{p_k}{q_k}$, uses the first k denominators of α's infinite continued-fraction expansion.

It can be shown that the rational approximants p_k/q_k belonging to α are the *best* rational approximations of the irrational number, in the sense that no rational number p/q with $q < q_k$ is closer to α than p_k/q_k. It can also be shown that

$$\left| \alpha - \frac{p_k}{q_k} \right| < \frac{1}{q_k q_{k-1}}.$$

There is a special class of irrationals whose simple continued-fraction denominators n_k form a periodic sequence. The simplest example of such an irrational is the golden mean, $\frac{\sqrt{5}+1}{2} = [1; 1, 1, 1, ...]$—that is, $n_k = 1$ for all k. The rational approximants belonging to the golden mean are ratios of successive Fibonacci numbers—that is, $p_k/q_k = F_{k+1}/F_k$, where F_k is the kth Fibonacci number (defined by the recursion $F_{k+1} = F_k + F_{k-1}$, where $F_0 = F_1 = 1$).

An irrational whose simple continued-fraction expansion's denominators are all equal to the integer m is a *quadratic irrational*, as it satisfies the quadratic equation $\alpha^2 - m\alpha - 1 = 0$. More generally, any real number whose continued-fraction denominators form a periodic sequence is an irrational solution of a quadratic equation with rational coefficients, such as $\sqrt{14} = [3; 1, 2, 1, 6, 1, 2, 1, 6...]$. On the other hand, we note that the irrational number π has a rather erratic simple continued-fraction expansion, namely: $\pi = [3; 7, 15, 1, 292, 1, 1, 1, 2...]$.

Appendix D: Nearest-integer continued fraction expansion

Below we give a simple algorithm for generating the nearest-integer continued fraction for any real number.

Given x, take the closest integer n. If $x > n$, set $y = x - n$; conversely, if $x < n$, then set $y = n - x$. By construction, y lies between 0 and 1/2, and $x = n \pm y = n \pm 1/(1/y)$. (Choose the appropriate sign—that is, + if $x = n + y$, an d − if $x = n - y$.) So n is the zeroth denominator of the continued fraction (and the zeroth sign has also been determined). Now set $x' = 1/y$. Since y is between 0 and 1/2, x' is at least 2. Now do the same thing for x' as was just done for x. That will give the first denominator (and first sign). Repeat this process *ad infinitum*; that will yield all the signs and all the denominators.

Appendix E: Farey paths and some comments on universality

An elegant way to express some universal aspects of the butterfly graph is to define a butterfly hierarchy via a sequence of letters "L" and "R" (for "left" and "right"). The resulting Farey path takes the center of generation l to the left edge, then to the right edge and finally to the center of generation $(l+1)$. If the sequence of letters is *infinite*, we will call it an "infinite Farey path". The term "finite Farey path" will denote a *finite* series of "L"s and "R"s, representing jumps in the Farey tree that

Table 2.1. Convergence of the ϕ_c for four *initial* flux intervals, $\Delta\phi(1)$. In all cases, the centers of the fixed-point butterflies are found to be irrational numbers whose simple continued fractions have tails that oscillate between 1 and 2, and the butterfly flux scaling is given by the constant ζ^*. We invite readers to verify this connection between figure 2.13 and table 2.1 by using either a simple continued-fraction expansion or the butterfly recursions discussed above.

Panel of figure 2.13	$\Delta\phi(1)$	ϕ_c^*
(a)	[2/5–1/3]	[2, 1, 2, 1..] = $\frac{\sqrt{3}-1}{2}$
(b)	[1/3–2/7]	[3, 3, 1, 2, 1, 2...]
(c)	[2/9–1/5]	[4, 1, 2, 1, 2, 1, 2..]
(d)	[3/7–2/5]	[2, 2, 2, 1, 2, 1, 2...]

connect two successive generations of butterflies. An infinite repetition of a finite Farey path (thus a *periodic* Farey path) gives rise to a self-similar butterfly nesting (i.e., a "butterfly hierarchy"). The infinite repetition can also be preceded by an arbitrary finite sequence of "L"s and "R"s, which we will call a "starting segment". An infinite Farey path made of a starting segment followed by a repeating finite Farey path is said to be *eventually periodic*. Four examples of eventually periodic Farey paths are shown in figure 2.13. Each one corresponds to a different butterfly hierarchy.

Table 2.1 shows the relationship between the four infinite Farey paths shown in figure 2.13 and simple continued-fraction expansions whose kth rational approximants, for even values of k, form the centers of butterflies. As can be seen in the table, the centers of all such butterflies converge to irrational numbers whose tails are characterized by the period-2 simple continued-fraction expansion:

$$\phi_c = [n_1, n_2, ...,1, 2, 1, 2, 1, 2...]. \tag{E.1}$$

Our discussion above illustrates the importance of eventually periodic Farey paths in characterizing the universal features of the butterfly, which are intimately related to the eventually periodic sequence of denominators of the simple continued-fraction expansion. In other words, the *starting segment* in an eventually periodic Farey path (or the *opening* of the simple continued-fraction expansion) plays no role in determining the fixed point and the scaling factor that characterize a butterfly hierarchy.

A nice example of this type of universality emerges if we look at the zooms associated with the entire set of level-1 butterflies centered at $\phi_c = \frac{1}{2n}$, where $n = 1, 2, 3...$. Following the recursion relations above, it can be shown that, asymptotically, the centers of these nested butterflies converge to $\phi_c^*(n)$:

$$\phi_c^*(n) = \lim_{l \to \infty} \frac{1}{2(n + \alpha_{2l-2})}, \quad \alpha = \frac{\sqrt{3}-1}{2} = [1, 2, 1, 2, 1, 2...] = \lim_{l \to \infty} \alpha_l. \tag{E.2}$$

Furthermore, this entire set is characterized by the scaling ratio R_ϕ given by equation (2.19).

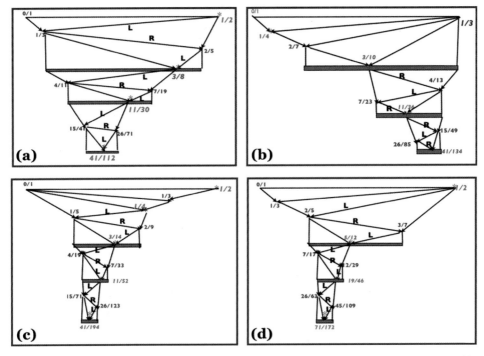

Figure 2.13. Four examples of eventually periodic Farey paths, each of which starts out, at the top, with a finite starting segment (whose "L"s and "R"s are not indicated), and later (i.e. lower down) becomes periodic, by repeating forever a finite Farey path (whose "L"s and "R"s are indicated). For example, case (c) starts off with "LL" and then becomes periodic, as follows: "LL–LRL–LRL–LRL–...". A finite Farey path is a finite sequence of "L"s and "R"s that connects two generations of butterfly through Farey addition. In the figure, the finite Farey paths that repeat have been made easily visible by having them always link a heavy horizontal gray line with a shorter one below it; moreover, to make matters maximally simple, we chose repeated finite Farey paths that are very short: either "LRL" or "RLR". The displayed Farey paths correspond to the butterfly hierarchy characterized by a periodic continued-fraction expansion of the form $[n_1, n_2, 1, 2, 1, 2, ...]$. What assures convergence to fixed-point behavior in all four of these cases is the *eventual periodicity* of the infinite Farey path taken.

References

[1] Ford L R 1938 Fractions *Am. Math. Mon.* **39** 586
[2] Kaku M 1995 *Hyperspace* (New York: Anchor) 330
[3] Hofstadter D R 1976 Energy levels and wave functions of Bloch electrons in rational and irrational magnetic fields *Phys. Rev.* B **14** 2239

IOP Concise Physics

Butterfly in the Quantum World
The story of the most fascinating quantum fractal
Indubala I Satija

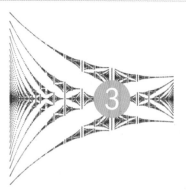

Chapter 3

The Apollonian–butterfly connection ($\mathcal{ABC}$)

Sometimes I write drunk and revise sober, and sometimes I write sober and revise drunk. But you have to have both elements in creation—the Apollonian and the Dionysian, or spontaneity and restraint, emotion and discipline.

—Peter De Vries

Integral Apollonian gaskets ($\mathcal{IAG}$s) and butterfly fractals are two classic types of fractals that are made up of integers. Could they possibly be related? In other words, is this yet another marvelous example of a physical incarnation of apparently abstract mathematics? This chapter addresses this fascinating topic, which is still in its infancy. The discussion below suggests a deep and intricate relationship between these two fractals. However, a complete understanding of this problem remains elusive at present.

An integral Apollonian gasket is an intricate hierarchical structure consisting of an infinite number of mutually kissing (i.e. tangent) circles that are nested inside each other, growing smaller and smaller at each level. Each hierarchical level is made of a finite number of quadruples of mutually tangent circles, and associated with each quadruple of circles is a quadruple of integers that are their curvatures (the reciprocals of their radii). The table in figure 3.1 lists some examples.

As was discussed in chapter 1 and also in chapter 0, the curvatures of the four kissing circles κ_i, with i running from 0 through 3, satisfy the following equation:

$$\kappa_0(\pm) = \kappa_1 + \kappa_2 + \kappa_3 \pm 2\sqrt{\kappa_1\kappa_2 + \kappa_2\kappa_3 + \kappa_3\kappa_1}. \tag{3.1}$$

It is easy to see that $\kappa_0(+)$ and $\kappa_0(-)$ satisfy this simple linear equation:

$$\kappa_0(+) + \kappa_0(-) = 2(\kappa_1 + \kappa_2 + \kappa_3). \tag{3.2}$$

If the first four circles have integer curvatures, then every other circle in the packing does too, thus forming a fractal landscape ruled by integers alone.

The above equations encode the subtle mathematical rules that permit certain integers to belong to an Apollonian quadruple while barring all other integers from this elite club. A key question is whether these mathematical subtleties also define

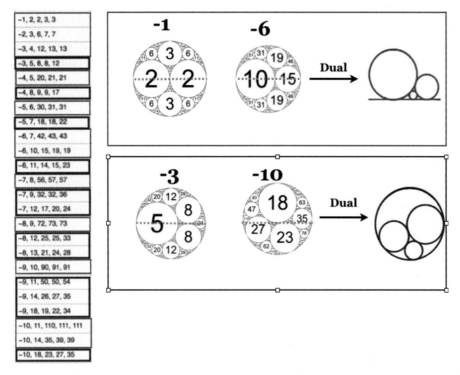

Figure 3.1. The table on the left side lists the curvatures of the circles comprising some members of the $\mathcal{IAG}$ family, with a minus sign attached to the curvature of the outermost bounding circle. The red and blue colors correspond, respectively, to Apollonian gaskets whose duals *are* and *are not* Ford Apollonian gaskets. These two categories of $\mathcal{IAG}$ unfold as one defines their *duals*, which belong to Descartes configurations of either type (b) or type (c), described in chapter 2 and shown on the right with two examples from each type. Of the five curvatures displayed on each line of the table, only the first three are needed to describe the corresponding quadruple. By including all five, however, we highlight the gaskets' mirror symmetry (the horizontal axis of symmetry is shown as a dotted brown line).

or encode the rules that form the butterfly fractal. Appendix A describes a Diophantine quadratic equation with constraints that determine all integral Apollonian gaskets.

3.1 Integral Apollonian gaskets ($\mathcal{IAG}$) and the butterfly

In chapter 2, we discussed a possible connection between the butterfly fractal and integer Apollonian gaskets in view of the fact that butterfly configurations can be associated with Ford circles. In the Hofstadter landscape, there are three rational numbers that define, respectively, the left edge, the center, and the right edge of a butterfly, and these three rationals form a Farey triplet ($\frac{p_L}{q_L}, \frac{p_c}{q_c}, \frac{p_R}{q_R}$). Such a triplet obeys the Farey sum condition: $\frac{p_c}{q_c} = \frac{p_L + p_R}{q_L + q_R}$. These three rationals can also be represented in terms of three mutually kissing Ford circles, sitting on a horizontal axis (a circle of infinite radius), and having curvatures $2q_L^2$, $2q_c^2$, and $2q_R^2$.

Such a quadruple of (generalized) circles is a Descartes configuration of type (c), as described in chapter 2. We will refer to such an Apollonian gasket as a "Ford–Apollonian gasket", meaning a set of four mutually kissing circles that have curvatures that make up a quadruple ($q_c^2, q_R^2, q_L^2, 0$), which we will denote as ($\kappa_c, \kappa_R, \kappa_L, 0$). (Here we have dropped the common factor of 2.) In the $\{l, k, n, m\}$ representation described in appendix B, Ford–Apollonian gaskets correspond to $m = 0$, and hence they obey the simple quadratic equation $l^2 = kn$.

However, Ford–Apollonian gaskets do not belong to the family of $\mathcal{IAG}$s, as can be seen from the table in figure 3.1, which consists solely of quadruples of *non-zero* integer curvatures of four kissing circles.

3.1.1 A duality transformation

Interestingly, it turns out that Ford–Apollonian gaskets are related to $\mathcal{IAG}$s by a *duality* transformation—that is, an operation that is its own inverse (also called an "involution"). This transformation gives us a bridge connecting the butterfly fractal, which is made up of Ford–Apollonian quadruples, with the world of integer Apollonian gaskets. We now proceed to describe this self-inverse transformation both geometrically and algebraically [1].

If we write the curvatures of four kissing circles as a vector A with four integer components, we can use matrix multiplication to obtain another such 4-vector $\bar{A}$. In particular, consider the matrix $\hat{D}$:

$$\hat{D} = \frac{1}{2}\begin{pmatrix} -1 & 1 & 1 & 1 \\ 1 & -1 & 1 & 1 \\ 1 & 1 & -1 & 1 \\ 1 & 1 & 1 & -1 \end{pmatrix}$$

$$\bar{A} = \hat{D}A.$$

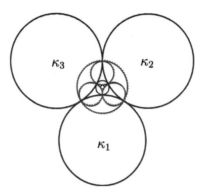

Figure 3.2. An illustration of a duality (a transformation that is its own inverse) involving $\mathcal{IAG}$s: four mutually tangent circles (shown in red) and their dual image (shown in blue). Each circle in the dual set passes through three of the kissing points of the original set of circles, and the reverse holds as well.

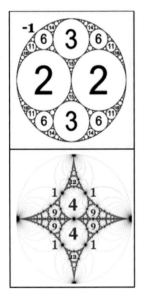
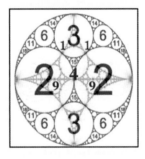

Figure 3.3. On the upper left, we see the $(-1, 2, 2, 3)$ Apollonian gasket. Underneath it is its dual. Note that $(-1, 2, 2, 3) = \hat{D}(4, 1, 1, 0)$. The black and blue integers are the curvatures of the set of larger circles in the $\mathcal{IAG}$ and of those in its dual set. On the right, we have overlaid the $\mathcal{IAG}$ and its dual. Here one easily sees how each circle in the dual configuration passes through the tangency points of three mutually kissing circles in the original configuration.

The matrix $\hat{D}$ is its own inverse. As is shown above, if we multiply A (the 4-vector of curvatures) by $\hat{D}$, we obtain its dual 4-vector $\bar{A}$. Since $\hat{D}^2 = 1$, this same transformation maps the dual gasket back onto the original gasket:

$$A = \hat{D}\bar{A}.$$

In sum, multiplication by the matrix $\hat{D}$ yields the desired duality transformation, mapping an arbitrary Ford–Apollonian gasket onto its dual gasket.

Figure 3.2 illustrates this duality relation geometrically. Given an $\mathcal{IAG}$ (the four red circles in the figure), there exists a related Descartes configuration (the blue circles). Each blue circle is defined by the intersection points of a subset—specifically, a trio—of the four red circles. Since there are four such trios in any quadruple, this construction gives us four new circles. This correspondence is further illustrated in figure 3.3, showing the case of the $\mathcal{IAG}$ (−1, 2, 2, 3).

3.1.2 Illustrating the Apollonian–butterfly connection

Given a butterfly and its representation in terms of Ford circles as discussed above, we can now map the butterfly onto an $\mathcal{IAG}$ using the duality transformation. Figures 3.4 and 3.5 illustrate the Apollonian–butterfly connection—that is, what we have called "$\mathcal{ABC}$". In general, every butterfly whose center and edges obey the Farey triplet rule (and is hence described by a configuration of four kissing Ford circles) can be mapped to an $\mathcal{IAG}$. In the general case, we write this mapping as follows:

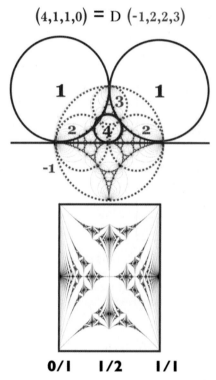

Figure 3.4. Illustrating the correspondence between an $\mathcal{IAG}$ (−1, 2, 2, 3) (top) and the butterfly centered at flux-value $\phi = 1/2$, and with edges at 0/1 and 1/1. The blue circles are Ford circles, representing the butterfly's center and edges, with reduced curvatures (4, 1, 1), all tangent to the horizontal line (whose curvature is zero). These four circles with curvature 4-vector (4, 1, 1, 0) form a Ford–Apollonian gasket that is dual to the (−1, 2, 2, 3) $\mathcal{IAG}$, which is shown in red. As usual, a colored integer inside a circle tells us the circle's curvature.

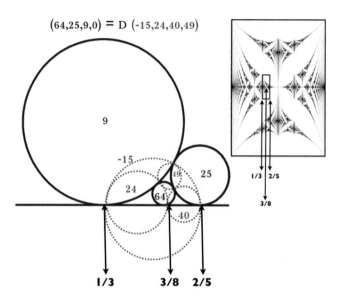

Figure 3.5. This figure, similar to figure 3.4, shows an $\mathcal{IAG}$ and the corresponding butterfly centered at flux-value $\phi = 3/8$ and having edges at 1/3 and 2/5. The four circles with curvature 4-vector (64, 25, 9, 0) form a Ford–Apollonian gasket that is dual to the $(-15, 24, 40, 49)$ $\mathcal{IAG}$.

$$\left(q_c^2, q_R^2, q_L^2, 0\right) \equiv \left(\kappa_c, \kappa_R, \kappa_L, 0\right) = \hat{D}\left(-\kappa_0, \kappa_1, \kappa_2, \kappa_3\right)$$
$$= \hat{D}\left(-q_L q_R, q_c q_L, q_c q_R, q_L q_R + q_R^2 + q_L^2\right).$$

The above equations, which determine a trio of denominators $\{q_L, q_c, q_R\}$ as a function of the four curvatures, appear to imply that the corresponding butterfly is not uniquely determined, as none of the equations involve the trio of numerators $\{p_L, p_c, p_R\}$. However, the requirement that p_x and q_x must be relatively prime leads to a unique set of of rationals—a unique Farey triplet $\{\frac{p_L}{q_L}, \frac{p_c}{q_c}, \frac{p_R}{q_R}\}$—given an $\mathcal{IAG}$.

3.2 The kaleidoscopic effect and trefoil symmetry

We next revisit the question of the hidden trefoil symmetry in the diamond hierarchy, discussed earlier in chapter 2, within the context of $\mathcal{ABC}$. This symmetry came to light when we realized that there is a set of Apollonian gaskets whose successive levels tend asymptotically towards perfect trefoil symmetry, corresponding to the scaling ratio of the butterflies in the diamond hierarchy. We will examine the relationship between butterflies in the diamond hierarchy and Apollonian gaskets that asymptotically exhibit trefoil symmetry. This subject is intimately tied to one of the most fascinating properties of Apollonian gaskets—namely, the "kaleidoscopic effect".

3.2.1 Seeing an Apollonian gasket as a kaleidoscope

An Apollonian gasket is like a kaleidoscope in which the image of the first four circles is reflected again and again through an infinite collection of curved mirrors. In particular, $\kappa_0(+)$ and $\kappa_0(-)$ are "mirror images" through a circular "mirror" that passes though the tangency points of κ_1, κ_2 and κ_3, as is shown in figure 3.7. The curvature of this circular mirror, which we denote as δ, can be shown to be equal to the quantity

$$\delta = \frac{1}{2}\sqrt{\kappa_1\kappa_2 + \kappa_2\kappa_3 + \kappa_3\kappa_1}. \tag{3.3}$$

In the case of Ford–Apollonian gaskets, where the quadruple of curvatures is $\{\kappa_c, \kappa_R, \kappa_L, 0\}$, the curvature of the mirror simplifies down to $\delta = \sqrt{\kappa_L\kappa_R}$.

Before discussing the kaleidoscopic effect in Apollonian gaskets and its relation to butterfly nesting, we shall briefly review the geometric operation called *inversion*, which can be thought of as reflection in a circle. Just as ordinary reflection exchanges what is on the two sides of the mirror, inversion in a circle maps everything in the circle's interior to its exterior, and vice versa—that is, inversion exchanges what is on the two sides of the "mirror".

Three key properties of inversion are:
- *Circularity is preserved* (i.e. circles are carried into circles).
- *Tangency is preserved.*
- *Any circle orthogonal to the "mirror" maps exactly onto itself.* (This happens not because each point on the given circle is individually preserved, but because the two arcs that together make up the given circle—one *inside* the "mirror" and one *outside* of it—are mapped onto each other.)

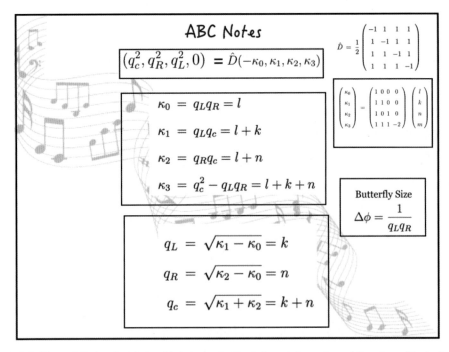

Figure 3.6. Here we display the relationship between the flux-values at the center and the edges of the butterfly, the curvatures κ_i, with $i = 0, 1, 2, 3$, of the corresponding $\mathcal{IAG}$, and the quadruple of integers $\{l, k, m, n\}$. Note that the radius (that is, the inverse of the curvature) of the outermost circle of the $\mathcal{IAG}$ that represents a butterfly is equal to the horizontal size $\Delta\phi$ of the butterfly.

All these properties of inversion are visible in figure 3.7, where both the object circle (outermost red circle) and its reflected image (innermost red circle) are tangent to the same set of three black circles, and where each of the three circles remains invariant under inversion in the dotted (blue) circle [2].

3.2.2 How nested butterflies are related to kaleidoscopes

In our discussion of butterfly hierarchies, the kaleidoscopic aspect of Ford–Apollonian gaskets (figure 3.7(b)) takes on a special new meaning, as the object and the mirror represent two successive generations of a butterfly. From our discussion in chapter 2, we know that the object and the mirror in figure 3.7(b) can be identified as κ_- and κ_+, which correspond to the curvatures of the Ford circles representing two successive levels of butterfly centers:

$$\kappa_+ = \left(\sqrt{\kappa_L} + \sqrt{\kappa_R}\right)^2 \equiv \kappa_c(l)$$
$$\kappa_- = \left(\sqrt{\kappa_L} - \sqrt{\kappa_R}\right)^2 \equiv \kappa_c(l-1).$$

In terms of butterfly coordinates, the ratio of the object and the image for the Ford–Apollonian gasket becomes:

$$\frac{\kappa_+}{\kappa_-} = \left(\frac{1 + q_R/q_L}{1 - q_R/q_L}\right)^2. \tag{3.4}$$

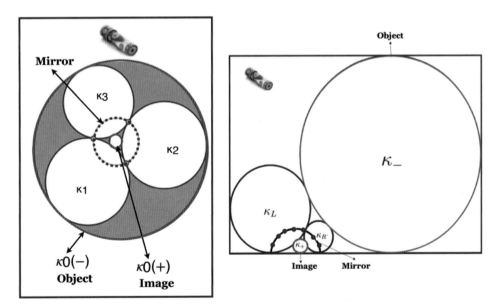

Figure 3.7. This figure shows the kaleidoscopic nature of Apollonian gaskets, where the outermost and the innermost circles (shown in red) are mirror images of each other, reflected through the (blue) circle that passes through the three tangency points of the three other circles. The figures on the left and right show, respectively, Descartes configurations of types (b) and (c), discussed earlier, in chapter 2.

The corresponding ratio for the $\mathcal{IAG}$ that is related to the above Ford–Apollonian gasket by a duality transformation is given by:

$$\frac{k_0(+)}{k_0(-)} = \frac{7q_L q_R + 4q_L^2 + 4q_L^2}{q_L q_R} = 7 + \frac{q_L}{q_R} + \frac{q_R}{q_L}. \tag{3.5}$$

In general, the ratios $\frac{\kappa_+}{\kappa_-}$ and $\frac{k_0(+)}{k_0(-)}$ are not equal. For example, for the diamond hierarchy, which is the most dominant hierarchy in the butterfly landscape, $\frac{q_R}{q_L} \to \sqrt{3}$ (see equation (2.14)).

$$\frac{\kappa_+}{\kappa_-} \to 7 + 4\sqrt{3} = \left(2 + \sqrt{3}\right)^2$$

$$\frac{\kappa_{0(+)}}{\kappa_{0(-)}} \to 7 + \frac{4}{\sqrt{3}}.$$

Therefore, although $\frac{\kappa_+}{\kappa_-}$ gives the correct scaling for the magnetic flux interval for the diamond hierarchy, its corresponding $\mathcal{IAG}$ does not.

However, in chapter 2, it was shown that the scaling ratio for the diamond hierarchy is described by an $\mathcal{IAG}$ that asymptotically has trefoil symmetry. Figure 3.8 shows an example of a butterfly Apollonian gasket (a Ford–Apollonian gasket), its corresponding $\mathcal{IAG}$, and also its trefoil-symmetric partner, which encodes its nesting property, as described in chapter 2.

Intriguingly, the two Apollonian gaskets shown on the bottom left of figure 3.8 are intimately related. We shall now discuss this correspondence for the diamond hierarchy.

3.2.3 $\mathcal{ABC}$ and trefoil symmetry

Given a butterfly represented by $\{\kappa_c, \kappa_R, \kappa_L, 0\}$ and its dual partner $\{\kappa_0, \kappa_1, \kappa_2, \kappa_3\}$, there exists another Apollonian gasket that encodes the nesting characteristics of the butterfly. This "conjugate" Apollonian gasket, which we will denote by $\{-\kappa_0^s, \kappa_1^s, \kappa_2^s, \kappa_3^s\}$ will be referred to as the *symmetric-dual* Apollonian gasket associated with the butterfly. For the diamond hierarchy, it is as follows:

$$\left(-\kappa_0^s, \kappa_1^s, \kappa_2^s, \kappa_3^s\right) = \left(-\kappa_0, \frac{\kappa_1 + \kappa_2}{2}, \frac{\kappa_1 + \kappa_2}{2}, \frac{\kappa_1 + \kappa_2}{2} + d\right) \tag{3.6}$$

$$= \left(-q_L q_R, \frac{(q_L + q_R)^2}{2}, \frac{(q_L + q_R)^2}{2}, \frac{(q_L + q_R)^2}{2} + d\right) \tag{3.7}$$

where $d = \frac{3q_L^2 - q_R^2}{2}$. Here, d reflects a deviation from perfect trefoil symmetry and is invariant for a given "set of zooms" corresponding to different generations of the butterfly. Since $d \to 0$ asymptotically (as $\frac{q_R}{q_L}$ approaches $\sqrt{3}$ in the diamond hierarchy), we see that these Apollonian gaskets evolve towards perfect trefoil symmetry.

Clearly, the ratio $\frac{k_0^s(+)}{\kappa_0^s(-)}$ determines the scaling ratio for the diamond hierarchy, as we saw earlier in equation (2.16). Therefore, for a given butterfly characterized by a

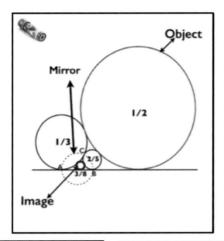
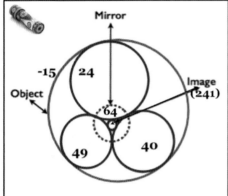
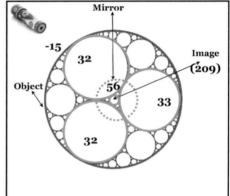

Figure 3.8. The upper figure shows the Ford-circle representation of the butterfly centered at flux-value 3/8. Reflection of the red circle in a circular mirror (dotted) gives another circle (blue), and of course reflecting the blue circle yields the red circle. The figure on the lower left shows the corresponding dual Apollonian gasket. The figure on the lower right is the trefoil-symmetric partner of the Apollonian gasket (−15, 24, 40, 49), with curvature 4-vector (−15, 32, 32, 33). The two partners have the same curvature value for their outermost circles (red) and in addition, their inner circles (shown in blue and green) obey the relation $\kappa_1 + \kappa_2 = \kappa_1^s + \kappa_2^s$.

Farey triplet ($\frac{P_C}{Q_C}$, $\frac{P_L}{Q_L}$, $\frac{P_R}{Q_R}$), and its associated $\mathcal{IAG}$ (−κ_0, κ_1, κ_2, κ_3), the size of the next-generation butterfly for the diamond hierarchy is determined not by the mirror image of the dual, but by its close cousin—the symmetric-dual Apollonian. In other words, the parent butterfly "rejects" the mirror image of its dual $\mathcal{IAG}$ to describe the next-generation daughter butterfly. Instead, it "chooses" the $\mathcal{IAG}$ with the highest form of symmetry possible for an $\mathcal{IAG}$—namely, trefoil symmetry.

The example below (see figure 3.9) illustrates the phenomena described above, explicitly elucidating the relationship between the $\mathcal{IAG}$ representing the butterfly and its trefoil-symmetric partner. We consider the diamond hierarchy as we

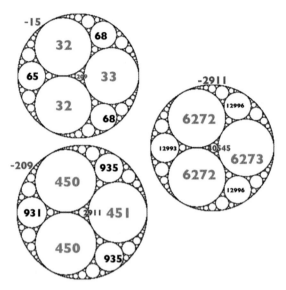

Figure 3.9. Evolution of Apollonian gaskets having approximate trefoil symmetry, and tending asymptotically towards perfect trefoil symmetry, showing how they shed light on the diamond hierarchy. The relation $\kappa_1 + \kappa_2 = \kappa_1^s + \kappa_2^s$ is satisfied at all levels of the hierarchy.

zoom into the butterfly sequence in magnetic-flux intervals $[\frac{0}{1}, \frac{1}{1}] \to [\frac{1}{3}, \frac{2}{5}] \to \ldots$ (see figure 3.5). The equations below list the corresponding dual $\mathcal{IAG}$ and also the symmetric-dual $\mathcal{IAG}$:

$$(0/1, 1/2, 1/1) \to (-1, 2, 2, 3)$$
$$(1/3, 3/8, 2/5) \to (-15, 24, 40, 49) \to (-15, 32, 32, 33)$$
$$(4/11, 11/30, 7/19) \to (-209, 330, 570, 691) \to (-209, 450, 450, 451)$$
$$(15/41, 41/112, 26/71) \to (-2911, 4592, 7952, 9633) \to (-2911, 6272, 6272, 6273).$$

By referring back to the four distinct sequences of butterflies that all belong to the diamond hierarchy (as shown in figure 2.13 in chapter 2), readers can check that their symmetric-dual $\mathcal{IAG}$ is the quadruple $\{-q_L q_R, \frac{q_c^2}{2}, \frac{q_c^2}{2}, \frac{q_c^2}{2} + d\}$, where d quantifies the deviation from perfect trefoil symmetry. Although d takes on different values—namely, 1, 11, 3 and 13 in these four cases (table 3.1)—it approaches a limit as one zooms further and further into the butterfly. In other words, κ_3^s differs from $\kappa_1^s = \kappa_2^s$ by a constant, irrespective of the butterfly generation. This implies that all such configurations evolve towards exact trefoil symmetry, satisfying the relations $d = \frac{3q_L^2 - q_R^2}{2}$ or $\frac{3q_R^2 - q_L^2}{2}$ and $\kappa_3^s = \kappa_4^s = (\kappa_2 + \kappa_3)/2$, as described in equation (3.7). These equations illustrate an important point regarding $\mathcal{ABC}$—namely, given a butterfly and its representation as a Ford–Apollonian gasket, there exists a unique pair of $\mathcal{IAG}$s: the dual of the Ford–Apollonian gasket and its corresponding symmetric partner defined by equation (3.7). We emphasize again that $\mathcal{ABC}$ has

Table 3.1. Three levels of four hierarchies a, b, c and d displayed in figure 13 of chapter 2, labeled here by $d = |\kappa^s - \kappa_1^s|$ (which does not depend upon l). We note that $\kappa_0^s(l)$ and $\kappa_+^s(l)$ are mirror images and the kaleidoscopic relation between two consecutive generations, where $\kappa_0(l)$ and $\kappa_0(l+1)$ are also mirror images, is obeyed, since $\kappa_+^s(l) = k_0(l+1)$ (highlighted in red). This table illustrates the universality as the scaling ratio R_ϕ (shown in bold) converges to a constant value $13.93.. = (2 + \sqrt{3})^2$. Also see table 2.1 in chapter 2.

l	$\left(\frac{p_L}{q_L}, \frac{p_c}{q_c}, \frac{p_R}{q_R}\right)$	$(-\kappa_0, \kappa_1, \kappa_2, \kappa_3, \kappa_+)$	$(-\kappa_0^s, \kappa_1^s, \kappa_2^s, \kappa_3^s, \kappa_+^s)$, $\mathbf{R_\phi} = \frac{\kappa_+^s(l)}{\kappa_0^s(l)} = \frac{\kappa_0(l+1)}{\kappa_0(l)}$
(a) $d = 1$			
1	$\left(\frac{0}{1}, \frac{1}{2}, \frac{1}{1}\right)$	$(-1, 2, 2, 3, 15)$	$(-1, 2, 2, 3, 15)$, **15**
2	$\left(\frac{1}{3}, \frac{3}{8}, \frac{2}{5}\right)$	$(-15, 24, 40, 49, 241)$	$(-15, 32, 32, 33, 209)$, **13.93**
3	$\left(\frac{4}{11}, \frac{11}{30}, \frac{7}{19}\right)$	$(-209, 330, 570, 691, 3391)$	$(-209, 450, 450, 451, 2911)$, **13.93**
(b) $d = 11$			
1	$\left(\frac{2}{7}, \frac{3}{10}, \frac{1}{3}\right)$	$(-21, 30, 70, 79, 379)$	$(-21, 39, 50, 50, 299)$, **14.23**
2	$\left(\frac{7}{23}, \frac{11}{36}, \frac{4}{13}\right)$	$(-299, 468, 828, 997, 4885)$	$(-299, 637, 648, 648, 4165)$, **13.93**
3	$\left(\frac{26}{85}, \frac{41}{134}, \frac{15}{49}\right)$	$(-4165, 6566, 11390, 13791, 67637)$	$(-4165, 8967, 8978, 8978, 58011)$, **13.93**
(c) $d = 13$			
1	$\left(\frac{2}{5}, \frac{5}{12}, \frac{3}{7}\right)$	$(-35, 60, 84, 109, 541)$	$(-35, 72, 72, 85, 493)$, **14.09**
2	$\left(\frac{7}{17}, \frac{19}{46}, \frac{2}{29}\right)$	$(-493, 782, 1334, 1623, 7971)$	$(-493, 1058, 1058, 1071, 6867)$, **13.93**
3	$\left(\frac{26}{63}, \frac{71}{172}, \frac{45}{109}\right)$	$(-6867, 10836, 18748, 22717)$	$(-6867, 14792, 14792, 14805, 95645)$, **13.93**
(d) $d = 3$			
1	$\left(\frac{1}{5}, \frac{3}{14}, \frac{2}{9}\right)$	$(-45, 70, 126, 151, 739)$	$(-45, 95, 98, 98, 627)$, **13.93**
2	$\left(\frac{4}{19}, \frac{11}{52}, \frac{7}{33}\right)$	$(-627, 988, 1716, 2077, 10189)$	$(-627, 1349, 1352, 1352, 8733)$, **13.93**
3	$\left(\frac{15}{71}, \frac{41}{194}, \frac{26}{123}\right)$	$(-8733, 13774, 23862, 28903, 141811)$	$(-8733, 18815, 18818, 18818, 121635)$, **13.93**

been rigorously established for butterflies whose centers lie on the symmetry axis, $E = 0$. These butterflies are related by a duality transformation to the subset of the infinite $\mathcal{IAG}$ family characterized by outermost circles whose curvatures are odd integers. A deeper understanding of this relationship and its extension to off-centered butterflies remains elusive and is a work in progress.

We close this section by pointing out that the trefoil symmetry hidden in the butterfly scaling reveals nature's attempt to restore a left–right symmetry to all the butterflies in the landscape. The broken symmetry of butterflies, whose origin lies in the fact that $q_L \neq q_R$, is asymptotically recovered as we zoom more and more deeply into these butterfly hierarchies. The curvatures of the circles that make up the Apollonian gaskets representing the nested butterflies are given by the symmetrization of q_L and q_R (as can be seen in equation (3.7)). This is another example of the many types of mathematical magic that can be found in the Hofstadter landscape.

3.3 Beyond Ford Apollonian gaskets and fountain butterflies

The red-framed entries in table 3.1 show duals of the Ford–Apollonian gaskets that map to butterfly configurations described by the Ford circles. They constitute only a subset of the entire set of $\mathcal{IAG}$s. Furthermore, Ford–Apollonian gaskets themselves describe only some of the butterflies in the full Hofstadter landscape. Below, we describe some other parts of the butterfly plot that can be related to various members of the $\mathcal{IAG}$ family. The ultimate question of whether one can associate each and every butterfly configuration with an $\mathcal{IAG}$ remains an open challenge.

We recall that Ford–Apollonian gaskets correspond to butterfly configurations whose centers and edges obey the Farey triplet rule. It appears, however, that the fine structure at the center and the boundaries of every one of these butterflies can be viewed as a hierarchy of different kinds of butterflies that do not follow the Farey triplet rule.

A close inspection of the Hofstadter landscape suggests that the entire butterfly graph can be viewed as consisting of two distinct classes of butterflies. This distinction is illustrated in figure 3.10. Depending on whether a given butterfly's

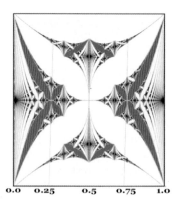

Figure 3.10. The upper two graphs show $\mathcal{K}$-type butterflies (red), whose four wings *kiss* at $\phi = 1/2$, and $\mathcal{C}$-type butterflies, also called "fountain butterflies" (blue), whose four wings *cross* at $\phi = 1/2$. In the graph on the right, the two pairs of wings are displaced in energy. The lower graph displays $\mathcal{K}$-type butterflies (pink), whose four wings *kiss* at $\phi = 1/4$ and $\phi = 3/4$, and $\mathcal{C}$-type butterflies (blue) centered at $\phi = 1/2$, illustrating the fact that fountain butterflies can be viewed as a continuation in the magnetic flux ϕ of $\mathcal{K}$-type butterflies.

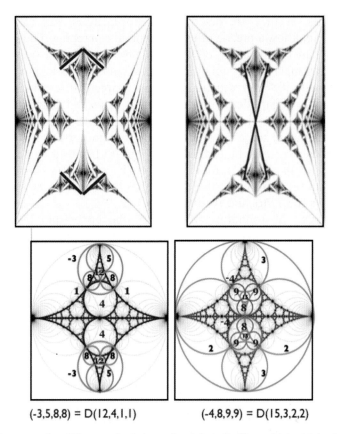

Figure 3.11. Two examples of the mapping between fountain butterflies and integral Apollonian gaskets.

wings *kiss* or *cross*, we will say that the butterfly belongs to the "kissing type" or "$\mathcal{K}$-type" (shown in red), or to the "crossing type" or "C-type" (shown in blue). Only the $\mathcal{K}$-type butterflies obey the Farey triplet rule. C-type butterflies, which we will also call *fountain butterflies*, appear to map onto non-Ford–Apollonian gaskets.

Figure 3.11 shows some examples of fountain butterflies. As can be seen in figure 3.10, fountain butterflies are a continuation of the kissing butterflies as one varies ϕ. Without going into further detail, we simply mention that chapter 10 will present a topological labeling of the gaps that provides a quantitative way of distinguishing $\mathcal{K}$-type butterflies from C-type butterflies. Incidentally, it is not necessary to view the butterfly landscape as consisting of two types of butterflies. The $\mathcal{K}$-type alone is sufficient to describe the butterfly fractal, since fountain butterflies can be regrouped by ripping apart their four wings, so that they become part of the $\mathcal{K}$-type structures that are off-centered.

Figure 3.11 shows fountain butterflies centered at $\phi = 1/2$ with $E = 0$ or at $E_{\max}$ or at $E_{\min}$ and their corresponding $\mathcal{IAG}$s. The actual process involved in identifying a given butterfly with the appropriate Apollonian gasket is rather straightforward for

$\mathcal{K}$-type butterflies, as they are described by Ford circles. For C-type butterflies, however, we do not know a precise rule that defines the mapping. The key point is that C-type butterflies are continuations of $\mathcal{K}$-type butterflies, as can be seen in the butterfly graph. A clear rationale for defining the fountain butterflies is tied to the topological aspects of the butterflies. As will be shown in chapter 10, every butterfly can be associated with a set of gaps determined by the flux value that defines the center of the butterfly and whose topological labeling is controlled by the main butterfly.

Below we summarize some key facts of the fountain butterflies.

- Hierarchical fountain butterflies at $E = 0$ are believed to be related to the sequence $(-4n, 4(n + 1), (n + 2)^2, (n + 2)^2)$ of $\mathcal{IAG}$s—the sequence that begins with the Ford circle that determines the center of the butterfly. Figure 3.11 shows the $\mathcal{IAG}$ and its dual for $n = 1$.
- Analogously, the fountain butterflies at the upper and lower boundaries appear to be related to the sequence $(-(2n + 1), (2n + 3), 2(n + 1)^2, 2(n + 1)^2)$. The $\mathcal{IAG}$ and its dual that corresponds to the $n = 1$ sequence is shown in figure 3.11.
- We note that both sequences mentioned above have outer circles whose curvatures are either even or odd integers and which encode the topological numbers of the gaps that form these fountains. (See figure 10.6 in chapter 10.)

In conclusion, various results described above point toward a deep and beautiful link between the Hofstadter butterfly and the infinite family of integral Apollonian gaskets. Among many other things, nature has found a way to use beautiful symmetric integral Apollonian gaskets in the quantum mechanics of the two-dimensional electron gas problem. Clearly, the preceding discussion has raised many more questions than it has answered. Finding a systematic mathematical framework that relates the Hofstadter landscape to the set of $\mathcal{IAG}$s is a work in progress, and we hope eventually to solve this problem.

> *I hope that posterity will judge me kindly, not only as to the things which I have explained, but also as to those I have intentionally omitted so as to leave to others the pleasure of discovery.*
>
> — René Descartes

Appendix: Quadratic Diophantine equations and $\mathcal{IAG}$s

An $\mathcal{IAG}$ is an integer wonderland that has intricate rules for admitting which integers belong to the set. The rules and regulations determining who is and who is not allowed entry into this exotic fractal land make for a fascinating topic. However, delving into such matters here would carry us too far afield; instead, we refer interested readers to the relevant literature. Rather, our focus here is to discuss the relationship between $\mathcal{IAG}$s and the butterfly fractal, while not worrying too much about the kind of rigor that mathematicians prefer. Below we state the rules that determine these Apollonian gaskets.

The following simple quadratic equation involving four non-negative integers $l, k, n,$ and m [3] determines all integral Apollonian gaskets:

$$l^2 + m^2 = kn. \qquad (A.1)$$

Here, $\{l, k, n, m\}$ is an irreducible quadruple of non-negative integers, with $2m \leqslant k \leqslant n$ and $l \geqslant \sqrt{3}m$. (By "irreducible" is meant that these numbers have no common factor.) These four integers are related to the quintet of $\mathcal{IAG}$ curvatures κ_i, with i running from 0 through 4, in the following way:

$$\kappa_0 = l; \quad \kappa_1 = l + k; \quad \kappa_2 = l + n; \quad \kappa_3 = l + k + n - 2m; \quad \kappa_4 = l + k + n + 2m.$$
$$(A.2)$$

The above set of linear equations relating the k_i (with i running from 0 to 3) with the four integers $\{l, k, n, m\}$ can be expressed more compactly as a matrix equation:

$$\begin{pmatrix} \kappa_0 \\ \kappa_1 \\ \kappa_2 \\ \kappa_3 \end{pmatrix} = \begin{pmatrix} 1 & 0 & 0 & 0 \\ 1 & 1 & 0 & 0 \\ 1 & 0 & 1 & 0 \\ 1 & 1 & 1 & -2 \end{pmatrix} \begin{pmatrix} l \\ k \\ n \\ m \end{pmatrix} \equiv \hat{L} \begin{pmatrix} l \\ k \\ n \\ m \end{pmatrix}. \qquad (A.3)$$

Here κ_0 is the curvature of the outer bounding circle. We remind readers that the curvature κ_0 is positive. However, it must be taken with a negative sign when one is using Descartes' theorem.

Conversely, given the four curvatures κ_i, with i running from 0 to 3, the four integers $l, k, n,$ and m can be calculated as follows:

$$l = \kappa_0; \quad k = \kappa_1 - \kappa_0; \quad n = \kappa_2 - \kappa_0; \quad 2m = \kappa_1 + \kappa_2 - \kappa_0 - \kappa_3. \quad (A.4)$$

The quadruple of integers $\{l, k, n, m\}$ is a good alternative to the quadruple of curvatures $\{\kappa_0, \kappa_1, \kappa_2, \kappa_4\}$ as a label that uniquely identifies an Apollonian gasket. For a geometrical interpretation of these integers, we refer readers to the paper by Kocik [3].

References

[1] Graham R L, Lagarias J C, Mallows C L, Wilks A R and Yan C H 2005 Apollonian circle packings: geometry and group theory I. The Apollonian group *Discrete Comput. Geom.* **34** 547–85
[2] http://www.artofproblemsolving.com/blog/34543
[3] Kocik J 2012 On a Diophantine equation that generates all Apollonian gaskets *ISRN Geom.* **2012** 348618

IOP Concise Physics

Butterfly in the Quantum World
The story of the most fascinating quantum fractal
Indubala I Satija

Chapter 4

Quasiperiodic patterns and the butterfly

Quasiperiodic tiles near Potsdamer Platz, Berlin.

In many years, flu sweeps the world. The actual strain varies from year to year; some years it has been Hong Kong flu, some years swine flu. In 1981, it was the almost periodic flu.

—Barry Simon

This chapter transports readers to another wonderland—the landscape of *quasiperiodic* or *almost-periodic* patterns. These non-repeating patterns are intimately related to the butterfly's fractal aspects. However, our aim here is broader—namely, to provide an introduction to the subject of quasiperiodicity and quasicrystals, thus taking a little detour in our journey, and revealing a world that possesses a new type of order, in spite of the lack of periodicity.

4.1 A tale of three irrationals

1. One of the most famous of all integer sequences is the Fibonacci sequence 0, 1, 1, 2, 3, 5, 8, 13, 21, 34, ..., defined by two very simple initial conditions and a recursion relation:

$$F_0 = 0; \quad F_1 = 1 \tag{4.1}$$

$$F_{l+1} = F_l + F_{l-1}. \tag{4.2}$$

The ratios of successive terms of the Fibonacci sequence converge to the golden mean:

$$\lim_{l \to \infty} \frac{F_{l+1}}{F_l} = \frac{1 + \sqrt{5}}{2} = [1; 1, 1, 1, 1, ...]. \tag{4.3}$$

The first person to study this sequence extensively was the Italian mathematician Leonardo of Pisa, more often called Leonardo Fibonacci, in the early 13th century, although the first several terms of the sequence had been known for centuries in India, thanks to their connection with Sanskrit versification. Among those who had commented on the sequence was the scholar Acharya Hemachandra, a poet and polymath who wrote on philosophy, history, grammar, and prosody[1]. For this reason, the Fibonacci numbers are also sometimes known as the Hemachandra–Fibonacci numbers.

The even entries in the Fibonacci sequence—0, 2, 8, 34, 144, ... (which we denote as F_l^e)—appear every third term, and they satisfy the following recurrence relation:

---

[1] The Fibonacci sequence turned up in Indian mathematics (around 200 BC) in connection with the Sanskrit tradition of prosody. It was of interest to scholars of Sanskrit to enumerate all patterns composed of long syllables, lasting two time units, and short syllables, lasting just one time unit. Counting the number of different patterns of 2's and 1's adding up to a given total time-length yields the Fibonacci numbers. In the West, the Fibonacci sequence first appears in the book *Liber Abaci* by Leonardo of Pisa, also known as Leonardo Fibonacci, in a study of rabbit breeding.

$$F^e_{l+1} = 4F^e_l + F^e_{l-1}. \tag{4.4}$$

The ratios of successive terms of this sequence converge to a limit, as follows:

$$\lim_{l \to \infty} \frac{F^e_{l+1}}{F^e_l} = 2 + \sqrt{5} = \left[\frac{1+\sqrt{5}}{2}\right]^3 = [4; 4, 4, 4, 4, 4, ...]. \tag{4.5}$$

This limiting ratio, the irrational number $2 + \sqrt{5}$, is a close cousin of the golden mean, and is sometimes referred to as the *fourth metallic mean*.

2. A sequence similar to the Fibonacci sequence is given by the recursion relation

$$\mathscr{P}_{l+1} = 2\mathscr{P}_l + \mathscr{P}_{l-1}, \tag{4.6}$$

using the initial conditions

$$\mathscr{P}_0 = 1, \quad \mathscr{P}_1 = 2, \tag{4.7}$$

to get it off the ground. These numbers are called the Pell numbers, named after the English mathematician John Pell, and they start out 1, 2, 5, 12, 29, 70, 169, ..., alternating between odd and even integers.

The ratio of consecutive Pell numbers tends to the so-called *silver mean*, $1 + \sqrt{2}$, whose continued fraction is $[2; 2, 2, 2, 2, ...]$. Thus,

$$\mathscr{P}_{l+1} = 2\mathscr{P}_l + \mathscr{P}_{l-1}, \quad \lim_{l \to \infty} \frac{\mathscr{P}_{l+1}}{\mathscr{P}_l} = 1 + \sqrt{2} = [2; 2, 2, 2, 2, ...]. \tag{4.8}$$

3. In our exploration of the butterfly hierarchy in the previous chapter, we came across a sequence of integers that obeyed the following recursion relation, and whose terms' ratios had the following limit:

$$\mathscr{D}_{l+1} = 4\mathscr{D}_l - \mathscr{D}_{l-1}, \quad \lim_{l \to \infty} \frac{\mathscr{D}_{l+1}}{\mathscr{D}_l} = 2 + \sqrt{3} = [3; 1, 2, 1, 2, 1, 2, ...]. \tag{4.9}$$

In analogy to the previously discussed limits of ratios of recursively defined integer sequences, we will refer to $2 + \sqrt{3}$ as the *diamond mean*.

Quadratic irrational numbers such as the foregoing three—the golden mean, the silver mean, and the diamond mean—are examples of what are known as *Pisot–Vijayaraghavan numbers*, or "PV numbers", for short. A PV number is a quadratic irrational (the solution to a quadratic equation with rational coefficients) that has the property that the product of the number itself with its *quadratic conjugate* (obtained by changing the sign of the square root) is equal to ± 1. A quadratic irrational always has the property that the sequence of denominators in its simple continued fraction is periodic, as we have already noted for $1 + \sqrt{2}$, the golden ratio $\frac{1+\sqrt{5}}{2}$, and so forth.

Here is a summary of some of the number-theoretical properties of these special quadratic irrationals:

- the diamond mean, $2 + \sqrt{3}$, is a solution of $x^2 - 4x + 1 = 0$, and its continued-fraction expansion is [3; 2, 1, 2, 1, ...];
- the golden mean, $\frac{1+\sqrt{5}}{2}$, is a solution of $x^2 - x - 1 = 0$, and its continued-fraction expansion is [1; 1, 1, 1, 1, ...]. Closely related to this is the fourth metallic mean, $2 + \sqrt{5}$, which is a solution of $x^2 - 4x - 1 = 0$, and whose continued-fraction expansion is [4; 4, 4, 4, 4, ...].;
- the silver mean, $1 + \sqrt{2}$, is a solution of $x^2 - 2x - 1 = 0$, and its continued-fraction expansion is [2; 2, 2, 2, 2, ...].

In our discussion of the butterfly hierarchies in chapter 2, the relation $\mathcal{D}_{l+1} = 4\mathcal{D}_l - \mathcal{D}_{l-1}$ emerged as the key recursion determining the numerators as well as the denominators of the flux values corresponding to the centers and left and right edges of central butterflies. (See equations (2.10)–(2.12).) That is, the sequences of integers $\mathcal{D}_l = p_x(l)$, $q_x(l)$—the numerators and denominators of the rational numbers $\frac{p_x(l)}{q_x(l)}$, where $x = C, L, R$—determine the magnetic flux values (ϕ) of the center, left edge, and right edge of the infinitely nested butterflies. Here the subscript l labels the lth generation, or the "zoom", of the butterfly.

It turns out that self-similar fractal hierarchies are described by irrational numbers with periodic continued-fraction expansions. In fact, in describing the phenomenon of asymptotic self-similarity, all that matters is the *tail* of the continued-fraction expansion. Therefore, associated with the golden, silver, and diamond means, there are three *classes* of irrational numbers, which we will denote by $\zeta_{1,2}$, ζ_1, and ζ_2, respectively:

$$\zeta_{1,2} \equiv [n_1, n_2, ...,2, 1, 2, 1, 2, 1...] \tag{4.10}$$

$$\zeta_1 \equiv [n_1, n_2, ...,1, 1, 1, 1...] \tag{4.11}$$

$$\zeta_2 \equiv [n_1, n_2, ...,2, 2, 2, 2...]. \tag{4.12}$$

Here $n_1, n_2, \ldots$ can be any positive integers; they form the initial entries of the continued-fraction expansion but they are irrelevant to the asymptotic scaling properties. This was illustrated in chapter 2, where we saw that the butterfly hierarchies were described by the scaling ratio equal to the diamond mean, for the entire set of fixed-point (asymptotic) butterflies centered at $\phi_c^* = \zeta_{12}$ (see table 2.1).

Aperiodic sequences and the discovery of the Hofstadter butterfly

Taking a historical perspective for a moment, we note that aperiodic sequences played a central role in the discovery of the Hofstadter butterfly. Linked to the tales of our three irrationals is the tale of what Douglas Hofstadter dubbed "η-sequences" when he first came across them in early 1961, and whose nature he explored for years thereafter. As he states in the prologue—"The grace of Gplot"—it was η-sequences that launched his study of certain areas of number theory in the 1960s. Because Hofstadter was plunged for so long into the study of η-sequences and related phenomena, especially certain self-similar two-dimensional graphs that very naturally came out of them, he developed a deep intuition for these things, and that intimacy was what allowed him to make his discovery of the butterfly and its recursive structure. And as we will see in chapter 10, for irrational magnetic flux values, the quantum numbers associated with the bands of the butterfly fractal are examples of η-sequences.

4.2 Self-similar butterfly hierarchies

To describe the self-similar properties of the butterfly fractal, we now proceed to determine the relationships between the butterfly and various quadratic irrationals. It will be useful to keep in mind the following important points related to the emergence of the diamond hierarchy as described in chapter 2:

- The diamond-mean scaling and the recursion relation given in equation (4.9) emerged from two types of tangency conditions—namely, the "inter-kiss precise" and "intra-kiss precise" conditions—when we examined the relationship between *equivalent* sets of butterflies obtained by zooming inwards in the original landscape. This set was found to correspond to a fixed path in the Farey tree, described by the symbol sequences "LRL" or "RLR".
- The nested sets of butterflies converged to a single center at ϕ_c^*, corresponding to the even-numbered rational approximants of $\zeta_{1,2}$, shown in table 2.1.

Two very natural analogous questions that we now address are: are there similar butterfly hierarchies that are related to other quadratic irrationals, such as the golden and silver means, and how do we track them?

To construct a hierarchy characterized by a given quadratic number, we follow these three rules:

(1) The rational approximants to a quadratic irrational having *even* denominators $f_c(l)$, $l = 1, 2, 3...$ are the fractions located at the centers of the butterflies.
(2) For each center $f_c(l)$, there exists a unique pair of fractions $(f_L(l), f_R(l))$ (with *odd* denominators), which give the left and the right edges of the butterfly. These fractions are also the left and right neighbors of $f_c(l)$ in the Farey tree.
(3) The butterfly spectrum in the flux interval $f_L(l) \leq \phi \leq f_R(l)$ is obtained numerically, where the energy interval is fixed by the central band of f_L.

Figure 4.1 shows three levels of the butterfly hierarchy for the diamond, silver, and golden means, and table 4.1 summarizes their key characteristics. Figure 4.2 shows Ford-circle representations of the magnetic-flux intervals for these hierarchies. A graphical illustration of the Farey sums that yield these nested sets of butterflies is shown in figure 4.3.

As we zoom into any given butterfly, we will find inside it a nested set of butterflies. The nesting relations between various generations of the butterfly

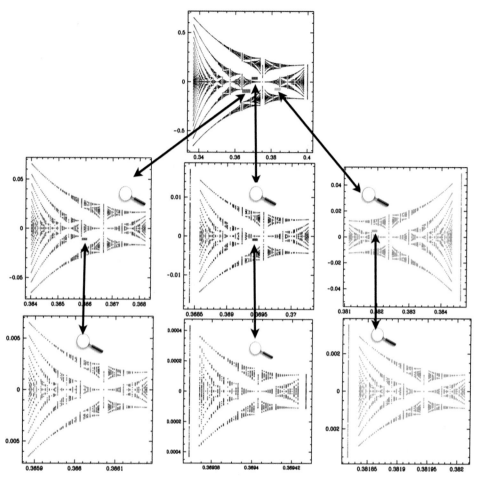

Figure 4.1. Butterfly plots for three levels of the diamond (red), silver (blue), and golden (green) hierarchies, where the uppermost graph (black) shows the first level of all three hierarchies (with the short horizontal lines indicating the sites of zooming-in). Close inspection of both the vertical scale (energy) and horizontal scale (flux) shows that $\zeta_{1,2}$ is the dominant hierarchy, where the scale factors along the *energy* axis are approximately 10, 38, and 14^2, for the diamond and the golden hierarchies. The corresponding scale factors along the *flux* axis are $(3.73)^2$, $(5.83)^2$ and $(4.24)^4$. (Also see table 4.1.) We note that the need to square the scale factors for the golden case is due to the flipping of the left and right edges between two successive generations.

depend on the type of hierarchy. For all hierarchies, the triplet (ϕ_L, ϕ_c, ϕ_R) is a friendly triplet—that is, any two members of this triplet are Farey neighbors—but different hierarchies have different relationships between two successive generations of butterflies. In our studies of three different hierarchies, which we have called the "diamond", "golden", and "silver" hierarchies, we have noted the following facts:

1. In the diamond hierarchy, the center of the lth generation butterfly is friendly with both the left and the right edges of the $l + 1$st generation butterfly, and

Table 4.1. Comparing the scaling ratios R_ϕ and R_E for three quadratic values of the flux, whose even-denominator approximations are the centers of the butterfly. Each of these three quadratic values represents the scaling of an infinite set of irrational numbers whose continued-fraction expansions all share the same tail, each associated with its unique Farey path.

ϕ_c^*	Recursion relations for $q_x, p_x \equiv s$	Farey pathway	$\sqrt{R_\phi}$	$\approx R_E$
$\zeta_{1,2} = [\ldots 1, 2, 1, 2 \ldots]$	$s(l+1) = 4s(l) - s(l-1)$	LRL	$2 + \sqrt{3} = 3.73205$	10
$\zeta_1 = [\ldots 1, 1, 1, 1 \ldots]$	$s(l+1) = 4s(l) + s(l-1)$	LRLRLR	$(\frac{1+\sqrt{5}}{2})^3 = 4.236068$	14
$\zeta_2 = [\ldots 2, 2, 2, 2 \ldots]$	$s(l+1) = 6s(l) - s(l-1)$	LRRL	$(1+\sqrt{2})^2 = 5.82843$	38

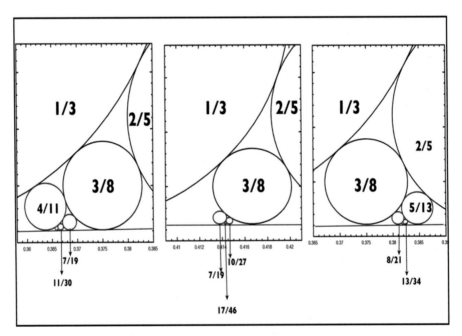

Figure 4.2. The images on the left, center, and right, respectively, show Ford-circle representations of the diamond, silver, and golden hierarchies. In each case, the circles represent the center, left, and right edges of the butterfly, corresponding to three levels, shown in black, blue, and red, respectively. The figure shows the flipping of the butterfly between successive levels for the golden hierarchy. Furthermore, in the silver hierarchy, the Ford circles for two consecutive generations do not form a close-packing of circles.

their corresponding Ford circles kiss. In addition, the left (or right) edge of the lth generation is friendly with the left (or right) edge of the $l+1$st generation, with kissing Ford circles.

2. In the golden hierarchy, the center of the lth generation butterfly is friendly with both the left and the right edges of the $l+1$st-generation butterfly, and

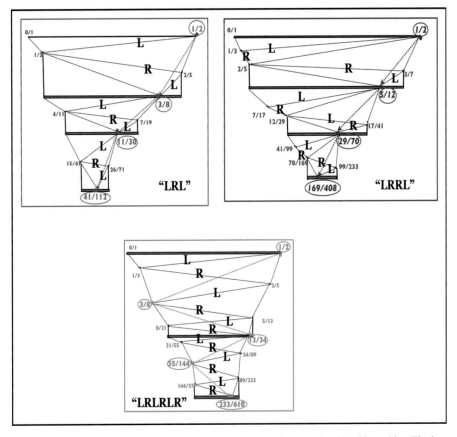

Figure 4.3. Farey paths and butterfly intervals for the diamond, silver, and golden hierarchies. The butterfly edges and centers obtained using Farey sums are explicitly shown. In the diamond hierarchy shown here, the butterfly edges and centers always converge towards the left, while in the golden hierarchy they oscillate, as is shown by the dashed lines connecting the butterfly centers (circled-ϕ values). This leads to flipping of the butterfly between successive levels, for the golden hierarchy, as is shown in figure 4.1. The black lines connecting Farey neighbors show Farey paths that can be symbolically represented using Ls and Rs. The figure shows three different hierarchies corresponding to different sequences of Ls and Rs. More generally, any butterfly hierarchy corresponds to a sequence of Ls and Rs.

their corresponding Ford circles kiss. However, the left (or right) edge of the lth generation is friendly to the right (or the left) edge of the $l + 1$st generation with kissing Ford circles.

3. In the silver hierarchy, the center of the lth generation butterfly is friendly with both the left and the right edges of the $l + 1$st generation butterfly, and their corresponding Ford circles kiss. However, neither the left nor the right edge at the lth generation is friendly with the edges of the $l + 1$st-generation butterfly.

Therefore, the simplest and most dominant hierarchy appears to be the diamond hierarchy, defined by the shortest possible Farey pathway—"LRL".

4.3 The diamond, golden, and silver hierarchies, and Hofstadter recursions

In his thesis and his *Physical Review* article [1], Douglas Hofstadter, in order to explain the nature of the infinite nesting of the butterfly inside itself, defined families of what he called "cells", "subcells", "subsubcells" (etc), and gave equations connecting the "inner variable" of any subcell with the "outer variable" of the cell containing it (see chapter 2). With the aid of these recursion relations, he was able to explain the hierarchical clustering-pattern of bands for any flux-value ϕ (see chapter 6), even including the Cantor-dust spectra belonging to irrational values of ϕ.

Here we will exploit Hofstadter's recursions to illustrate the self-similar scaling properties of various hierarchies (such as the diamond, silver, and golden hierarchies discussed above). In other words, we will solve for the fixed points of these recursions, as well as the scaling ratios R_ϕ that characterize the scaling along the ϕ-axis of the butterflies making up various hierarchies. Although establishing an equivalence between Hofstadter's recursions involving continuous variables and this chapter's integer recursions using s-values (the numerators p and denominators q of the edges and centers of butterflies making up a given hierarchy, as shown in table 4.1) remains an open problem, both ways of looking at the phenomenon lead to the scaling behavior described above.

The trajectories of the flux-values ϕ_l connecting two successive generations of butterflies fall into two distinct categories of C and L cells. Since here we are focusing on *central* butterflies—namely, those butterflies whose centers are located on the line $E = 0$—we will first discuss the recursions for the C-subshells. As is stated below, the ϕ trajectories or recursions are sensitive to the relative location of ϕ_{l+1} with respect to the "butterfly center" at generation l. Consequently, the prescription for describing the golden hierarchy, where centers of the butterflies at various generations follow a zigzag pattern, oscillating to the left and to the right (as can be seen in figure 4.3), differs from the rules for the diamond and silver hierarchies, where the butterfly centers at all generations are located to the left of the centers from the previous generation.

Without further ado, here is Hofstadter's rule relating the local variables for two C-cells belonging to successive generations:

C-cells

$\phi_{l+1} = \dfrac{1}{2+\frac{1}{\bar{\phi}_l}}$, if ϕ_{l+1} is to the left of the butterfly center of generation l, and

$1 - \phi_{l+1} = \dfrac{1}{2+\frac{1}{\bar{\phi}_l}}$, if ϕ_{l+1} is to the right of the butterfly center of generation l.

Here, $\bar{\phi}_l = [\bar{\phi}_l] + \{\bar{\phi}_l\} \equiv [\bar{\phi}_l] + \phi_l$, where the notations "$[x]$" and "$\{x\}$" denote the *integer part* and the *fractional part* of an arbitrary real number x, respectively.

If we denote the integer part of $[\bar{\phi}_l]$ by N, we obtain the C_N cells shown in figure 4.4. To describe the diamond, silver, and golden hierarchies, we consider two distinct cases.

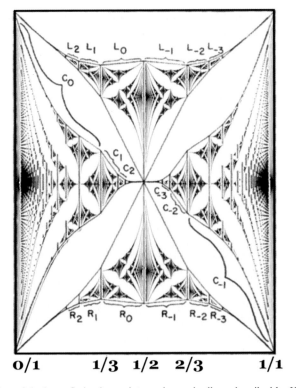

Figure 4.4. Partitioning of the butterfly landscape into various subcells as described by Hofstadter in his 1976 *Physical Review* article [1]. Figure adapted with permission from [1] copyrighted by the American Physical Society.

1. For butterflies whose centers remain on the same side for all generations, such as the diamond and the silver hierarchies, the ϕ recursion relation is:

$$\phi_{l+1} = \cfrac{1}{2 + \cfrac{1}{N + \phi_l}}. \tag{4.13}$$

To determine the asymptotic scaling for this subshell, we first solve the fixed-point equation:

$$\phi_{l+1} = \phi_l = \phi_{C_N}^* = \cfrac{1}{2 + \cfrac{1}{N + \phi_{C_N}^*}}.$$

This is a quadratic equation, and it determines the fixed point $\phi_{C_N}^*$:

$$\phi_{C_N}^* = \frac{\sqrt{N^2 + 2N} - N}{2}.$$

This irrational number $\phi_{C_N}^*$ has the following period-2 continued-fraction expansion:

$$\phi_{C_N}^* = \cfrac{1}{2 + \cfrac{1}{N + \cfrac{1}{2 + \cfrac{1}{N + \cdots}}}} \equiv [2, N, 2, N, \ldots].$$

The asymptotic butterfly scaling between two successive generations is the derivative of ϕ_{l+1} with respect to ϕ_l, evaluated at ϕ^*:

$$\lim_{l \to \infty} \frac{\Delta \phi_{l+1}}{\Delta \phi_l} \bigg|_{\phi = \phi^*} = \frac{1}{\left(N + 1 + \sqrt{N^2 + 2N}\right)^2}. \tag{4.14}$$

The $N = 1$ butterfly hierarchy is the most dominant C_N hierarchy, characterized by the least amount of shrinking of butterflies between successive generations. Furthermore, the case $N = 0$ describes a set of butterflies where successive generations always share their left edge. Such a hierarchy of butterflies is characterized by the trivial scaling factor of unity.

2. When the centers of successive butterflies zigzag back and forth from left to right (this case includes the golden hierarchy), then the ϕ recursion needs to be modified as discussed above. Below, we illustrate this case for the golden hierarchy, whose centers are given by $\phi_l = 1/2, 5/8, 21/34, 89/144, \ldots$. Readers can easily see that this corresponds to $\bar{\phi}_{2l-1} = 1 + \phi_{2l-1}$ and $\bar{\phi}_{2l} = -2 - \phi_{2l}$, resulting in the following recursions:

$$1 - \phi_{2l} = \cfrac{1}{2 + \cfrac{1}{1 + \phi_{2l-1}}}$$

$$\phi_{2l+1} = \cfrac{1}{2 - \cfrac{1}{2 + \phi_{2l}}}.$$

The above equations can be rewritten as follows:

$$\phi_{2l} = \cfrac{1}{2 - \cfrac{1}{2 + \phi_{2l-1}}} \tag{4.15}$$

$$\phi_{2l+1} = \cfrac{1}{2 - \cfrac{1}{2 + \phi_{2l}}}. \tag{4.16}$$

The fixed-point equation $\lim_{l \to \infty} \phi_{2l-1} = \phi_{2l+1} = \phi_g^*$ gives $\phi_g^* = \frac{\sqrt{5}-1}{2}$. This gives $R_{\phi^*} = (8\phi_g^* + 13)^2 = \phi_g^{*6}$, in agreement with the result shown in table 4.1.

The recursions for the rational $\phi = \frac{p}{q}$ (as discussed by Hofstadter) and the recursions for the associated integers p and q (discussed here and in chapter 2) yield exactly the same scaling exponents for the butterfly.

And now here is Hofstadter's rule connecting the local variables in L-cells that belong to successive generations:

L-Cells

$$\phi_{l+1} = \frac{1}{2 + N + \phi_l}.$$

Its fixed point is $\phi_{L_N}^* = \frac{\sqrt{(N+2)^2 + 4} - (N+2)}{2} \equiv [N+2, N+2, N+2, N+2, \ldots]$. The corresponding scaling exponent is determined by this equation:

$$\lim_{l \to \infty} \frac{\Delta\phi_{l+1}}{\Delta\phi_l}\bigg|_{\phi = \phi^*} = -\frac{4}{\left(N + 2 + \sqrt{(N+2)^2 + 4}\right)^2}. \quad (4.17)$$

In parallel with our discussion for the C-cells, the recursion for the L-cells needs to be modified when the butterfly centers zigzag back and forth.

4.4 Symmetries and quasiperiodicities

It turns out that these three quadratic numbers—the golden, silver, and diamond means—are associated, respectively, with 5-fold, 8-fold, and 12-fold symmetries. An illustration of this is given in figure 4.5. Note that these quadratic numbers are

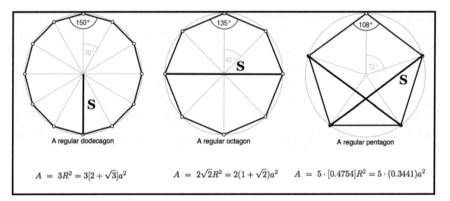

Figure 4.5. With S/a equal to (from left to right) the diamond mean, the silver mean, and the golden mean, this figure shows regular polygons whose angles relate to the corresponding quadratic numbers. Here R is the radius of the circumscribed circle, a is the length of the polygon's side, and A is the polygon's area. The polygons have vertex angles of 30, 45, and 72 degrees respectively, or equivalently, $\frac{2\pi}{12}$, $\frac{2\pi}{8}$, and $\frac{2\pi}{5}$ radians. This reveals the relationship between these quadratic irrationals and 12-fold, 8-fold, and 5-fold symmetries. The diamond mean stands out here, because of the simplicity of its relationship to both A and R.

related to the angles at the vertices of a regular pentagon, octagon, and dodecagon, respectively. In the preceding discussion, we saw that 12-fold symmetry is the dominant symmetry underlying the Hofstadter butterfly, and this intriguing result seems to be related to the trefoil symmetry of an Apollonian gasket. In chapter 10, we will see that these symmetries are also shared by the topological characterization of the butterfly. The significance of this result remains elusive.

Quasiperiodic patterns associated with quadratic irrationals that underlie the self-similar butterfly are examples of one-dimensional "quasicrystals", a new kind of crystalline structure that lacks the exact periodicity associated with classical crystal lattices. Experimentally discovered in 1982, and honored with a Nobel Prize in Chemistry in 2011, quasicrystals constitute a new paradigm at the frontiers of physics.

One-dimensional quasiperiodic patterns are projections of higher-dimensional periodic lattices. Recent studies have argued that quasiperiodic systems possess nontrivial topological properties, thanks to their relation to higher-dimensional entities. In view of this connection between quasicrystals and higher-dimensional structures, it is conceivable that the study of these sorts of mathematical objects may have some bearing on other areas of physics, such as string theory [2].

In the appendix below, we give an overview of some of the properties of quasicrystals, including a brief history of their discovery.

Appendix: Quasicrystals

Modern crystallography started in 1912, with the seminal work of Max von Laue, who performed the first x-ray diffraction experiment [6]. The crystals von Laue studied were ordered and periodic, and all the hundreds of thousands of crystals studied during the 70 years from 1912 through 1982 were also found to be ordered and periodic. From these observations emerged a paradigm that all crystals are periodic, and this paradigm was accepted by the community of crystallographers and by the scientific community in general.

Crystalline structures are in general characterized by sharp diffraction patterns that reflect the crystal structure and its underlying symmetry. It was an (unproven) assumption that the existence of sharp peaks in a diffraction pattern from any material proved the crystalline nature of the material.

Diffraction patterns of crystals are periodic patterns in Fourier space, more commonly known as "reciprocal space". (See figure 4.6.) Reciprocal space is a concept in crystallography that is invaluable for the interpretation of diffraction patterns. While the atoms are located at sites belonging to the crystal lattice itself, the reflections arising from certain atomic planes are located at sites belonging to the reciprocal lattice. There is a fundamental and symmetrical duality that links the crystal lattice and the reciprocal lattice.

If the structure of a crystal lattice is known, one can easily derive the possible positions of reflections from it, and the Fourier transform of its electron density (structure) gives us the intensities that one can expect to see in an experiment. This technique for simulating diffraction patterns is very useful, as it allows physicists to compare experimentally obtained patterns with theoretically predicted ones.

In two-dimensional crystals, the only possible rotational symmetries are 2-fold, 3-fold, 4-fold, and 6-fold. Rotations of higher order (e.g. 5-fold, 7-fold, and so on) are disallowed ([3]). Therefore, the diffraction patterns found in experiments should reflect only these four basic symmetries.

Nonetheless, in the year 1982, the 70th anniversary of crystallography, Israeli materials scientist Dan Schechtman discovered diffraction patterns having a 5-fold symmetry, a pattern that was inconsistent with any known periodic lattice. In other words, the systems he was studying defied the long-established canons of crystallography that had been derived more than two centuries earlier, which restricted crystalline symmetries to 2-fold, 3-fold, 4-fold, and 6-fold rotational symmetry axes.

This led to a paradigm shift at the frontiers of physics. The recently found structures, soon christened "quasicrystals", exhibited long-range order and sharp diffraction peaks, despite the lack of a periodic lattice. In 1992, in order to include quasicrystals, the International Union of Crystallography changed the definition of a crystal, retaining only the criterion of an essentially sharp diffraction pattern. Schechtman was awarded the Nobel Prize in Chemistry in 2011.

A.1 One-dimensional quasicrystals

The non-repeating sequences shown in figure 4.7 are examples of *one*-dimensional quasicrystals. The sequence of B's and S's is perfectly ordered (predictable) but is aperiodic. The same sequence can be obtained by projecting a regular two-dimensional lattice onto a line. If the line is at an irrational slope that avoids any lattice plane, then

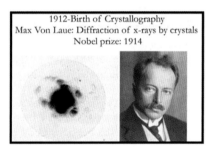

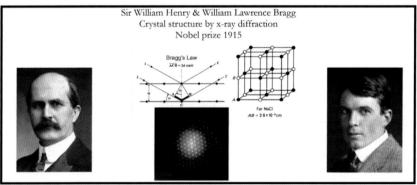

Figure 4.6. The birth of crystallography, in which scattered waves reveal periodic arrangements of atoms.

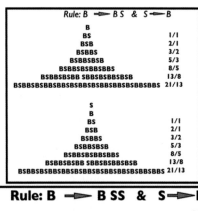
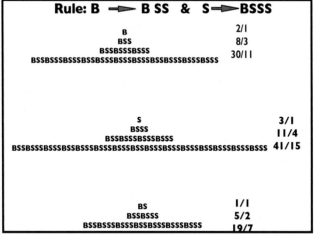

Figure 4.7. The figures inside the top and bottom boxes, respectively, show the construction of the Fibonacci sequence (associated with the golden mean) and the analogous sequence associated with the diamond mean, starting with the letters B and S. The figure illustrates an important distinction between the golden hierarchy and the diamond hierarchy in the butterfly fractal. The ratio of the number of Ss to the number of Bs represents the magnetic-flux values for the butterfly discussed above. We note that, unlike the golden-mean case, the diamond hierarchy of the butterfly requires different initial conditions to generate the flux-values corresponding to the center and to the left and right edges, as can be seen in the three rows on the right side of the figure.

the nearest lattice points projected onto the line will slice it into a quasicrystal sequence. This is illustrated in figure 4.8.

A.2 Two-dimensional quasicrystals: Quasiperiodic tiles

Two-dimensional quasicrystals come in three varieties:
- octagonal quasicrystals, with local 8-fold symmetry;
- decagonal quasicrystals, with local 10-fold symmetry;
- dodecagonal quasicrystals, with local 12-fold symmetry.

Two-dimensional lattices whose diffraction patterns exhibit 8-fold, 10-fold, and 12-fold symmetry are shown in figure 4.9. These diffraction patterns encode the three

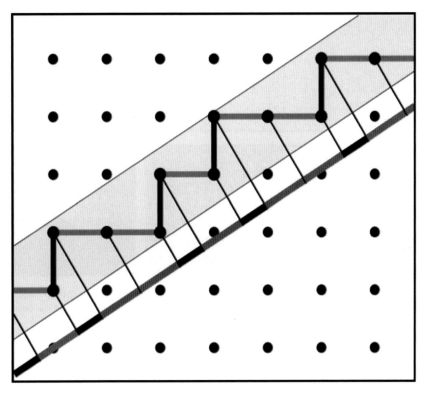

Figure 4.8. The "slice-and-project" method for obtaining sequences associated with the golden mean. To do this, one first draws a "slice" (a slanted line whose slope equals the chosen irrational number—say, the golden mean), and then a second line parallel to the first one. The shaded gray region between the two slices encloses some of the lattice points of the two-dimensional lattice. When perpendiculars are dropped from those lattice points onto the lower slice, the "feet" (i.e., the red points thus created) form a non-repeating sequence of two different-sized intervals, red representing "*B*" (for "big") and black representing "*S*" ("small"). (Such a pattern of letters, incidentally, forms an example of the η-sequences studied by Douglas Hofstadter.) Some of the lines intersect the "real world"—that is, lattice sites of the two-dimensional crystal—thereby allowing observation of the real quasiperiodic structure.

quadratic numbers discussed above, in the sense that they are characterized by wave vectors that are collinear but are incommensurate in length. In other words, the ratio of distances between the bright spots for the 5-fold lattice is equal to the golden mean, while the analogous ratio in the 8-fold lattice is equal to the silver mean, and in the 12-fold lattice it is equal to the diamond mean. The fact that $\cos \frac{2\pi}{Q}$ equals a quadratic irrational only for $Q = 5, 8, 10$, and 12 distinguishes these three special quadratic numbers from all other irrationals.

The only rotational symmetries that have been observed to date in real materials are icosahedral, decagonal (or perhaps pentagonal), octagonal, and dodecagonal. Analysis of quasicrystal tilings of the plane suggests that these may be the only symmetries that support an ordered quasicrystalline phase.

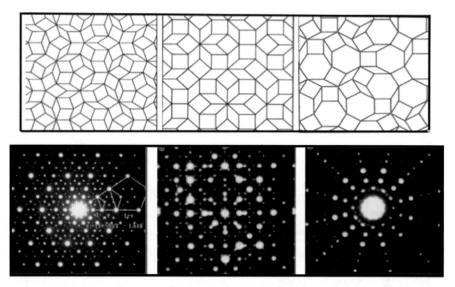

Figure 4.9. 5-fold, 8-fold, and 12-fold lattices (left to right) and the diffraction patterns corresponding to them. In the diffraction pattern coming from a 10-fold symmetry (bottom left), the golden mean has been marked explicitly on the pattern.

A.3 A brief history of the discovery of quasicrystals

Figure 4.10 summarizes the history of this discovery in chronological order. To this pictorial history, one should add the discovery of a *natural* quasicrystal in 2009. Before 2009, all known quasicrystals were synthetic alloys produced in the laboratory under controlled conditions, but in 2009, after a dedicated search, a new mineral, icosahedrite ($Al_{63}Cu_{24}Fe_{13}$), was found, which suggested that quasicrystals can form and remain stable under geological conditions [4]. However, how icosahedrite was actually formed in nature remains an open question.

Quite surprisingly, aperiodic structures lacking the translational symmetry of crystals have been known and admired for quite some time. Medieval Islamic artisans developed intricate geometric tilings to decorate their mosques, mausoleums, and shrines. Some of these patterns, called *girih tilings*, appeared as early as the 12th century AD [5].

The following excerpts from the 2011 Nobel Prize ceremony convey a sense of the importance of the discovery of quasicrystals. Figure 4.10 gives a pictorial summary of the discovery of quasicrystals.

A.4 Excerpts from the ceremony of the Nobel Prize in chemistry in 2011 [6]

For three millennia we have known that five-fold symmetry is incompatible with periodicity, and for almost three centuries we believed that periodicity was a prerequisite for crystallinity. The electron diffraction pattern obtained by Dan Shechtman on April 8, 1982 shows that at least one of these statements is flawed, and it has led to a revision our view of the concepts of symmetry and crystallinity alike.

The objects he discovered are aperiodic, ordered structures that allow exotic symmetries and that today are known as quasicrystals. Having the courage to believe in his observations and in himself, Dan Shechtman has changed our view of what order is and has reminded us of the importance of balance between preservation and renewal, even for the most well established paradigms. Science is a theoretical construction on an empirical foundation. Observations make or break theories ...

Your discovery of quasicrystals has created a new cross-disciplinary branch of science, drawing from, and enriching, chemistry, physics, and mathematics. This is in itself of the greatest importance. It has also given us a reminder of how little we really know and perhaps even taught us some humility. That is a truly great achievement ...

From the presentation speech by Professor Sven Lidin, Member of the Royal Swedish Academy of Sciences.

Figure 4.10. A summary of the discovery of quasicrystalline order, based on the 2011 Nobel ceremony [6]. © The Royal Swedish Academy of Sciences.

References

[1] Hofstadter D R 1976 Energy levels and wave functions of Bloch electrons in rational and irrational magnetic fields *Phys. Rev.* B **14** 2239
[2] http://physics.aps.org/articles/v5/99
[3] Kittel C 2004 *Introduction to Solid State Physics* (New York: Wiley)
[4] Bindi L, Steinhardt P J, Yao N and Lu P J 2009 Natural quasicrystals *Science* **324** 1306
[5] Tennant R 2009 Medieval Islamic architecture, quasicrystals, and Penrose and Girih tiles: questions from the classroom *Symmetry: Cult. Sci.* **19** 113–25 http://home.earthlink.net/~mayathelma/sitebuildercontent/sitebuilderfiles/medieval.islamic.arch.girih.tiles.pd
[6] http://www.nobelprize.org/nobel_prizes/chemistry/laureates/2011/popular-chemistryprize2011.pdf

Part II

Butterfly in the quantum world

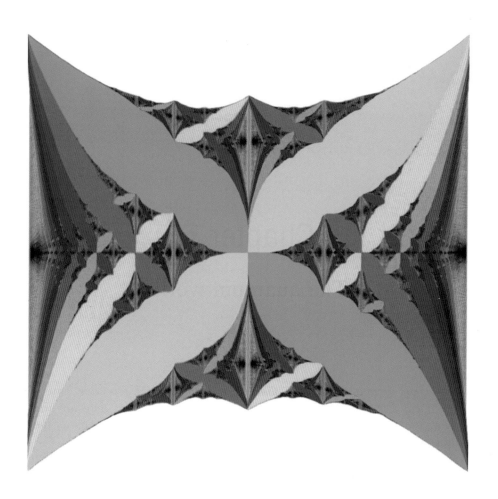

IOP Concise Physics

Butterfly in the Quantum World
The story of the most fascinating quantum fractal
Indubala I Satija

Chapter 5

The quantum world

> *One cannot escape the feeling that these mathematical formulae have an independent existence and an intelligence of their own, that they are wiser than we are, wiser even than their discoverers, and that we get more out of them than was originally put into them.*
>
> —Heinrich Hertz

The quantum revolution is usually considered to have started in 1900, with the proposal, made by German physicist Max Planck, that the vibration energies of atoms forming the walls of a black body are restricted to a set of discrete energy values—a very bold and wild-seeming yet very carefully worked-out hypothesis in which Planck himself did not fully believe, but which he nonetheless published since it was the first theory anyone ever proposed that fully agreed with the experimentally observed spectra of black-body radiation. It took another 25 years for others to develop this first very tentative foray into a mature theory—the theory of quantum mechanics that was then, and that still remains today, a profound challenge to our

Figure 5.1. Starting with Max Planck at the top, this figure displays some of the pioneers who created the quantum revolution. In counterclockwise (but chronological!) order, they are: Albert Einstein (Credit: Bangkokhappiness/Shutterstock.com), Niels Bohr, Louis de Broglie, Werner Heisenberg, and Erwin Schrödinger.

everyday experiences and ingrained intuitions. Perhaps the most inspired period ever of scientific creativity, the quarter-century of the quantum revolution was jointly created by many thinkers of extraordinary brilliance and imagination, some of whom are shown in figure 5.1.

This chapter cannot claim to be a genuine introduction to quantum science; it offers merely a quick peek into the quantum world[1]. In it we highlight and partially explain some of the key quantum phenomena and quantum concepts that are indispensable to an understanding of the butterfly landscape. The aim of this chapter is thus twofold:

- Beginning with the wave nature of matter, we present a simple picture of the quantization of energies in the physical world. It will be useful for readers to internalize this first image of quantization before learning about the quantization of Hall conductivity, which is an entirely different kind of thing. The integer quanta that make up the butterfly fractal, related to the quantization

---

[1] For a relatively simple guide to the quantum world, see [1, 2].

of Hall conductivity, have their roots in phenomena that are quite distinct from those that give rise to the quantization of energies in an atom. Keeping this distinction in mind will help one to appreciate the exotic nature of the quantum Hall effect that will be presented in chapters to come.
- This chapter also serves as a bridge between parts 2 and 3 of the book, as it explains the crucial concepts that are necessary to make a transition from the quantum world of a solitary atom to that of a solid body—a pristine crystalline structure made up of millions of atoms.

5.1 Wave or particle—what is it?

Try telling someone that an electron can sometimes act like a particle and sometimes like a wave. This news is bound to cause confusion and skepticism. When we imagine a particle, we tend to think of a tennis ball or a pebble, perhaps even an infinitesimal ball bearing, while the mention of waves makes us dream of huge long swells of water drifting in, every ten or twenty seconds, and breaking on beaches. Waves are gigantic and spread-out, while particles are tiny and local. No two things could be more different than a particle and a wave! Perhaps the only familiar connection between particles and waves is the fact that tossing a pebble in a pond generates ripples that gracefully spread out in perfect circles. Richard Feynman once summed up the situation by saying that although we do not know what an electron is, there is nonetheless something simple about it: *it is like a photon*.

The quantum world is very counterintuitive, for it is inhabited by species that have dual personalities: they can behave *both* as particles *and* as waves. These species are, however, not necessarily alien or extraterrestrial beings. Some of them are very familiar entities. Indeed, they can be baseballs or even baseball players themselves, as well as more exotic entities, such as electrons, quarks, or Higgs particles. Interestingly, they can also be light waves or x-rays or waves vibrating on the string

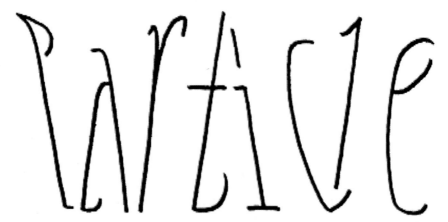

"WAVE/particle" ambigram by Douglas Hofstadter, reproduced here with his permission.

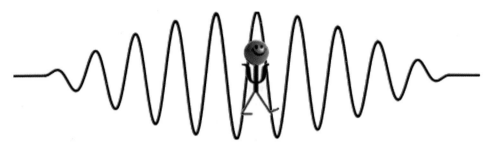

Matter–wave cartoon.

of a violin. The revolutionary idea of *wave–particle duality* was first proposed by Louis de Broglie in his doctoral thesis in 1924, for which he was awarded the Nobel Prize in Physics in 1929.

5.1.1 Matter waves

In the quantum world, any particle possesses a wavelength and a frequency (like waves on the ocean or ripples on a pond), in addition to having usual particle-like properties, such as mass, size, position, velocity, and electric charge. The quantum waves that de Broglie postulated, sometimes called "matter waves", reflect the intrinsic wave–particle duality of matter. According to de Broglie's hypothesis, the wavelength λ associated with any particle is inversely proportional to the particle's momentum p, while the frequency ν associated with a particle is proportional to the particle's total energy E. The constant of proportionality in both cases is the fundamental constant of nature h, called Planck's constant, named after Max Planck, who, in his pioneering 1900 work on the black-body spectrum, discovered this number and its central role in nature, and who, for those discoveries, was awarded the 1918 Nobel Prize in Physics. The following two equations constitute the core of Louis de Broglie's hypothesis:

$$\lambda = \frac{h}{p}; \qquad \nu = \frac{E}{h}. \tag{5.1}$$

In most situations, it turns out to be simpler and more natural to use the *angular* frequency $\omega = 2\pi\nu$ than to use the simple frequency ν. In the case of a rotating body, the quantity ν is the number of full rotations made by the body per second. Thus a body that rotates exactly once per second (360 degrees per second) has a simple frequency of 1 Hz, while its *angular* frequency is 2π Hz. The angular frequency thus equals the number of radians turned per second by the body (one radian equaling $360°/2\pi \approx 57°$). Since the angular frequency is generally more natural, whether we are talking about rotating bodies or wave phenomena, it is often useful to write quantum expressions not in terms of the simple frequency ν and Planck's constant h, but in terms of the angular frequency ω and the *reduced* Planck's constant, which equals Planck's constant divided by 2π, and is denoted by the symbol "$\hbar$" (pronounced "h-bar"):

$$\hbar = \frac{h}{2\pi} = 1.054571726(47) \cdot 10^{-34} \text{ Joule} - \text{sec}. \tag{5.2}$$

Readers will notice the explicit presence of h or $\hbar$ in all equations involving quantum aspects of a particle. These constants, which have the units of angular momentum, lie at the heart of quantum science. Since they are extremely small compared to familiar amounts of angular momentum (such as the angular momentum of an ice skater doing a spin, or that of a frisbee flying through the air), quantum effects are usually completely unobservable in the macroscopic world.

Three years after de Broglie announced his hypothesis about the wave nature of particles, the idea was confirmed in two independent experiments involving the observation of electron diffraction. At Bell Labs in New Jersey, Clinton Davisson and his assistant Lester Germer sent a beam of electrons through a crystalline grid and, to their amazement, observed interference patterns. They were utterly baffled at first, but when they heard about de Broglie's wave–particle hypothesis, they realized that what they were seeing confirmed de Broglie's predictions exactly. A few months later, at the University of Aberdeen in Scotland, George Paget Thomson passed a beam of electrons through a thin metal film and also observed the predicted interference patterns. For their independent contributions, Davisson and Thomson shared the 1937 Nobel Prize for Physics.

5.2 Quantization

The idea that atoms might be like small solar systems, with negatively charged particles in orbits around a positively charged central particle, was first proposed by French physicist Jean Baptiste Perrin in 1901, and two years later in a far more detailed manner by Japanese physicist Hantaro Nagaoka. Unfortunately, almost no physicists took their ideas seriously. However, just a few years later, very careful scattering experiments performed by the New Zealander Ernest Rutherford in Manchester, England showed that some kind of planetary model was in fact correct.

In 1910, Australian physicist Arthur Haas tried valiantly to incorporate Planck's constant into a planetary-style atom but ran into roadblocks. Two years later, British physicist John William Nicholson went considerably further, but still, his theories did not match known data. Then in 1913, Danish physicist Niels Bohr entered the scene. Bohr knew well that a particle in a circular orbit is constantly undergoing accelerated motion (change of direction being a type of acceleration), and that, according to classical electromagnetic theory, an accelerating charged particle must radiate energy away, so it will quickly lose all of its energy and the system will collapse. In other words, according to classical physics, any planetary model of an atom is unstable. To explain the stability of atoms, Bohr introduced some counterintuitive and highly revolutionary concepts. For simplicity, we will deviate slightly from Bohr's original way of presenting his model, and will instead present it using de Broglie's wave–particle hypothesis described above. (Our presentation is anachronistic, as de Broglie first proposed that particles are wavelike roughly ten years after Bohr developed his model of the atom.)

Figure 5.2 illustrates the key idea of the Bohr model, using de Broglie's idea. We imagine an electron moving in a circular orbit, and on that circle we superimpose a sinusoidal de Broglie wave. If, after one trip around the nucleus, the periodic waving pattern returns exactly in phase with itself (i.e., if the circumference equals an integral number of de Broglie wavelengths), then we will get a constructive interference pattern, where the wave reinforces itself on each new swing around the atom's center. Metaphorically speaking, by returning in phase with itself, the electron "reinforces its existence", whereas by returning out of phase with itself, it "undermines its own existence". The classical story of the electron orbit gradually decaying and finally going out of existence, as in panel A of figure 5.2, corresponds to the impossible image shown in panel C, while the image in panel B is deeply quantum-theoretical, and corresponds to no classical story at all. Alternatively, you can just accept the idea that an electron in an atom, being a wave at the same time, is constrained to vibrate with specific frequencies for much the same reason as a guitar string is.

Whichever way you choose to see it, this very simple idea results in the following elegant quantization condition for the de Broglie wavelength λ associated with the electron:

$$2\pi R = n\lambda, \qquad (5.3)$$

where R is the radius of the orbit and n is a positive integer, $n = 1, 2, 3, 4, \ldots$. Each different value of n corresponds to a different possible orbit of the electron around

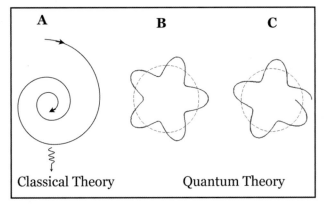

Figure 5.2. Panel A depicts the behavior of a negatively charged electron in orbit around a positive charge, according to classical theory. Rather than remaining in a stable circular orbit, the electron would constantly radiate energy away and thus would spiral ever inwards, until the system collapsed. Quantum theory, however, allows stable orbits. Panel B shows how a wavelike electron in a circular orbit can interfere constructively with itself, resulting in a stable orbit. Panel C shows that if the electron's wavelength does not fit exactly into the circumference, then the electron interferes destructively with itself, and therefore cannot exist in such an orbit.

the proton. Using the above relation, along with the classical equation for circular motion and the expression for the total energy, we get:

$$\frac{mv^2}{R} = \frac{Ke^2}{R^2}; \qquad E = \frac{1}{2}mv^2 - \frac{Ke^2}{R}. \tag{5.4}$$

Here, m is the mass of the electron, v is its velocity, e is its charge, and K is a universal constant that defines the strength of the electromagnetic attraction between two given charges (here a proton and an electron).

The last equation above leads to the following quantization condition for the orbital energy levels of the electron:

$$E = -\frac{(Ke^2)^2 m}{2\hbar^2 n^2} \approx \frac{-13.6}{n^2} \text{ electron} - \text{volts}. \tag{5.5}$$

The smallest value of n—namely, $n = 1$—corresponds to the tightest orbit, which is the orbit of lowest energy, and is thus called the "ground state". There are infinitely many other possible orbits of higher energies. Bohr postulated that an electron can "jump" from any H-atom orbit to a lower-energy H-atom orbit, and in so doing will release an electromagnetic wave whose energy must (because of the law of conservation of energy) equal the energy difference between the two H-orbits. This allowed Bohr to calculate what all the spectral lines of hydrogen should be, and to his enormous satisfaction, his predictions coincided perfectly and precisely with the observations of spectral lines coming from the Sun.

Moreover, the mathematical expression that yielded the values of these spectral lines, and which Bohr had rigorously deduced from his "magical" quantum hypothesis, coincided precisely with the mysterious empirical formula that had been guessed in 1885 by Johann Jakob Balmer, a Swiss high-school teacher who was 60 years old at that time. For all his life, Balmer had been obsessed with numerical patterns in nature and had sought beautiful formulas that matched them, but never before had he hit such a jackpot. Over many years of careful observation after 1885, Balmer's miraculously simple formula had always been totally confirmed, but no one had ever been able to say what secrets of nature lay behind it. It was just a wonderfully lucky guess. But now, all of a sudden, the world understood *why* the Balmer formula was the way it was, and with that, the profound mysteries of the atom were starting to be uncloaked.

Given all these reasons, the Bohr model of the hydrogen atom was immediately accepted by the world physics community, and in 1922, Niels Bohr, for his path-breaking explanation of the hidden quantum phenomena that lay behind the spectral lines of hydrogen, was honored with the Nobel Prize in Physics.

At a conference in 1998 in honor of the great Dane, Douglas Hofstadter pointed out that the name "Niels Bohr" resonates profoundly with the phenomena that its bearer so beautifully explained:

NIELS BOHR = H-ORB LINES.

It's a marvelous coincidence. Or is it a coincidence? As the Romans said, *Nomen est omen*, which could roughly be translated as "What one's name conceals, one's life reveals."

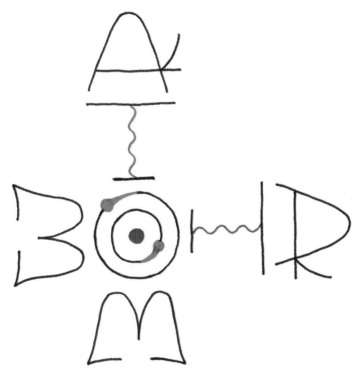

"Bohr–Atom" ambigram by Douglas Hofstadter, reproduced here with his permission.

The Bohr model was deeply revolutionary when it was first proposed, but just a few years later it started to seem rather primitive, since it was far from being a complete theory of atoms. In fact, it was unable to explain the spectral lines of any atom other than hydrogen! From 1913 on, intense efforts were made by physicists all over the world to develop a complete quantum description of atoms—an analogue to Newton's equations, describing the behavior of particles, or to Maxwell's equations, describing the behavior of electromagnetic waves. At long last, in the mid-1920s, German physicist Werner Heisenberg and Austrian physicist Erwin Schrödinger succeeded. They independently devised two radically different frameworks, each of which connected the final dots and provided precise quantum rules to describe a physical system. For a while, the two theories were considered to be rivals, but eventually it was proven—indeed, by Schrödinger himself—that although on the surface they involved very different images and very different mathematical ideas, they were nonetheless totally equivalent at a deeper level. Interestingly enough, both of these theories gave exactly the same results as the Bohr model had given for the spectrum of the hydrogen atom. Below we will briefly discuss the Schrödinger theory.

5.3 What is waving?—The Schrödinger picture

If a particle acts like a wave, then a natural question is: *what is it that is waving?*

In 1925, Erwin Schrödinger came up with an equation that predicted the behavior of a quantum particle, just as Newton's famous equation $F = ma$ predicts the behavior of a classical particle. Schrödinger's equation is a partial differential equation that describes how the quantum state of a physical system—that is, the system's "wave function"—changes across space and over time. The wave function, usually denoted $\Psi(\mathbf{r}, t)$, encodes all the information there is about the state of the particle at spatial location $\mathbf{r}$ and time t. In the standard interpretation of quantum mechanics, this function is the most complete description that can be given to a physical system.

The Schrödinger equation for a non-relativistic particle of mass m moving in a potential-energy field V runs as follows (here i is $\sqrt{-1}$):

$$i\hbar \frac{\partial}{\partial t}\Psi(\mathbf{r}, t) = \left[\frac{-\hbar^2}{2m}\nabla^2 + V(\mathbf{r}, t)\right]\Psi(\mathbf{r}, t). \tag{5.6}$$

To solve this equation for Ψ, one often uses a method called "separation of variables", a standard technique for solving partial differential equations. This transforms the time-*dependent* Schrödinger equation into the following time-*independent* equation:

$$\left[\frac{-\hbar^2}{2m}\nabla^2 + V(\mathbf{r})\right]\psi(\mathbf{r}) = E\psi(\mathbf{r}). \tag{5.7}$$

The letter E on the right side of this equation represents the energy of the quantum state Ψ. However, as Schrödinger soon realized, not all values of E will work. To be more specific, for some special values of E, there will be a solution Ψ to this equation, but for other values of E, there will be *no* solution. Those very special values of E for which there exists a solution Ψ to the Schrödinger equation are called the *eigenvalues* of the equation. Also, $\psi(\mathbf{r})$ is called the *eigenfunction* belonging to the eigenvalue E (which is also sometimes called the "eigen-energy" of the particle).

When $V(\mathbf{r})$ is set equal to the potential-energy field due to the positive electric charge of a proton, then this equation becomes a full quantum-mechanical description of the hydrogen atom. In 1926, Schrödinger solved this equation, determining its eigen-energies, and he found that they coincided exactly with the quantized energy levels that Niels Bohr's very early "semiclassical" model had predicted for electrons in the hydrogen atom. This confirmation of Bohr's model by the Schrödinger equation, analogous to the earlier confirmation of Balmer's formula by Bohr's model, was a remarkable event in physics, and showed how deep Bohr's intuitions in 1913 had been, and how amazingly precise Balmer's aesthetics-based numerological guess in 1885 had been.

The fact that only certain special values of E will allow Schrödinger's equation to be solved—a fact that we might call the *eigenvalue constraint* on the solvability of the Schrödinger equation—explains many deep phenomena in physics. For example, as we have just pointed out, this constraint explains why electrons in atoms can only move in orbits having certain precise energy values (the values that Bohr found in 1913, roughly a dozen years before Schrödinger dreamt up his wave equation).

These special energy levels are the eigenvalues of the Schrödinger equation for the hydrogen atom.

The eigenvalue constraint also explains why a free electron in a magnetic field can only take on the so-called "Landau levels" of energy, rather than any arbitrary amount of energy at all (as would be possible in classical electromagnetic theory). The Landau levels are the eigenvalues of the Schrödinger equation for an electron moving solely under the influence of a magnetic field.

The eigenvalue constraint also explains why electrons in crystals are limited to having their energies in Bloch bands (regions along the E-axis such that the Schrödinger equation has solutions), which are separated by energy gaps (complementary regions along the E-axis such that the Schrödinger equation has no solutions).

Last but not least, it is the eigenvalue constraint that explains why electrons in crystals in magnetic fields can only take on energies shown in the black bands that make up the butterfly graph, and cannot take on energies in the white gaps.

Unlike classical waves, whose amplitudes are always real numbers, the wave function Ψ that obeys the Schrödinger equation takes on *complex* values, and is not itself a physically measurable or observable quantity. In other words, unlike water waves, or waves on a string, or sound waves, or light waves, where what is waving is always a perfectly familiar, observable entity—for example, the height of the water at some spot in a lake as circular ripples spread out after a stone has been tossed in, or the displacement from equilibrium of a plucked violin string as it vibrates back and forth, or the fluctuating value of the air pressure at a chosen spot in space, or the rapidly oscillating strengths of the electric and magnetic fields at some fixed point in space—what is waving in a quantum-mechanical situation is an abstract, physically unobservable, complex-valued quantity. This fact about the wave function was deeply bewildering to physicists, who didn't know what to make of these invisible, unobservable quantum-mechanical complex numbers floating about, filling up every point in space, and oscillating with time, like ghostly ripples ubiquitously undulating in the void.

It was Heisenberg's mentor Max Born who first successfully interpreted the wave function Ψ as giving what he called the "probability amplitude" associated with the particle. According to Born, it is only the *absolute square* of this amplitude—namely, $|\Psi(\mathbf{r}, t)|^2$—that is observable (in some sense). That is, the absolute square of the complex wave function is a *real* number that tells the probability of finding the particle at location $\mathbf{r}$ and at time t. For having made this discovery in 1926, Max Born was awarded the Nobel Prize in Physics many years later (1954).

While the complex probability amplitude encodes all the information about the state of the particle, the act of taking its absolute value (or "modulus") and squaring it means that one loses some information (namely, the *phase*). This subtle loss of information is the ultimate source of all quantum-mechanical "weirdness".

5.4 Quintessentially quantum

We will now discuss two different experiments that illustrate "quantum-mechanical weirdness", and that continue to challenge our basic intuitions even today, in spite of the almost universal acceptance of the laws of quantum physics.

5.4.1 The double-slit experiment, first hypothesized and finally realized

In 1801, the English physicist Thomas Young introduced the double-slit interferometer. In such a device, a light wave spreads outward from a point source and is allowed to pass through two slits in an opaque barrier; once the wave is beyond those slits, it interferes with itself (in a way, it is more like two waves interfering with each other, one emanating from each slit), and what results is an *interference pattern* on a distant two-dimensional surface. Even today, this type of device remains one of the most versatile tools for demonstrating interference phenomena for waves of any imaginable sort.

Here, we will consider a variation on this theme, involving electrons rather than light. Our hypothetical two-slit experiment (see figure 5.3) was originally dreamt up by Richard Feynman in volume 3 of his famous *Feynman Lectures on Physics* [3]. In Feynman's thought experiment, deeply inspired by Young's interferometer, an electron (which, from many experiments over the past century, we have every reason to conceive of as a microscopic dot carrying electric charge) is released at point A towards a screen, and somewhere between point A and the screen there is an impenetrable wall that has two slits in it, at, say, points P and Q. We immediately see that the rightward-moving electron can reach the screen and leave a little mark on it only if it passes through either slit P or slit Q.

This simple conclusion is not merely commonsensical, but totally obvious and not worth giving a moment's thought to. Or at least that is what classical thinking would tell us. But quantum mechanics violates this "obvious" fact, because quantum mechanics tells us that particles—tiny points moving through space like tiny pebbles flying through the sky—do not act like pebbles in the sky but like ripples on water. However, this fact, when first encountered, is very disorienting, to say the least, so let

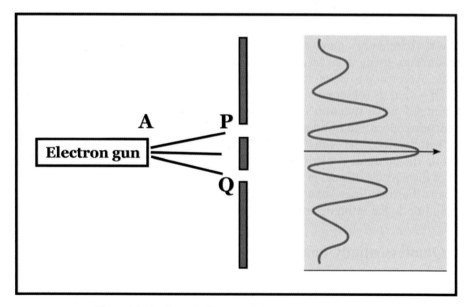

Figure 5.3. A schematic diagram of the double-slit experiment.

us first spell out what our deep-seated classical intuitions would predict for such a setup.

If we were to send a broad stream of electrons from point A toward the screen, we would expect to find two dark splotches building up on the screen as more and more electrons came in for landings, each point-like electron leaving a tiny mark where it landed. This seems absolutely straightforward and obvious. More specifically, we would expect to see splotches gradually building up at *exactly two predictable places* on the screen—namely, at (or very near) the two points on the screen that are determined by drawing a straight line first from A to P and extending it all the way to the screen, and then from A to Q and likewise extending it to the screen. These two straight lines, determined by the point of release and the two slits in the wall, are the only conceivable trajectories that could carry an electron from point A to the screen, since the wall, aside from slits P and Q, is impenetrable.

But commonsensical and even watertight though this conclusion may seem, what we have just described is *not* what is actually observed on the screen. What is seen on the screen, instead of two isolated splotches where electrons land, is an *interference pattern*—that is, a blurry pattern all over the screen, which is darker in certain areas and lighter in others—and *in no way* does it look like two splotches! In fact, oddly enough, the pattern is darkest exactly halfway between the two hypothetical splotches that classical thinking gave us, and a short distance from there it fades to zero, and then a little further away it again becomes dark, and then it fades away to zero again, then darkens and lightens again, and so forth and so on. This pattern of alternating lighter and darker zones—the trademark of an interference pattern, just like those observed by Thomas Young in the early 1800s—is what is symbolized by the wavy line shown to the right side of the screen in figure 5.3. The peaks of the wavy graph are the areas where the screen is darkest, and the troughs are where it is lightest.

To further reveal the mysteries of the wave–particle duality intrinsic to quantum mechanics, Feynman invited his readers to imagine firing just one single electron toward the screen (rather than a beam comprised of many electrons), and then marking the position where it strikes the screen, and then repeating this one-electron experiment over and over again. After many electrons have been fired, the marks on the screen will *still* comprise an interference pattern, which shows that each electron *on its own* was interfering with itself. In other words, each electron on its own somehow went, in a ghostly manner (or at least in a wavy manner!), through *both* slits, rather than through just one or the other of the slits (which is what we would expect of a particle that manifests itself as a tiny dot wherever it hits the screen).

If we now cover up, say, slit A, so that each electron can pass only through slit B, then no interference pattern will appear on the screen—just a splotch directly behind slit B will build up over time. This agrees with our classical intuitions, and shows us that the intuition-defying interference pattern arises only when we give each electron the chance to pass through both slits. When an electron is given that chance, it will always take it, and so, as one electron after after gets released from point A, the interference pattern gradually takes shape on the screen!

Before Feynman dreamt up his thought experiment (in the early 1960s), experiments of this sort using double-slit setups had been done, and they indeed showed the interference pattern we have just described, but they all used a *beam* of electrons rather than just one electron at a time. Because of this, these experiments did not establish a crucial point of Feynman's thought experiment—namely, that an individual electron traveling by itself will behave like a wave. Single-electron double-slit diffraction was first demonstrated in 1974 by Giulio Pozzi and colleagues at the University of Bologna in Italy, who passed single electrons through a biprism—an electronic optical device that serves the same function as a double slit—and they observed the predicted build-up of an interference pattern. A similar experiment was also carried out in 1989 by Akira Tonomura and colleagues at the Hitachi research lab in Japan. The actual Feynman-style double-slit experiment, in which the arrivals of individual electrons in a double-slit situation were recorded one at a time, was finally realized only in 2012 by Pozzi and colleagues. Perfecting the double-slit experiment with a single electron continues to obsess many physicists even today.

The double-slit interference pattern with a single electron makes one dizzy irrespective of whether we imagine the electron to be a particle or a wave. If we ask, "Did the electron pass through slit P or slit Q?", the answer is, "Neither—it passed through them both." This is because an electron is a wavelike entity, and we have to imagine it spreading through space like ripples moving on the surface of a pond—or if you wish to have a three-dimensional image, then like sound waves propagating through the air (of course, since sound waves are invisible, they are harder to imagine than ripples).

The weird thing is that although each electron wears its "wave hat" while propagating through space (that is, while moving away from point A, passing through the slits, and approaching the screen), it doesn't keep that hat on at the very end. Instead, when it finally lands on the screen, it doffs its "wave hat", puts its "particle hat" back on, and deposits a little dot in just one single point on the screen. Why and how does this weird hat-trick take place? No one can say. This unfathomable mystery lies at the very heart of quantum mechanics. As Richard Feynman said, "Nobody can explain quantum mechanics." Or as Albert Einstein once wistfully remarked, toward the end of his life, "I have been trying to understand the nature of light for my entire life, but I have not yet succeeded."

5.4.2 The Ehrenberg–Siday–Aharonov–Bohm effect (ESAB) [4]

Two examples of quantum phenomena that have no analogues in classical physics are Heisenberg's uncertainty principle and quantum tunneling, both discovered in the early days of quantum mechanics, and both quite famous, even outside physics. There are also less famous quantum phenomena that were discovered later, such as the so-called *Aharonov–Bohm effect*, dating from 1959, and the *Berry phase*, dating from 1984, both of which were discovered in Bristol, England, although 25 years apart. In chapter 9, we will discuss the Berry phase, but here we will discuss the Aharonov–Bohm effect, published in 1959 by David Bohm and his student Yakir Aharonov. Shortly after their article was published, Bohm and Aharonov learned that Raymond Siday and Werner Ehrenberg had published exactly the same result a decade earlier. This must have been a great shock to them, but David Bohm, to his credit, always referred to the discovery thereafter as the "ESAB effect", and in this book we shall follow his lead, although "Aharonov–Bohm effect" is the usual name.

The ESAB effect involves a setup closely related to the double-slit experiment described above. However, in this case, when an electron is released at point A, rather than having the possibility of passing through slits in a wall, it has the option of taking various pathways around an obstacle, after which it lands on a screen, leaving a mark showing where it hit. The collective pattern built up on the screen by the landings of many electrons is what we are interested in. Such a setup, since it involves different pathways quantum-mechanically interfering with each other (just as in the two-slit experiment) is called an *interferometer*.

In this case, the obstacle around which the electron must move (wearing its "wave hat", of course) will be a tightly wound coil of wire (technically called a *solenoid*), which is surrounded by a very intense repulsive electric field (technically called a *potential barrier*), which is so strong that it prevents the electron from entering the solenoid. Thus the electron's only option is to go *around* the solenoid, either to its left or to its right—or more accurately, to pass by it on *both* sides at once, like a ripple rippling around a stone jutting up in the middle of a pond. When the electron hits the screen, it will be wearing its "particle hat", and will deposit a mark on it. All the marks together will add up to an interference pattern. The close analogy between this setup with a solenoid and the earlier-described two-slit experiment should be clear.

We first imagine that no current is flowing in the wires of the solenoid. In that case, we will get an interference pattern on the screen. Being an old quantum hand by now, you of course are not surprised by this at all. It's self-evident! And thanks to the equations of quantum mechanics, the exact interference pattern can be calculated precisely in advance, although nobody in the world can explain *why* it happens. So far so good.

Now let us turn on the current. When a current flows in the wires of any solenoid, a non-zero magnetic field is produced *inside* the solenoid; however, everywhere *outside* the solenoid, the magnetic field remains exactly zero. As a result, we "old quantum hands" would not expect any change in the interference pattern on the screen, since the electrons passing outside the solenoid experience exactly the same magnetic field (namely, zero!) as they did before the current was turned on. Before 1959, most physicists would have bet their bottom dollar that the interference pattern on the screen would be unaffected by turning on the current.

But in 1949 and then in 1959, first Siday and Ehrenberg and then Aharonov and Bohm realized that turning on the current causes a perfectly observable and precisely measurable *shift* of the interference pattern on the screen. This visible shift in space corresponds to an invisible *phase shift* between pathways that pass *one way* around the solenoid (say, to its left side) with respect to pathways going *the other way* around the solenoid (to its right side). This phase shift equals $\gamma = 2\pi \frac{\phi}{h/e}$, where ϕ is the amount of magnetic flux enclosed between the two pathways—that is, ϕ equals the magnetic field inside the solenoid times the area of the solenoid. This is said to be a "geometric phase", in the sense that it does not depend on the velocity or energy or momentum of the traveling electron, nor on the time taken by it to travel from the emitter to the detection screen—just on the geometry of the situation.

Let us try to summarize what is truly weird here—weird not just to classical physicists, but even to highly experienced "old quantum hands". In classical mechanics, the trajectory of a charged particle is not affected by the presence of a magnetic field in regions of space that the particle never visits. For a particle to be affected by a nonexistent field is unimaginable. What the ESAB effect teaches us, however, is that for a charged particle in a *quantum* situation, there can be an observable effect produced by a magnetic field even though the particle never "feels" that field. In figure 5.4, we see that the particle, after passing near the solenoid, while

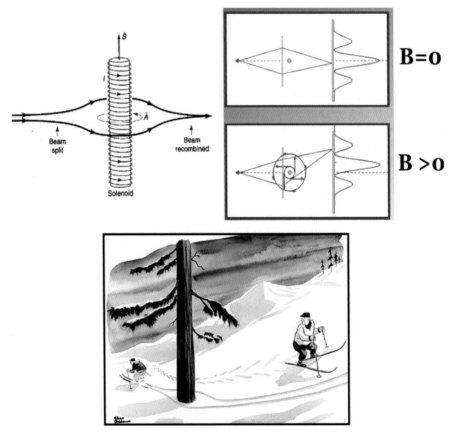

Figure 5.4. The left side of the upper figure shows a schematic diagram of the ESAB effect. To its right, we first see (the $B = 0$ case) an interference pattern on the screen, which is due to quantum-mechanical interference of the two alternate routes around the solenoid, and below it we see (the $B > 0$ case) a *displacement* of the interference pattern, which is due to the presence of a non-zero magnetic field inside the solenoid, even though the particle, while following its trajectory, never "feels" the magnetic field. The *New Yorker* cartoon by Charles Addams depicts a phenomenon highly reminiscent of the ESAB effect. © 1940 Charles Addams, renewed 1967. With permission Tee and Charles Addams Foundation.

always remaining in a region of exactly zero magnetic field, produces an interference pattern on a screen. When there is no current and thus no magnetic field, the peak of the interference pattern is precisely centered (the $B = 0$ picture on the right). When there is a current and thus a magnetic field (but only *inside* the solenoid), the peak of the interference pattern *shifts away* from the center (the $B > 0$ picture on the right), and in fact the whole pattern shifts.

This shift is due to the fact that although the magnetic field is zero at all points of space that are accessible to the particle (that is, at all points outside the solenoid), there is another field—a more abstract field—associated with the magnetic field, called the *magnetic vector potential* $\vec{A}$, whose value is *not* zero at those points. The vector potential $\vec{A}$, invented by the great Scottish physicist James Clerk Maxwell, and today also known as a "gauge field", mathematically determines the magnetic field $\vec{B}$ at all points in space (through the equation $\vec{B} = \vec{\nabla} \times \vec{A}$), just as the electric scalar potential (invented a few decades earlier by the French physicist Siméon Denis Poisson) mathematically determines the electric field at all points in space. However, whereas the electric and magnetic fields themselves are observable and measurable, the scalar and vector potentials are *not* observable or measurable (or at least in *classical* electromagnetic theory, this is the case). Thus quantum mechanics makes us confront the most peculiar situation of a totally non-observable and non-measurable (and thus arguably non-physical) variable (the magnetic vector potential) affecting the observable physical state of a particle. This is troubling, to say the least.

A *gauge transformation* is a special kind of transformation of the "non-physical" scalar and vector potentials that changes their values everywhere in space while leaving the "physical" electric and magnetic fields completely invariant throughout space. (Specifically, a gauge transformation modifies the the electric scalar potential by adding a constant to it, and it modifies the magnetic vector potential by adding the gradient of a scalar to it.)

In the ESAB effect, the phase shift that brings about the shift of the interference pattern on the screen depends on the magnetic flux enclosed between the two different pathways, and this is determined by the vector potential (or more precisely, on its gauge-independent part). What this tells us is that in quantum mechanics, the electric and magnetic *potentials* take on just as fundamental a reality as did, in classical physics, the electric and magnetic *fields*. To state this in another way, in classical physics only the electric and magnetic fields were believed to be fundamental, observable quantities, while the scalar and vector potentials, though useful for calculations, were seen as merely mathematical aids, which, on their own, could be neither observed nor measured—not even in principle. But with the advent of quantum mechanics, it was discovered that the scalar and vector potentials are *not* merely mathematical aids, but are themselves first-class physical entities.

In summary, the ESAB effect has taught us that the magnetic vector potential (or at least its gauge-invariant part) is a *real physical field*. This discovery shattered the old way of thinking about electromagnetic fields, and brought about a totally new vision of the nature of *potentials*.

The ESAB effect is just one example of a far more general phenomenon. As we will see in chapter 8, in certain situations the wave function of a particle acquires a phase having a purely geometric origin. This phase, known as the "geometric phase" or "Berry phase", is due to an abstract analogue of the magnetic flux and the vector potential $\vec{A}$ inside the solenoid in the ESAB setup described above. Moreover, in situations such as that of electrons in a two-dimensional crystal in a magnetic field, the Berry phase itself may be quantized. It is these "phase quanta", which are lurking behind the scenes in the geometry and topology of a purely abstract space, that are key to understanding the quantum numbers associated with the Hofstadter butterfly. More generally, an understanding of the Berry phase deepens our appreciation of the underlying geometrical nature of quantum mechanics.

5.5 Quantum effects in the macroscopic world

Quantum mechanics provides a deep understanding of the properties of solid materials, just as it provides a deep understanding of a single atom, such as the hydrogen atom. For example, phenomena like superconductivity—the complete absence of any resistance to flowing current in certain solid materials—cannot be explained by classical physics. Even normal electrical conduction, which arises in metallic and insulating materials, cannot be explained without quantum theory. The range of resistivity of different materials, running all the way from the very best of conductors to the very worst of conductors (that is to say, the very best of insulators), covers more than thirty orders of magnitude! There is simply no way that classical physics could even begin to explain anything like this. As far as classical physics is concerned, all of these materials are simply quite similar combinations of positive nuclei and negative electrons. We are thus led to ask a most natural question: *when many atoms aggregate together to make a whole, what kinds of new properties can emerge at the level of the whole, and how can quantum mechanics account for these properties?*

In our discussion of solids, we will focus on the way that the electrons in them determine their properties. In other words, we will discuss only *electronic* states of matter[2]. Electrons in materials can organize themselves in many different ways, and these various electronic patterns define what are called *phases* (not to be confused with the geometrical Berry phases mentioned above, and even less with phases of the Moon!). Condensed-matter physics is the study of these phases of matter, which can roughly be classified as *metals, semi-metals, insulators/semiconductors, topological insulators, superconductors, magnets, charge-density wave systems,* and *spin liquids*. Each of these phases has a common set of characteristics (e.g. all liquids flow). In this sense, each of the above is an electronic phase.

In what follows, we will limit our discussion to the *metallic state* and the *insulating state* of matter. This will set the stage for our next key topic—namely, the quantum Hall states of matter, also known as *topological insulators*, which will be covered in

---
[2] For a very accessible lay-level discussion of solid-state physics, see [2].

the coming chapters. We note that the insulating states described here are called "band insulators". They can be classified as ordinary (trivial) insulators or as topological insulators. In addition, there are Anderson insulators (which are the result of impurities) and Mott insulators (which require interaction effects, and are beyond the scope of this book).

We provide a brief summary of this subject by listing some of the key concepts of condensed-matter physics.

5.5.1 Central concepts of condensed-matter physics [5]

- *Periodic potentials*
  A periodic electric potential $V(\mathbf{r})$ arises automatically in a crystal because the positively charged nuclei in the crystal are arranged in a perfect lattice—an endlessly repeating spatial pattern. The periodicity of the electric potential is expressed by the equation $V(\mathbf{r}) = V(\mathbf{r} + \mathbf{a})$, where $\mathbf{a}$ is a lattice vector. The *primitive unit cell* of the crystal is the smallest building block that, when periodically placed next to itself in space, yields the full crystal lattice.
- *Non-interacting electrons and electron gases*
  In crystals, electrons move under the influence of the just-described periodic potential, and their behavior and properties are described by the laws of (non-relativistic) quantum mechanics. It is rather surprising that many aspects of matter can be understood by assuming that the electrons in such a system never interact with one another, but that is the case, and in this book only such phenomena are discussed. A system of non-interacting electrons is known as an "electron gas".
- *Pauli's exclusion principle*
  In our description of non-interacting electrons, we need to take into account a very deep quantum effect that arises due to the fact that we are dealing with particles (electrons in particular) that are all *perfectly identical*. This feature has no classical analogue, since in classical physics one can always distinguish identical-seeming particles simply by following their distinct trajectories. In other words, in classical physics, identical particles simply do not exist; any two particles can always be distinguished. But in the quantum world, this deeply intuitive property fails to hold.

  Electrons, which are the main actors in solid-state systems, are literally indistinguishable from each other, and they obey what is known as the "Pauli exclusion principle", named after the Austrian-born physicist Wolfgang Pauli (who, over the course of his life, held Austrian, German, Swiss, and American citizenships). In 1945, Pauli was awarded the Nobel Prize in Physics for the discovery of this central quantum-mechanical principle, which states that no two electrons can ever occupy the same quantum state simultaneously. Actually, Pauli's principle is more general than this, as it applies not just to electrons but to every type of *fermion*. A fermion is a particle that inherently possesses *half-integer* spin—that is, an angular momentum that is an odd multiple of $\frac{\hbar}{2}$. Electrons are fermions, since they have spin $\frac{1}{2}$. Pauli's full exclusion principle states that two identical fermions can never occupy the same quantum state simultaneously.

On the other hand, there are identical particles that *can* occupy the same quantum state, and such particles are called *bosons*. A boson is a particle that inherently possesses an *integral* amount of spin—that is, an angular momentum that is an integer multiple of $\hbar$. Photons are bosons, since they have spin 1. As such, they are not subject to the exclusion principle; any number of photons can occupy the same quantum state, and indeed they tend to do exactly that.

- *Bands and band gaps*
The term "electronic band structure" (or just "band structure" for short) denotes the set of energy values that electrons in a solid may take on. The allowed energy levels are limited to certain intervals of the energy axis called "Bloch bands", "energy bands", "allowed bands", or simply "bands". The ranges of energy values that an electron may *not* take on are called "band gaps" or "forbidden bands" (see figure 5.6). Band theory derives these bands and band gaps by solving the Schrödinger equation to determine the quantum-mechanical eigenvalues for an electron in a periodic lattice of atoms or molecules.

The reader, having seen that in many circumstances quantum mechanics gives rise to a set of isolated, discrete energy levels, and nothing like the continuous range of values comprising an energy band, might well ask: why do solids have *bands*, as opposed to *discrete levels*, of energy? This is an excellent question, and a sketch of the answer is as follows (see figure 5.5).

The electrons in a single isolated atom have discrete energy levels. When multiple atoms join together to form into a molecule, their wave functions overlap in space, and because of the Pauli exclusion principle, their electrons cannot occupy the same state (meaning they cannot have the same energy). Therefore, each discrete eigenvalue (energy level) splits into two or more new levels, all clustered close to the original level. As more and more atoms are brought together, the allowed energy levels have to split into more and more new levels (again because of the exclusion principle), and thus the cluster of energy eigenvalues becomes increasingly dense and widens. Eventually, when there are so many atoms periodically spaced together that they constitute a macroscopic crystal, the allowed energy levels are so astronomically numerous and are clustered so densely along certain portions of the energy axis that they can be considered to form continua, or bands. Band *gaps* are essentially the leftover ranges of energy not covered by any band (see figure 5.5).

If electrons completely fill one or more bands (i.e., occupy all the levels in them), leaving other bands completely empty, the crystal will be an insulator. Since a filled band is always separated from the next higher band by a gap, there is no continuous way to change the momentum of an electron if every accessible state is occupied. A crystal with one or more *partly filled* bands is a metal. These ideas are schematically depicted in figure 5.6.

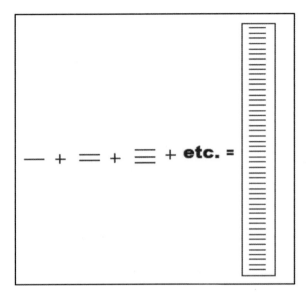

Figure 5.5. This schematized diagram shows a pure atomic energy level splitting into two when two identical atoms are brought together, and into three when three identical atoms are brought together. (The splitting is due to the Pauli exclusion principle, which forbids electrons to occupy the same state.) When huge numbers of atoms (10^{24} of them, say) come together to form a periodic crystal lattice, then what was originally a single energy level splits into a cluster of enormously many energies, which are so close to each other that they essentially constitute a continuum. This continuum of energies is called a "Bloch band".

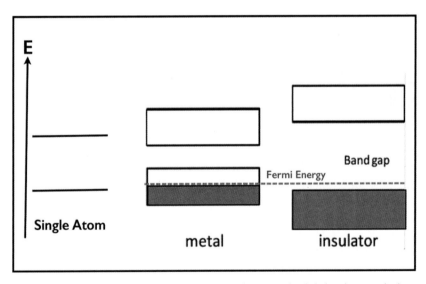

Figure 5.6. A schematic depiction of the difference between the energy levels belonging to a single atom, to a metal, and to an insulator. The *discrete energy levels* belonging to an isolated atom evolve into *energy bands* belonging to a crystal, as each atom's structure is modified by the close approach of other atoms. Inside the bands, the allowed energies take on a continuum of values. Two neighboring bands are separated by a *band gap* or simply a *gap*—a region of forbidden electron energies. The shaded regions in the figure represent levels that are occupied. In a metal, the Fermi energy lies inside a partially filled band. In an insulator, the Fermi energy lies inside the energy gap.

- *Fermi energy*
  A key notion in band theory is the *Fermi level* or *Fermi energy*, denoted E_F. This is the highest energy level occupied by a crystal electron when the crystal is at a temperature of absolute zero (see figure 5.6). The Fermi level can thus be thought of as the "surface" of the "sea" of electrons in a crystal—but it must be remembered that this "sea" lies in the abstract space of energy levels, not in a physical space.

  The position of the Fermi level relative to a crystal's band structure is a crucial factor in determining the crystal's electrical properties (e.g. its propensity to conduct electric current). At absolute zero, the electrons in any solid systematically fill up all the lowest available energy states, one by one by one. In a metal, the highest band that has electrons in it is not completely full, and hence the Fermi level lies *inside* that band. By contrast, in an insulating material or a semiconductor, the highest band that has electrons in it is completely filled, and just above it there is an energy gap, and then above that, a band that is completely empty. In such a case, the Fermi level lies somewhere between the highest filled band and the empty band above it. In a semiconductor at zero degrees Kelvin, no electrons can be found above the Fermi level, because at absolute zero, they lack sufficient thermal energy to "jump out of the sea". However, at higher temperatures, electrons can be found above the Fermi level—and the higher the temperature gets, the more of them there will be.

  The *Fermi velocity* v_F is the velocity of an electron that possesses the Fermi energy E_F. It is determined by the equation $E_F = \frac{1}{2}mv_F^2$, which, when solved for the unknown quantity v_F, gives the formula $v_F = \sqrt{2E_F/m}$.

- *Bloch's theorem*
  Named after physicist Felix Bloch, Bloch's theorem states that the energy eigenstates of an electron moving in a crystal (a periodic potential) can be written in the following form:

  $$\Psi(\mathbf{r}) = e^{i\mathbf{k}\cdot\mathbf{r}}u(\mathbf{r}), \tag{5.8}$$

  where $u(\mathbf{r})$ is a periodic function with the same periodicity as that of the underlying potential—that is, $u(\mathbf{r}) = u(\mathbf{r} + \mathbf{a})$. The exponential preceding the periodic function u is a kind of helical wave, or "corkscrew", which multiplies the the wave function by a spatially changing phase that twists cyclically as one moves through space in a straight line.

  Inside a crystal, the noninteracting electrons are not free, but are Bloch electrons moving in a periodic potential. When the potential is zero, the solutions reduce to that of a free particle with $\Psi(\mathbf{r}) = e^{i\mathbf{k}\cdot\mathbf{r}}$ and with energy $E = \frac{\hbar^2 k^2}{2m}$ (see figure 5.7). The existence of Bloch states is the key reason behind the electronic band structure of a solid.

- *Crystal momentum*
  The vector $\mathbf{k}$ of a given eigenstate for a crystal electron is that state's *Bloch vector*. When multiplied by $\hbar$, the Bloch vector gives the so-called *crystal*

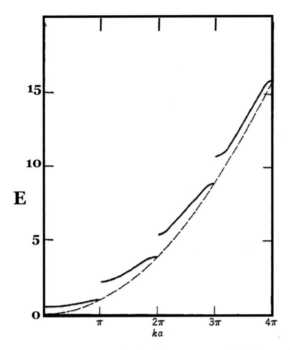

Figure 5.7. This figure illustrates how the parabolic plot of energy versus k of a free particle (the dashed curve) breaks up into a set of energy bands and energy gaps (the solid S-shaped curves that do not touch each other) when particles are constrained to move in a periodic potential.

momentum $\hbar \mathbf{k}$ of that state. Although crystal momentum has the same units as momentum, and although in some ways an electron's crystal momentum acts very much like a momentum, it should nonetheless not be conflated with the electron's momentum, because unlike momentum, crystal momentum is not a conserved quantity in the presence of a potential.

- *Brillouin zone*

    This notion was developed by the French physicist Léon Brillouin (1889–1969). For any crystal lattice in three-dimensional physical space, there is a "dual lattice" called the *reciprocal lattice*, which exists in an abstract space whose three dimensions are *inverse lengths*. This space lends itself extremely naturally to the analysis of phenomena involving wave vectors (because their dimensions are inverse lengths). If we limit ourselves to crystals whose lattices are perfectly rectangular (as has generally been done in this book), then given a lattice whose unit cell has dimensions $a \times b \times c$, the reciprocal lattice's unit cell will have dimensions $\frac{1}{a} \times \frac{1}{b} \times \frac{1}{c}$. This cell is called the *first Brillouin zone*. The various locations in the Brillouin zone—wave vectors—act as indices labeling the different Bloch states (since there is a one-to-one correspondence between wave vectors and Bloch states). Thus each point in the Brillouin zone is the natural "name" of a quantum state.

5.5.2 Summary

Both the quantized *energy levels* of an isolated atom and the quantized *energy bands* of a crystalline solid are rooted in the wave nature of electrons. In a macroscopic system, such as a two-dimensional sheet of a solid, there is an additional type of quantization where geometry and topology play a central role. This fascinating quantum effect will be the subject of the coming chapters.

References

[1] Lederman L M and Hill C T 2011 *Quantum Physics for Poets* (New York: Prometheus)
[2] Chandrasekhar B S 1997 *Why Things Are the Way They Are* (Cambridge: Cambridge University Press)
[3] http://www.feynmanlectures.caltech.edu/III_toc.html
[4] http://mafija.fmf.uni-lj.si/seminar/files/2010_2011/seminar_aharonov.pdf
[5] Kittel C 2004 *Introduction to Solid State Physics* (New York: Wiley)

IOP Concise Physics

Butterfly in the Quantum World
The story of the most fascinating quantum fractal
Indubala I Satija

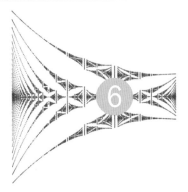

Chapter 6

A quantum-mechanical marriage and its unruly child

I want to know how God created this world. I am not interested in this or that phenomenon, in the spectrum of this or that element. I want to know his thoughts; the rest are details.

—Albert Einstein.

Having dipped into the quantum world and acquainted ourselves with the quantum Hall effect, we now return to the butterfly landscape. Whence came this wondrously strange shape? Who could better answer this question than Douglas Hofstadter, who came across the butterfly in the years 1974–5, and who still recalls those crucial events of his life almost as if they happened yesterday? He was very pleased to be asked to tell the tale as he remembers it, and so we will now enjoy a guest chapter from the pen of Douglas Hofstadter…

6.1 Two physical situations joined in a quantum-mechanical marriage

Among the most important and exciting findings of quantum mechanics in its early days were rigorous explanations of the behavior of electrons in two canonical but highly contrasting types of physical situation—on the one hand, electrons in a *crystal lattice*, and on the other hand, electrons in a (homogeneous) *magnetic field*. *Per se*, these two fundamental types of physical situation are totally unrelated, but of course there is nothing to prevent one from combining them in a metaphorical "marriage". But before introducing them to each other, let's briefly review the two partners separately.

As we saw earlier, Swiss physicist Felix Bloch, already in 1928, discovered that electrons inside a crystal (in the absence of a magnetic field) have energy values that lie in *Bloch bands*—that is, their energies range across a continuous swath of possible values, without any gaps. Bloch also found that such electrons have quantum-mechanical states (now called *Bloch states*, naturally) that can be described as standing waves inside the lattice. Such waves of course oscillated periodically, with a period depending on their energy.

We also saw earlier that Russian physicist Lev Landau, just two years later, in 1930, discovered that electrons immersed in a magnetic field (in vacuum, not in a crystal) have very sharp energy values—*Landau levels*—that are evenly spaced, like the rungs of a ladder, with wide energy gaps separating them. Landau also found that such electrons have quantum-mechanical states that correspond to classical circular orbits of various radii, yet all these different states share the same period of oscillation (the reciprocal of the "cyclotron frequency", which was the *classical* frequency of electrons making circular orbits in a magnetic field). In other words, this quantum-mechanical period is independent of the energy the Landau electron possesses.

Thus inside a Bloch band there are *no* energy gaps, but between Landau levels there *are* gaps. A fascinating question was therefore this: what would happen when these two contrasting situations—one with a continuous spectrum and an energy-dependent period, the other with a discrete spectrum and an energy-independent period—were physically combined? More concretely, what would happen to electrons in a crystal when a magnetic field was turned on? Would the band to which they belonged somehow split into a set of mini-bands—perhaps "narrow" Bloch bands, perhaps "thick" Landau levels—separated by mini-gaps? And how many such "narrow bands" or "thick levels" would there be? And how would they be spaced? And would the electron's behavior be periodic, and if so, with what period? How, in short, would nature manage to reconcile these profoundly opposing situations? None of this was at all obvious, and solving the Schrödinger equation for such a hybrid situation was, to say the least, mathematically very challenging.

6.2 The marvelous pure number ϕ

There is one extremely important new physical quantity that arises when the two situations are combined, and that is the *pure number* ϕ ("pure" in the sense that it has no units attached to it), which measures the magnetic field's strength in an extremely natural fashion. How can a pure number, with no units at all, tell how strong a magnetic field is? The answer is not too subtle and very revelatory; in fact, this all-important quantity ϕ can be thought of in (at least) three conceptually different but mathematically equivalent ways, described below.

It all hinges on the fact that there is a fundamental amount of magnetic flux—the *flux quantum*, equal to hc/e—that emerges intrinsically out of quantum mechanics. This minuscule quantity is an inherent fact about our universe, just as are the speed of light and the charge on the electron. Given that this tiny amount of flux is the natural chunk of flux, it is as if nature had handed us a measuring stick on a silver platter! This beautiful and generous favor on nature's part must not be ignored.

A first way of thinking about the variable ϕ, then, is as *how many flux quanta pass through a unit cell of the lattice*. After all, the flux passing through a unit cell is proportional to the magnetic field B (it equals $a^2 B$ in the case of a square lattice, which is what we are focusing on for the moment), and the flux quantum hc/e is the natural unit—the unit *par excellence*—with which to measure that flux. And indeed, counting the number of flux quanta passing through a unit cell gives a dimensionless answer—a pure number, a flux divided by another flux—namely, $\frac{a^2 Be}{hc}$. In truth, nothing could be more natural than this dimensionless way of measuring the strength of a magnetic field in a crystal. (By the way, this trick won't work to measure a magnetic field in a vacuum, because the trick involves flux, which involves area, and in a vacuum, unlike in a crystal, there is no scale defining a natural chunk of area.)

A second way to compute exactly the same number, yet conceptually quite different, vividly reflects the fact that ϕ is the fruit of the "marriage" of two unrelated "parents"—the isolated crystal and the isolated magnetic field. How can one find a pure number that combines the unrelated "genes" coming from each of ϕ's parents? Well, the idea is to *compute the ratio of two natural geometrical areas*, one coming from each of the two parents. These geometrical areas can be thought of as the "genes" to exploit. In particular, the salient area having to do with the *crystal* alone is the area of a lattice cell (a^2). The salient area having to do with the *magnetic field* alone is the area of a circle (a Landau–orbit) that intercepts exactly one flux quantum (this is $\frac{hc}{eB}$). Now take the ratio of these two natural areas, and *voilà!* You get a pure number, and a very simple calculation shows that it is equal to the previous number.

A third way to compute the same number reflects, once again, ϕ's dual origins. This time we again take a dimensionless ratio, but this new ratio involves two different "genes"—namely, *the ratio of two natural time intervals*, one coming from each of the two parents. The salient time interval having to do with the *crystal* alone is the time taken by an electron with momentum h/a to cross a unit cell (h/a is the canonical momentum for a crystal electron, and the associated velocity is $\frac{h}{am}$, with m being the mass of the electron); this time is therefore $\frac{a}{(h/am)} = \frac{a^2 m}{h}$. The salient time interval having to do with the *magnetic field* alone is the cyclotron period (the reciprocal of the cyclotron frequency $\frac{eB}{mc}$, which, as was mentioned before, is independent of the size of the Landau–orbit). As before, take the ratio of these two quantities, and *voilà* once again! The units cancel, you get a pure number, and another very simple calculation shows that it is equal to the previous two numbers.

Seeing ϕ as a ratio of two independent entities of the same sort (whether they are areas or time intervals), one coming from each "parent", is a fundamental and insight-giving way to look at what ϕ means. Also, seeing ϕ as the number of flux quanta threading a lattice cell is another deep way to look at what ϕ means, equally fundamental and insight-giving. Whichever may be your favorite way of conceiving of what ϕ means, in any case it is a pure number that naturally measures the magnetic-field strength in the hybrid situation. Indeed, ϕ is the key parameter at the very heart of the situation.

6.3 Harper's equation, describing Bloch electrons in a magnetic field

How can one figure out how to represent the hybrid situation—the marriage of a crystal lattice and a magnetic field—mathematically? In this section, we will go through the steps, one by one, that over a few decades led physicists to a deep equation that captured the essence of the hybrid situation, or at least the essence of an extremely idealized model of the hybrid situation. That equation—Harper's equation—is the very source, root, and wellspring of everything that this book is about, and so having a sense of this equation's origins is a crucial part of the story.

The steps that led up to Harper's equation involve some fascinating acts of virtuoso thinking by great physicists, which it will give me great pleasure to talk about. A number of these steps, however, will be mathematically quite advanced. For some readers, therefore, this will be a delicious gourmet meal, but for others, it will be exotic and alien cuisine. I hope, nonetheless, that all readers will try to plow through the whole section, even if at times it is forbidding. It is my hope that I have included enough thought-provoking historical and philosophical comments along the way to make it worthwhile even for readers without great mathematical knowledge.

The story starts out with one of the simplest and most famous equations of classical physics, relating two ubiquitous quantities in mechanics: kinetic energy and momentum. For a classical particle of mass m, the kinetic energy E is a quadratic function of the momentum p:

$$E = \frac{p^2}{2m}.$$

It would seem natural to try to carry this equation over to electrons in a crystal. (By the way, we'll assume we are dealing with a two-dimensional square lattice whose cells all have side a, as usual.) As was mentioned in previous chapters, an electron in a crystal has a *crystal momentum* $\hbar \mathbf{k}$, and the analogical link between that notion and ordinary momentum is so obvious and so strong that it almost inevitably leads to a naïve generalization of the above formula, in which crystal momentum is simply substituted for momentum. That analogical guess would yield this formula:

$$E = \frac{(\hbar \mathbf{k})^2}{2m}.$$

As a function of the wave vector $\mathbf{k}$, this is a parabola, and for small values of $\mathbf{k}$, it is accurate. However, as was mentioned in the previous chapter, all physical phenomena in a crystal must be *periodic* in the wave vector $\mathbf{k}$, and of course a parabola is not a periodic curve. As most readers will surely know, the quintessential periodic curves are sines and cosines. If, therefore, we make another tentative guess, and simply replace the parabola by a cosine function (which, after all, is a parabola for small arguments), we will get the following equation, hopefully defining the relation of electron energy E to wave vector $\mathbf{k} = (k_x, k_y)$ in a hypothetical Bloch band in a two-dimensional crystal:

$$E = E_0(\cos k_x a + \cos k_y a). \tag{6.1}$$

This humble equation, arrived at here by educated guesswork (although it can also be more rigorously derived) is the starting point for a quantum-mechanical treatment of the marriage of Bloch electrons to a magnetic field. (I call it "humble" because it embodies the most idealized model imaginable of a Bloch band, using the most rudimentary periodic function that exists. The formula reflects what is commonly called the "tight-binding approximation".) Over the next few pages, bit by bit, we will witness this equation turning into a Schrödinger equation, with its left side turning into an eigenvalue, and its right side turning into a differential operator acting on a wave function. But let us not jump the gun. One step at a time!

The next step in the process leading to a Schrödinger equation for this marriage was an extraordinarily bold and ingenious move devised first in the 1930s by English physicist Rudolf Peierls and then, in the 1950s, in a different form, by Norwegian physicist Lars Onsager Peierls (see figure 6.1). (As one might guess from his name,

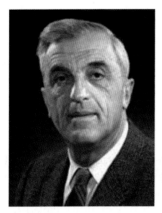

Figure 6.1. Felix Bloch, Lars Onsager, and Rudolf Peierls.

Peierls was born in Germany, but because of the rise of Nazism, he left Germany for Britain in the 1930s and remained there for the rest of his life.) Both of these seminal thinkers must be counted among the major pioneers of 20th century physics.

Onsager and Peierls had the inspiration of replacing the crystal momentum $\hbar\mathbf{k}$ in this formula by a quantum-mechanical *momentum operator*, a replacement that would convert the sum of cosines into a Hamiltonian operator which could act on a wave function Ψ; with that stroke, this purely *numerical* equation would become a *differential* equation—indeed, a form of the Schrödinger equation.

The motivation for this strange-seeming substitution is two-fold. First is the fact that a canonical way to turn a classical equation into a quantum-mechanical equation is to replace certain numerical variables (e.g. momentum) by corresponding differential operators. (Dirac, who discovered this trick, referred to it as the replacement of "c-numbers" by "q-numbers", where presumably "c" stood for "classical" and "q" for "quantum-mechanical".) Of course Onsager and Peierls, both being excellent physicists, were very aware of this method of turning a classical-physics formula into a quantum-physics formula. Second is the fact that crystal momentum has the same units as momentum and behaves in many ways like momentum. So if you were in a daring, risky mood, and if you happened to think of these two ideas at the same time, you just might dream up the idea of replacing the crystal momentum by a quantum-mechanical momentum operator. Nonetheless, it is a little like pulling a rabbit out of a hat. After all, crystal momentum is *not* momentum. Treating the two as if they were synonyms is an intuitive stab in the dark rather than a rigorous act of reasoning. There is nothing to guarantee that it will even be *meaningful* to make such a substitution, let alone that it will be *correct*. And yet this was the rather wild flash that came to Onsager and Peierls.

This was already a very clever idea, but carrying it out properly required not just great care but also some true touches of genius. How, indeed, does one understand what it means to have a differential operator acting as an argument of a cosine? This is not self-evident. However, before tackling that subtle mathematical question, let us figure out—either following in the footsteps of, or perhaps standing on the shoulders of, the aforementioned giants—just *which* operator we want to plug into the above equation.

A clever physics student's first reflex might be to plug in the "vanilla" momentum operator $-i\hbar\nabla = -i\hbar(\frac{\partial}{\partial x}, \frac{\partial}{\partial y})$, well known to all students of first-year quantum mechanics, but if one thinks about this for a moment, one realizes that doing this will not take into account the existence of a magnetic field in the situation. The crystal is represented by the cosine formula involving the crystal momentum $\hbar\mathbf{k}$, and quantum mechanics itself is represented by the act of substituting a q-number for a c-number, but as of yet, there is no representation of magnetism anywhere here.

Spurred on by this oversight, our imaginary physics student might realize that what is called for here is, so to speak, a "chocolate" momentum operator—that is, a fancier operator that somehow takes the magnetic field into account. From courses in classical mechanics, the student might recall that there is a quantity called the "canonical momentum" of an electron in a magnetic field, which involves both the electron's *ordinary* momentum $m\vec{v}$ and the vector potential $\vec{A}$ for the magnetic field that the electron is immersed in. To be specific, an electron's "canonical momentum" is the quantity $m\vec{v} - \frac{e}{c}\vec{A}$. (I interrupt the flow very briefly to mention that the rest of this book uses a slightly different system of units, in which $c = 1$, and so some of the equations in this chapter will have factors of c or of $1/c$ that are lacking in their counterparts in other chapters.) And now, with a bit of nonchalant hand waving, we will replace the *ordinary* momentum term $m\vec{v}$ in this expression by the "vanilla" momentum operator $-i\hbar\nabla$, so that we wind up with the "chocolate" momentum operator $-i\hbar\nabla - \frac{e}{c}\vec{A}$.

As you can probably tell, there is a lot of slippery conceptual playing-around going on here, but slippery or not, this is precisely the differential operator that Onsager and Peierls came up with. And now, for our next trick, we will blithely plug the chocolate momentum operator into the cosine function, letting it replace $\hbar\mathbf{k}$. (Actually, we first have to divide it by $\hbar$, since we are replacing $\mathbf{k}$, not $\hbar\mathbf{k}$.) What we will need to do after that is to figure out how to make sense out of the resulting expression—or if we can't make sense of it, then hopefully we can at least *manipulate* it as if it made sense, a bit like whistling in the dark.

Come to think of it, we are not quite ready to tackle that tricky step, because we haven't yet said what the vector potential $\vec{A}$ should be, if we wish it to capture the physical situation that we are concerned with. Luckily, finding such an $\vec{A}$ is not terribly hard. If we choose our magnetic field $\vec{B}$ to have constant magnitude along the z direction, then we want a vector potential $\vec{A}$ whose *curl* equals such a magnetic field. This means that $\vec{A}$ has to be a vector field such that $\vec{B} = \nabla \times \vec{A}$ is a constant vector pointing in the z direction.

There are, in fact, many such vector fields, all related to each other by gauge transformations, but to save time and to avoid details, we will skip straight to the chase and reveal the choice for $\vec{A}$ made by Onsager and Peierls. This is $\vec{A} = B(0, x, 0)$, usually called the *Landau gauge*. It is easily verified that $\nabla \times \vec{A} = B\hat{z}$, as desired.

If we plug this very simple formula for $\vec{A}$ into the definition of our chocolate momentum operator, we get the following equations for the two non-zero components of the operator:

$$p_x = -i\hbar\frac{\partial}{\partial x}$$

$$p_y = -i\hbar\frac{\partial}{\partial y} - \frac{e}{c}Bx.$$

And now we are cooking with gas!

The main remaining challenge is to interpret the meaning of a mysterious-looking expression like "$\cos(-ia\frac{\partial}{\partial x})$". We can get partway there by recalling that for any θ, $\cos\theta = (e^{i\theta} + e^{-i\theta})/2$. If, in place of θ, we plug the differential operator $-ia\frac{\partial}{\partial x}$ into this expression, we get the following equation (in which we have applied both sides to a two-dimensional wave function Ψ):

$$\cos\left(-ia\frac{\partial}{\partial x}\right)\Psi(x, y) = \frac{\left(e^{a\frac{\partial}{\partial x}} + e^{-a\frac{\partial}{\partial x}}\right)}{2}\Psi(x, y).$$

Now all we have to do is expand the two exponentials on the right side by exploiting the elegant power series for e^x, known for centuries:

$$e^x = 1 + x + \frac{x^2}{2!} + \frac{x^3}{3!} + \frac{x^4}{4!} + \cdots.$$

What this gives, for the first exponential, is:

$$e^{a\frac{\partial}{\partial x}}\Psi(x, y) = \Psi(x, y) + a\frac{\partial}{\partial x}\Psi(x, y) + \frac{a^2}{2!}\frac{\partial^2}{\partial x^2}\Psi(x, y) + \frac{a^3}{3!}\frac{\partial^3}{\partial x^3}\Psi(x, y) + \cdots.$$

(6.2)

Similarly,

$$e^{-a\frac{\partial}{\partial x}}\Psi(x, y) = \Psi(x, y) - a\frac{\partial}{\partial x}\Psi(x, y) + \frac{a^2}{2!}\frac{\partial^2}{\partial x^2}\Psi(x, y) - \frac{a^3}{3!}\frac{\partial^3}{\partial x^3}\Psi(x, y) + \cdots.$$

(6.3)

This does feel rather magical, like pulling a rabbit out of a hat, does it not? I say this in all seriousness, and indeed with a sense of profound wonder, because to me, even after decades of seeing such manipulations, it still feels mysterious, miraculous, and nearly mystical. After all, to spell it out very clearly, we started with a very basic high-school trigonometry function, $\cos\theta$, whose argument is an everyday angle, like 30 or 60 degrees. For many very highly intelligent people, this is already approaching their abstraction ceiling, where the oxygen is sparse; however, to a mathematician or physicist, nothing could be possibly more tangible and concrete. From there we nimbly leapt to a deeper and more sophisticated vision of the cosine function, involving a sum of two exponentials of complex numbers—but that nimble leap, although quite a feat and leading us into far more abstract, far less visualizable territory, was not nearly bold enough. The next step was to replace each exponential by an infinitely long polynomial (an idea that ought to make you smile), but even that level of wildness was *still* not wild enough, because we then replaced the power-series variable—a complex number, admittedly, but at least a "thing" that was more or less visualizable—by a dangling partial differential operator "$\frac{\partial}{\partial x}$", representing a pure *action*, rather than a *thing*. In the end, then, we wound up with a pair of very mysterious-looking power series equations (6.2) and (6.3) that would leave most cosine-savvy high-school students choking in the dust. This was a rather dizzying sequence of rapid-fire generalizations, and yet in a technical physics paper, such manipulations would be treated as if they were trivial and utterly self-evident. But I think that it's important, at least once in a while, to point out and to savor such magically fluid thinking, because it lies at the core of what theoretical physics is all about. But enough editorializing for the moment. Let us return to the story of where Harper's equation came from.

At this stage of the game, our hypothetical astute physics student would hopefully recognize the expressions on the right-hand sides of equations (6.2) and (6.3) as the Taylor expansions of $\Psi(x + a, y)$ and $\Psi(x - a, y)$, respectively. In other words, the effect of the *x*-component of the chocolate momentum operator has been to translate the first argument, *x*, of the wave function Ψ through the lattice distance *a*, first rightwards and then leftwards.

A very similar argument holds for the *y*-component of the chocolate momentum operator, except that there is also an extra numerical factor involved, which comes from the nonvanishing *y*-component of the magnetic vector potential $\vec{A}$. Instead of merely giving us $\Psi(x, y + a)$ and $\Psi(x, y - a)$, the operator will attach *phase factors* to both of these values of the wave function.

In short, our humble original equation relating energy to wave number in a highly idealized Bloch band (equation (6.1)) has bit by bit been turned into a Schrödinger equation that relates Ψ at point (*x*,*y*) of the lattice to four other values of Ψ on the lattice—namely, those located at the lattice sites $(x + a, y)$, $(x - a, y)$, $(x, y + a)$, and $(x, y - a)$, which are the four nearest neighbors of (*x*,*y*). Here, then, is the full Schrödinger equation that we have "derived" (or to put it more accurately, that we have arrived at via a series of slick sleights of hand):

$$E_0 \left[\Psi(x + a, y) + \Psi(x - a, y) + e^{-\frac{ie}{\hbar c} Bxa} \Psi(x, y + a) + e^{\frac{ie}{\hbar c} Bxa} \Psi(x, y - a) \right]$$
$$= E\Psi(x, y).$$

Like any self-respecting Schrödinger equation, this is an eigenvalue equation that admits of solutions for the wave function Ψ only for certain values of the energy E on the right side. Finding those values of E as a function of the magnetic field ϕ is our "Holy Grail", so to speak.

At this point, it will be useful to introduce some notational conventions. Firstly, we will divide both sides by E_0, giving us E/E_0 on the right side, but we will replace this ratio by the letter E (this is just a rescaling of the energy, and turns E into a dimensionless quantity). Secondly, since it seems that only lattice points (*ma*, *na*) are involved, with *m* and *n* running over integer values only, we can replace the *continuous* variables *x* and *y* by the *discrete* variables *ma* and *na*, or even by just *m* and *n*. Lastly, we can introduce our old friend, the pure number ϕ, into this equation. All of this will give us the following equation:

$$\Psi(m + 1, n) + \Psi(m - 1, n) + e^{-2\pi i m \phi} \Psi(m, n + 1) + e^{2\pi i m \phi} \Psi(m, n - 1) = E\Psi(m, n).$$

And just think: this is no longer a differential equation, but a *difference* equation, since it only involves discrete positions in a two-dimensional lattice. That's quite a surprise!

We have come a long way, but we can still simplify this equation quite a bit more if we guess that $\Psi(m, n)$ is a product of two independent functions ψ_m and χ_n, with χ_n being purely periodic along the *x*-axis (or the *n*-axis, if you prefer). (This is actually not just a guess but a necessary consequence of our prior assumptions.) If we set $\chi_n = e^{ik_y n}$ and plug it in, then the equation simplifies as follows:

$$\psi_{m+1} + \psi_{m-1} + \left[e^{2\pi im\phi - k_y} + e^{-2\pi im\phi + k_y} \right] \psi_m = E\psi_m.$$

All that's left of χ_n, which represented the wave function's behavior along the y-axis, is the number k_y in the two exponents. Other than that, there is no more trace of χ_n. Everything comes down to behavior along just the x-axis (i.e., the m-axis).

The sum of the two exponentials in the square brackets actually amounts to twice a cosine, so we can finish up our simplification work as follows:

$$\psi_{m+1} + \psi_{m-1} + 2\cos(2\pi\phi m - k_y)\psi_m = E\psi_m. \tag{6.4}$$

Et voilà!

This equation was first published in 1955 by P G Harper (see figure 6.2), a student of Rudolf Peierls, and thus it is usually called "Harper's equation", although it was independently discovered at about the same time by Gregory H Wannier. Note that what Peierls, Onsager, Harper, and Wannier did boils the whole problem of a Bloch electron in a magnetic field down to a *one-dimensional difference equation*.

I remind readers that ϕ represents the strength of the magnetic field, E represents the electron's energy, and k_y is just a constant that represents the periodicity of the wave function along the "boring" y-axis. All the *interesting* action takes place along the x-axis, which has here become the m-axis. Now who would ever have guessed that a one-dimensional difference equation could capture the essence of such a

Figure 6.2. Philip Harper. This image is reproduced with permission from the Heriot-Watt University Heritage and Information Governance.

complex physical situation? Well, perhaps someone with great gifts of clairvoyance might have been able to anticipate that the Schrödinger equation in this context would have to have more or less this form. The argument would go something like this: "The first two terms ($\psi_{m+1} + \psi_{m-1}$) represent the second spatial derivative of Ψ, the next term captures both the periodic potential due to the lattice (via the cosine) and the magnetic field (via ϕ), and the right-hand side represents the eigen-energy E." Well, this argument makes a bit of sense after the fact, but it would take an exceptionally brilliant mind to guess such a subtle equation in advance.

It's key to remember that equation (6.4) is an eigenvalue equation, meaning that for certain values of E, it will have solutions ψ_m, and for others it will not. What does "have solutions" mean here? It means that for some values of E, ψ_m is "well-behaved" when m gets large. To be more precise, for certain values of E—most, in fact—the wave function ψ_m will blow up as m becomes large (positively or negatively), while for a very special set of other values of E, the wave function ψ_m will remain bounded, no matter how large m grows. Unbounded explosions of the wave function as one moves out towards infinity cannot exist in nature, and thus, to make a long story short, what makes an E "good" is that the wave function ψ_m should remain bounded for all m. Finding all the "good" values of the energy E, given a fixed pure-number strength ϕ of the magnetic field, is the challenge.

6.4 Harper's equation as a recursion relation

Harper's equation can be thought of as a recursion relation that defines each new term ψ_{m+1} as a weighted sum of the previous two terms, ψ_m and ψ_{m-1}. To see this, all we need to do is rewrite it as follows:

$$\psi_{m+1} = C_m \psi_m - \psi_{m-1},$$

where "C_m" stands for the oscillating quantity $E - 2\cos(2\pi\phi m - k_y)$. The frequency with which the coefficient C_m makes a cycle and comes back to where it was is proportional to ϕ, the magnetic-field strength. In that sense, C_m might remind us of the cyclotron orbits of an electron in a magnetic field, whose frequency is also proportional to the magnetic-field strength.

If ϕ is an integer, then C_m is a constant. Harper's equation then becomes

$$\psi_{m+1} = C\psi_m - \psi_{m-1},$$

which, in its extreme simplicity, is reminiscent of the Fibonacci equation.

If ϕ is not an integer, but is rational—say it is 1/3—then C_m first takes on one value, then a second value, then a third value, and then (since the cosine function is periodic) it starts all over again, taking on those same values over and over again, cyclically. If ϕ were 2/5, then the values of C_m would repeat after going through a cycle of length five. In general, if ϕ equals p/q, the denominator q tells you how long you will have to wait for the sequence of values of C_m to come back, full circle, to where they started (courtesy of the cosine).

If ϕ is not rational, then the sequence of values taken on by C_m doesn't ever close back on itself exactly, but on the other hand, its values are *almost* periodic, in that after a while, they will *nearly* come back in a cycle, and if one waits a longer while, they will come even closer to cycling back—and if one has the patience to wait infinitely long, they will come back exactly.

6.5 On the key role of inexplicable artistic intuitions in physics

As we recall the origin of Harper's equation in a daring manipulation of the humble Bloch-band formula (equation (6.1)), we feel it is important to take our hats off to Peierls and Onsager, who made this reckless substitution of a quantum-mechanical operator for the crystal momentum purely intuitively. It was certainly an artistic move, absolutely indefensible in purely logical terms, and thus it almost feels like magical thinking.

Other physicists have occasionally pointed out the almost incomprehensible magic in the thinking processes of the rarest geniuses in physics. For example, Japanese Nobelist Sin-Itiro Tomonaga, in his marvelous book *The Story of Spin* [1], wrote of P A M Dirac's uncanny hypothesis of second quantization, "We ordinary mortals stand here bewildered." That memorable phrase (and it was uttered by someone who himself was a genius) could apply equally well to the rash and purely aesthetic leap of imagination made by Onsager and Peierls.

And Steven Weinberg, one of the most distinguished physicists of recent decades, in his book *Dreams of a Final Theory*, made a similar comment about Werner Heisenberg's astonishing thought processes in 1925, when Heisenberg first came up with what was eventually called "quantum mechanics". Weinberg wrote this:

If the reader is mystified at what Heisenberg was doing, he or she is not alone. I have tried several times to read the paper that Heisenberg wrote on returning from Heligoland, and although I think I understand quantum mechanics, I have never understood Heisenberg's motivations for the mathematical steps in his paper. Theoretical physicists in their most successful work tend to play one of two roles: they are either sages *or* magicians. *The sage-physicist reasons in an orderly way about physical problems on the basis of fundamental ideas of the way that nature ought to be. Einstein, for example, in developing the general theory of relativity, was playing the role of a sage; he had a well-defined problem—how to fit the theory of gravitation into the new view of space and time that he had proposed in 1905 as the special theory of relativity. He had some valuable clues, in particular the remarkable fact discovered by Galileo that the motion of small bodies in a gravitational field is independent of the nature of the bodies. This suggested to Einstein that gravitation might be a property of space–time itself. Einstein also had available a well-developed mathematical theory of curved spaces that had been worked out by Riemann and other mathematicians in the nineteenth century. It is possible to teach general relativity today by following pretty much the same line of reasoning that Einstein used when he finally wrote up his work in 1915.*

Then there are the magician-physicists, who do not seem to be reasoning at all but who jump over all intermediate steps to a new insight about nature. The authors of physics textbooks are usually compelled to redo the work of the magicians so that they seem like sages; otherwise no reader would understand the physics. Planck was a magician in inventing his 1900 theory of heat radiation, and Einstein was playing the part of a magician when he proposed the idea of the photon in 1905. (Perhaps this is why he later described the photon theory as the most revolutionary thing he had ever done.) It is usually not difficult to understand the papers of sage-physicists, but the papers of magician-physicists are often incomprehensible. In this sense, Heisenberg's 1925 paper was pure magic.

It is important to point out that, contrary to the general opinion of lay people and even of most physicists themselves, doing physics is an *art* that is deeply rooted in aesthetic intuitions and ineffable nuances of taste. What Onsager and Peierls did, for example, was a poetic act of interpretation of a mathematical expression (namely, the right side of equation (6.1)). They read into that expression a metaphorical meaning that no one thinking logically would ever have thought of. Their flash of replacing the crystal momentum by a generalized momentum operator was a piece of inspired guesswork, rooted in aesthetics, motivated by an intuitively sensed likeness. It was a kind of magical thinking.

If one wanted to be harsh, one might call their leap of faith "superstitious thinking" rather than "magical thinking", but that would be a gross exaggeration. In the first place, science is all about the *testing* of guesses, so that if Onsager and Peierls' speculative idea had yielded wrong predictions, they would have simply dropped it and gone back to the drawing board. That's the diametric opposite of superstitious thinking! But in the second place, there is no reason to describe intuitively motivated guesswork in physics as "superstitious thinking". The fact is that some physicists have an uncanny ability to peer behind the veils of nature and to spot hidden connections that no one else sees. Such artistic thinking is what doing physics can be, at its best, and it is important to point this out, rather than to misrepresent physical thinking as just a series of purely logical acts and rigorous manipulations of mathematical formulas.

It is also interesting to point out that although the "Onsager–Peierls ansatz" (as Gregory Wannier used to call their aesthetics-based heuristic substitution) gives perfectly accurate results when a crystal has only a single Bloch band, it is quite inaccurate when there is more than one Bloch band. So in a sense, Onsager and Peierls were very lucky in their guess that they could make this substitution. And decades later, *I* was lucky that *they* had been lucky, since it was their ansatz that led to the equation whose behavior I studied.

6.6 Discovering the strange eigenvalue spectrum of Harper's equation

We have now seen how Harper's equation—Schrödinger's equation for Bloch electrons in a magnetic field—was derived (or, if "derived" is too strong a term, then at least *discovered*), and we also understand that the condition defining its eigenvalues E is that ψ_m should not blow up as the variable m grows large. But given a specific value ϕ of the magnetic field, how can these special values of E be pinpointed?

Up to 1974, many physicists had studied this question and some progress had been made, but no one knew in detail what the eigenvalue spectrum—the concrete *graph* of eigenvalues E as a function of the magnetic field ϕ—looked like. If this seems strange to today's readers, it must be explained that at that time, computers, though quite advanced and used in many disciplines, were not commonly exploited by physicists—especially not by theoretical physicists, who believed that physics was done by coming up with equations and then manipulating them with all sorts of fancy techniques learned in graduate school. In the research group of Gregory Wannier, a matrix-based method had been devised to characterize the good values of E, but no one in the group had been able to take advantage of that result to derive important new facts about the nature of the spectrum.

The basic idea of the matrix-based method was to look at Harper's equation as a recipe telling how to compute the couple (ψ_{m+1}, ψ_m), given the couple (ψ_m, ψ_{m-1}). If we think of these couples as vectors of size 2, then we can rewrite Harper's equation using a 2×2 matrix:

$$\begin{pmatrix} \psi_{m+1} \\ \psi_m \end{pmatrix} = \begin{pmatrix} C_m & -1 \\ 1 & 0 \end{pmatrix} \begin{pmatrix} \psi_m \\ \psi_{m-1} \end{pmatrix} = \begin{pmatrix} E - 2\cos(2\pi\phi m - k_y) & -1 \\ 1 & 0 \end{pmatrix} \begin{pmatrix} \psi_m \\ \psi_{m-1} \end{pmatrix}.$$

Let us abbreviate the 2×2 matrix as H_m. Then the preceding equation becomes:

$$\begin{pmatrix} \psi_{m+1} \\ \psi_m \end{pmatrix} = H_m \begin{pmatrix} \psi_m \\ \psi_{m-1} \end{pmatrix}.$$

If we start out with the couple (ψ_1, ψ_0) and then multiply it by the matrix H_1, we will get the couple (ψ_2, ψ_1). If we multiply *this* couple by the matrix H_2, we will get the couple (ψ_3, ψ_2)—and so forth. Symbolically,

$$\begin{pmatrix} \psi_4 \\ \psi_3 \end{pmatrix} = H_3 \begin{pmatrix} \psi_3 \\ \psi_2 \end{pmatrix} = H_3 H_2 \begin{pmatrix} \psi_2 \\ \psi_1 \end{pmatrix} = H_3 H_2 H_1 \begin{pmatrix} \psi_1 \\ \psi_0 \end{pmatrix}.$$

So far we have not done much; we have just written out, using matrix notation, the idea that we can get new values of ψ from old values by using the coefficients in the recursion relation that is Harper's equation. But let us now imagine that ϕ is a rational number—say, 1/3. In this case, the matrices will start repeating after a cycle of length 3; that is to say, $H_4 = H_1$, and $H_5 = H_2$, etc. Thus the product $H_6 H_5 H_4$ will be equal to the product $H_3 H_2 H_1$, and likewise, $H_9 H_8 H_7$ will be exactly the same matrix once again. If we abbreviate the product $H_3 H_2 H_1$ as M, then if we want to get ψ_{3m} for a very large value of m, all we need to do is use the mth power of M:

$$\begin{pmatrix} \psi_{3m+1} \\ \psi_{3m} \end{pmatrix} = M^m \begin{pmatrix} \psi_1 \\ \psi_0 \end{pmatrix}.$$

All we have done is to take advantage of the periodicity of the cosine, given that $\phi = 1/3$. But of course the same idea will hold whenever ϕ is a rational number p/q. In the general case, the matrix M will be the product of q matrices $H_q H_{q-1} ... H_2 H_1$, instead of just three matrices.

So under what circumstances will the terms ψ_m given by Harper's equation blow up when m grows very large? The matrix M gives us a handle on this question. (Of course the existence of this matrix depends on ϕ being rational, as we have just indicated, but let's not worry about that for now.) If putting M to higher and higher powers results in matrices with larger and larger elements, then we are sunk. So the question boils down to when high powers of M themselves have bounded elements.

There are theorems about when a matrix put to higher and higher powers will blow up, and in the early 1970s, Gustav Obermair, while working on this problem with Wannier, had looked into those theorems, and with their aid he had demonstrated that what mattered, in this case, was whether the *trace* of M (the sum of its two diagonal elements) lay between -2 and $+2$. If yes, the matrix's powers would remain bounded, which was good; if no, the matrix's powers would blow up, which was bad. This meant that the condition for E to be an eigenvalue was implicit in the matrix M. If ϕ were given as, say, 2/5, one would merely need to calculate M, which in this case would be $H_5 H_4 H_3 H_2 H_1$, since the denominator of ϕ is 5, and then to take the trace of M and see if it is between -2 and $+2$. This is pretty simple!

Bear in mind, though, that the trace of this product of five matrices (or q matrices, in the general case) is not a numerical value, because there are two *variables* in all the 2×2 matrices being multiplied together—namely, E and k_y. Therefore, the trace of the matrix M will be a qth-degree polynomial in the variable E. (By the way, the dependence on the variable k_y turns out to be very simple—so simple that we can take all possible values of k_y into account in one fell swoop. All we do is set k_y to zero and relax the condition on the trace of M so that it can lie anywhere between -4 and $+4$, instead of between -2 and $+2$. That allows us to forget about k_y.) So in short, we have found, using this mode of analysis, that for any rational value p/q of ϕ, there will be a qth-degree polynomial in the variable E, and what we want to know is: *when will the values of this polynomial be less than or equal to 4, in absolute value?*

This is more or less where I came in, in the fall of 1974 in Regensburg, Germany. As I said in my prologue, I was very frustrated that I was unable to participate in the theorem-proving activity that the senior members of the Regensburg group (Wannier, Obermair, and Alexander Rauh) were eagerly engaged in, but at some point I had the good fortune to discover a desktop computer in the hallway, and since I knew how to program, I made the only contribution I knew how to make, which was to numerically calculate the "fat roots", as I called them, of many different qth-degree polynomials (a fat root being an interval along the x-axis, which is to say, the E-axis, in which the polynomial is within a distance of 4 of the axis).

These "Harper polynomials" are quite fascinating, as we shall see. Figure 6.3 shows the degree-3 and degree-5 Harper polynomials belonging to $\phi = 1/3$ and $\phi = 2/5$, respectively. Note that in these two cases, the subbands lie *inside* the Bloch band (which stretches from $E = -4$ to $E = +4$). Figure 6.4 shows the degree-6 Harper polynomial belonging to $\phi = 1/6$, and once again all the subbands are

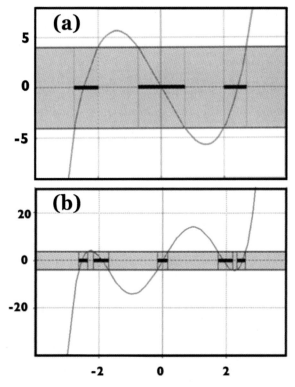

Figure 6.3. In panel (a), we see a cubic Harper polynomial with three "fat roots" along the E-axis. In fact, the axis itself has been "fattened", becoming a horizontal strip of thickness 8 located between the parallel lines $E = -4$ and $E = +4$. The three "fat roots" are the three intervals along the E-axis where the polynomial makes three quick passes through the "fat axis". And of course those intervals are the three mini-bands into which the original Bloch band splits when $\phi = 1/3$. Similarly, panel (b) shows the degree-5 Harper polynomial belonging to $\phi = 2/5$, with its five fat roots along the E-axis. As before, these five "fat roots" are the five mini-bands into which the original Bloch band splits when $\phi = 2/5$.

contained inside the Bloch band. Could this tendency for subbands to lie inside the Bloch band be relied on in general? The answer was yes. Alexander Rauh proved an important theorem to the effect that the "fat roots" of the Harper polynomial for any rational value of ϕ always lie between $E = -4$ and $E = +4$. This was an excellent piece of progress, but there were nonetheless dark clouds on the horizon.

6.7 Continued fractions and the looming nightmare of discontinuity

What worried everyone working on this problem, whether in Regensburg or elsewhere, was that if there was a qth-degree Harper polynomial for each rational value p/q of ϕ, then for that rational value of ϕ, there would be q fat roots, and thus q sub-bands of the original Bloch band. This may not sound too worrisome, unless you think about how rational numbers are distributed along the number line. Very close to 1/6, for instance, is the rational number 3/19. The former has six bands, while the latter has 19 bands. That sounds very different from six bands. We can get

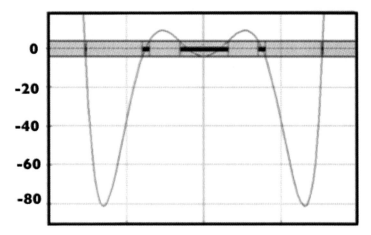

Figure 6.4. The Harper polynomial belonging to magnetic-field value $\phi = 1/6$. Note that because this polynomial grazes the "fat axis" at at $E = 0$, there are two "fat roots" that touch each other at that point, which gives the misleading impression that for $\phi = 1/6$, there are only five bands. In fact, there are six, as there must be, but two of them kiss at a point, so they look like just one.

yet closer to 1/6 by setting ϕ equal to 10/61, whose 61st-degree polynomial of course gives rise to 61 bands, which obviously sounds even less like six bands. Then there are ϕ-values of 100/601, 1000/6001, and so forth, which are clearly approaching 1/6 in the limit, yet they have more and more bands.

This is very worrisome if one believes that the phenomena of physics must be *continuous*, meaning that tiny changes in ϕ must be accompanied by tiny changes in the spectrum belonging to ϕ. It would seem, however, that precisely the reverse is happening—the closer a rational number gets to 1/6, the greater its denominator must be, and so the more subbands it will have in its spectrum. This doesn't sound like continuity, but like its diametric opposite! This paradox was deeply bewildering to all the people working on the problem in those days, but happily, there turned out to be a beautiful resolution to the mystery.

Incidentally, one might imagine that for $\phi = p/q$, there could be *fewer* than q bands, because the qth-degree Harper polynomial might enter one side of the "fat axis" and then turn around inside it, instead of exiting it on the other side. For instance, in the case of $\phi = 1/3$, one could imagine that the "S"-shaped double reversal made by the polynomial could have taken place entirely *within* the fat axis, so there would be only *one band*, rather than three. As it turns out, however, this kind of turnaround inside the fat axis never happens. Every time any Harper polynomial enters the fat axis, it always reemerges from it on the other side without turning around inside it. This theorem about the behavior of Harper polynomials associated with all rational values of ϕ was formally proven in a letter to me in the mid-1970s by an old friend of mine, the mathematician John Mather, with whom I had been in correspondence.

The only case that resembles the just-described turnaround scenario occurs when q is even, and it occurs only at the very middle of the E-axis. In those cases, the

Harper polynomial, after entering the fat axis, slows down, bends toward the horizontal, and just barely reaches the other side, at which point it turns around while just grazing its upper (or lower) edge from the inside, in a perfect tangency. This phenomenon is shown in figure 6.4, for the value $\phi = 1/6$. One sees that there are four normal bands away from the center (two on the left, two on the right), but that at the very center, where $E = 0$, the polynomial does *not* break out of the fat axis, but elegantly grazes its lower edge from the inside. This curious phenomenon makes the two middle bands belonging to 1/6 (and more generally, to any even-denominator rational value of ϕ) "kiss" each other, as was described in chapter 1. For such values of ϕ, this gives the appearance of only $q - 1$ bands, but in fact, there are still q bands—it's just that two of them gently nuzzle each other at one single point ($E = 0$). But that, it must be stressed, is the worst case (or the best, depending on how you feel about nuzzling). In all other cases, there are exactly q cleanly separated subbands.

The question remains, however, how those q bands are distributed. Consider, for the sake of concreteness, values of ϕ that lie close to 1/3, such as $\phi = 100/301$. It is rather confusing, if not downright mystifying, to think that this value's 301 bands are situated right next door to the set of just three bands belonging to $\phi = 1/3$. And then think of the 3001 bands belonging to $\phi = 1000/3001$, which lie even *closer* to the three bands of 1/3! The closer you get to $\phi = 1/3$, the more bands there are! From such simple considerations, it would seem inevitable that the spectrum of Harper's equation not only is discontinuous, but is *perversely* discontinuous!

And yet, it turns out that nature was clever enough to save continuity in a very elegant fashion. To spell this out, consider the seeming discontinuity just mentioned. It turns out that for $\phi = 100/301$, there are 100 extremely narrow subbands tightly clustered together very close to the *leftmost* subband belonging to $\phi = 1/3$. These 100 subbands collectively "act like" a *single* band. And symmetrically, there are also 100 extremely narrow subbands tightly clustered together very close to the *rightmost* band belonging to $\phi = 1/3$. This leaves 101 subbands still to be accounted for, and they are found, luckily, just where one would hope to find them: tightly clustered around the *middle* band belonging to $\phi = 1/3$. In other words, the 301 subbands belonging to $\phi = 100/301$ do a superb job of *emulating* the three bands belonging to $\phi = 1/3$, by clustering together in three tight groups. If one squints at the spectrum of $\phi = 100/301$, so that one isn't able to make out the fine details, one sees essentially the spectrum of $\phi = 1/3$. (Figure 6.5 shows this very clearly.) This is a beautiful trick that nature figured out, to ensure the continuity of the spectrum of Harper's equation despite what would seem, at first thought, to be a hopeless situation.

6.8 Polynomials that dance on several levels at once

Looking directly at the graphs of the Harper polynomials sheds considerable light on this fascinating emulation of one rational number by another rational number very close to it. Let us take the case of $\phi = 3/19$, which is very close to $\phi = 1/6$. The 19th-degree Harper polynomial in question is displayed at several different levels of detail in figure 6.6. Most people have probably never seen or imagined a polynomial that behaves anything like this. The first view, in panel (a), shows the Harper

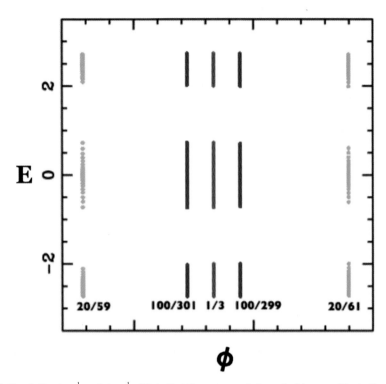

Figure 6.5. Bands for $\phi = \frac{1}{3}$ and $\phi \approx \frac{1}{3}$. (Note that the energy axis is vertical here, unlike in figures 6.1 and 6.3.) Although the bands are very numerous when the denominators are large, the fragmented band structure, when viewed from a distance, very closely resembles the band structure for $\phi = \frac{1}{3}$. Thus for $\phi = \frac{20}{61}$, only a few gaps are visible, and just barely so. For $\phi = \frac{100}{301}$, one instantly sees two large gaps, but the 298 remaining gaps between the fragments are far too tiny to make out here. The idea that making radical changes in the number of bands might nonetheless have no significant visible (or physical) effect is a lovely and subtle idea, showing how elegant are the solutions that nature finds to paradoxical-seeming situations.

polynomial as seen from very far away. Note the scale—the values on the *y*-axis are well over 1 000 000. This is a very strange graph, with cones on both sides (one pointing up, one pointing down), and what looks like a perfectly horizontal plateau separating them. Of course, we are seeing that plateau from very far away, so we can't be sure whether it is truly horizontal. We still have no idea how many times the graph crosses the fat *E*-axis. So we must zoom in on it.

Panel (b) shows the same polynomial but with the values along the *y*-axis being much smaller—only about 2000 or so. We have really zoomed in a lot! Now the two cones that were so salient in panel (a) have become gigantic spikes that sail off the top and bottom of the frame. But closer to the middle, we see two new cones that look very much like those old cones, only flipped. (Of course we have to keep in mind that they are about 1000 times smaller, vertically.) So let's zoom in once again, in panel (c).

Now the heights along the *y*-axis are about 50, and all sorts of structural details are starting to emerge. Once again, we see two cones (and once again flipped), but these are much smaller than the previous four—their height is only about 30,

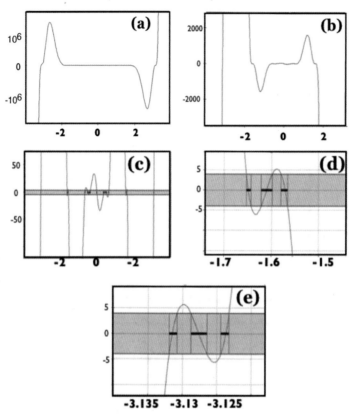

Figure 6.6. The Harper polynomial belonging to flux-value $\phi = \frac{3}{19}$, shown at five different levels of zooming-in.

compared with 1 000 000 and 2000—and they connect to each other at the very center of the graph. (At this scale, all that we can see of the four cones mentioned earlier are extremely steep spikes that shoot off the top and bottom of the frame.) But now, at last, we are getting a good glimpse of some of the "fat roots" of the polynomial.

For example, near $E = -1.6$, we see a very rapid "S"-shaped dance taking place just around the fat axis. This little dance is blown up in panel (d) of the figure, both vertically and horizontally, which allows us to view it very clearly. And in fact, one can't help but be reminded, by this dance, of the *cubic* polynomial belonging to $\phi = 1/3$. This is most curious!

And speaking of curiosity, what is going on near $E = -3.1$, where the Harper polynomial is almost perfectly vertical, dashing like crazy to cross the E-axis in as rapid an eyeblink as possible? Well, as before, let us zoom in on that area of the graph, expanding both the vertical and horizontal axes until the desired area comes into view in the desired manner. (See panel (e) of the figure.) Once again, this delicate little dance, producing three tiny "fat roots" or "bandlets" belonging to $\phi = 3/19$, is very reminiscent of the dance done by the cubic polynomial belonging to $\phi = 1/3$.

To make a long story short, the same can be said for all six of the zigzag dances done by this 19th-degree Harper polynomial—when appropriately zoomed in on, they all look very much like the zigzag dance done by the cubic polynomial belonging to $\phi = 1/3$. In sum, it is as if, six times in a row, very locally and very briefly, the 19th-degree polynomial belonging to $\phi = 3/19$ were somehow putting on a mask, assuming a different identity, and emulating the *cubic* polynomial belonging to $\phi = 1/3$. Is that just a pipe dream, or is it conceivable?

Yes, indeed—it is more than conceivable. This is exactly what is going on, and for a very good reason. To understand this phenomenon, we need to look at the continued-fraction expansion of the number 3/19. It is a very short and very simple expansion—namely:

$$3/19 = \cfrac{1}{6 + \cfrac{1}{3}}.$$

The fraction looks, at first glance, like 1/6 (since the denominator is 6 + 1/3, which is very close to 6). But the "correction term"—namely, that "+1/3"—reveals the secret of what is going on. On a *global* scale, the spectrum belonging to 3/19 is trying to emulate that of 1/6, but on a *local* scale, it is trying to imitate that of 1/3. In short, this 19th-degree polynomial is actually doing a marvelous job of emulating two different polynomials at once, on two different scales. It is doing a set of six "local dances", each of which looks like the dance for 1/3, and it is doing a "global dance" that looks like the dance for 1/6. To be more specific about the global dance, this polynomial makes six rapid triple-crossings of the E-axis in spots that are widely separated from each other, but that are very close to where the six bands belonging to $\phi = 1/6$ are located. In that sense, the Harper polynomial for $\phi = 3/19$ globally emulates the Harper polynomial for $\phi = 1/6$. (It also crosses the E-axis one time at the very middle, which it has to do, since 19 is odd.)

If we were to take a rational number with a longer continued-fraction expansion, then its Harper polynomial will do a very complex dance with even more levels of emulation. For example, consider the Harper polynomial for $\phi = 15/38$. The continued-fraction expansion for this rational number is as follows:

$$15/38 = \cfrac{1}{3 - \cfrac{1}{2 + \cfrac{1}{7}}}.$$

Here, we have three integer denominators—3, 2, and 7—representing three different hierarchical levels. Most globally, therefore, the Harper polynomial for $\phi = 15/38$ will try to emulate the Harper polynomial for $\phi = 1/3$. More locally, it will try to emulate the Harper polynomial for 1/2, and on the most local level, the 15/38 Harper polynomial will do its best to "dance" just the way the 1/7 Harper polynomial dances. Jumping through these three hoops at three different scales is a bit of a trick, but the 15/38 polynomial manages to do it.

Notice the minus-sign just after the first denominator ("3 − ...") in this continued fraction. This tells us that we are not dealing with the most standard variety of continued fraction, which features only plus-signs (such fractions are called "simple continued fractions"). What we have here, instead, is a *nearest-integer continued fraction*, which sometimes has plus-signs and sometimes minus-signs. What is the logic behind the sign choices?

I will explain with an example. In this case, if we try to expand 15/38 as a continued fraction, the first step is to take its reciprocal, and then we break that into an integer part and a fractional part. Doing so gives us:

$$15/38 = \frac{1}{38/15} = \frac{1}{2 + \frac{8}{15}} = \frac{1}{3 - \frac{7}{15}}.$$

Why did we replace the expression "$2 + \frac{8}{15}$" by "$3 - \frac{7}{15}$"? Because the quantity we are working on—namely, $2 + \frac{8}{15} = 2.533333333...$—is slightly *closer* to 3 than it is to 2 (in other words, because $\frac{7}{15}$ is smaller than $\frac{8}{15}$). Always go for the closer integer! In general, the trick for making a nearest-integer continued fraction is, *at each level of the fraction*, always to take the integer *nearest* to the quantity you are currently working on (instead of mechanically just taking the integer just below it, which would result in a *simple* continued fraction).

For example, consider the golden ratio, 1.61803..., whose simple continued fraction we have already referred to a number of times:

$$1 + \cfrac{1}{1 + \cfrac{1}{1 + \cfrac{1}{1 + ...}}}.$$

This is quite different from the corresponding nearest-integer continued fraction, because the golden ratio is closer to 2 than to 1, hence the desired expression necessarily begins with "2−" rather than "1+". Here, without further ado, is the nearest-integer continued fraction for the golden ratio:

$$2 - \cfrac{1}{2 - \cfrac{1}{2 - \cfrac{1}{2 - ...}}}.$$

It is continued fractions of this sort, as opposed to simple continued fractions, that are best suited for speaking about the recursive patterns that make up Gplot.

If the nearest-integer continued-fraction expansion for a rational value of ϕ has more denominators than the above two-level and three-level examples, then there are simply more hierarchical levels of Harper-polynomial dancing, but the patterns that these dances exhibit, on the most global and the most local scales, as well as on all intermediate scales, are still specified by the successive denominators in the nearest-integer continued fraction for ϕ. This idea, incidentally, was heuristically suggested by Russian physicist M Ya Azbel' in a remarkably prescient article, already in 1964. Although Azbel''s article was not able to describe the spectrum of Harper's equation in detail, and was wrong in some ways (for instance, it guessed that simple continued fractions, rather than nearest-integer continued fractions, were involved, and it did not state that central bands behave differently from all other bands, splitting up into a different number of sub-bands, and it thus implied the wrong number of bands for almost all rational flux-values), it nonetheless accurately anticipated certain key truths about the nature of the graph.

Like so many important pieces of progress in science, Azbel''s 1964 article [2] was a mixture of brilliant intuitive guesswork and intellectual rigor, and some of its central ideas were exactly right, while some were not. But being 60 percent right (say) and 40 percent wrong is very important, because overall it is a step in the right direction. Science is an art form that only converges very slowly on the truth, and sometimes even an article that is only 10 percent right is worth its weight in gold!

6.9 A short digression on INT and on perception of visual patterns

The representation of a real number as a nearest-integer continued fraction is closely related to the INT function, which I discussed in the prologue. They way I described INT(x) there, it is calculated by swapping two sequences (the **coun**-sequence and the **sep**-sequence) belonging to x. However, INT can be defined equivalently in terms of nearest-integer continued fractions. To show this in a pleasing case, I will demonstrate it for the famous constant $e = 2.718281828...$. First, here is how e is represented as a nearest-integer continued fraction:

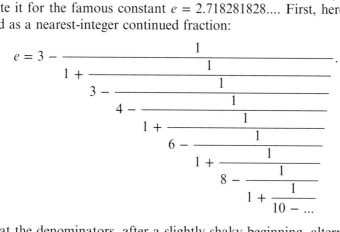

$$e = 3 - \cfrac{1}{1 + \cfrac{1}{3 - \cfrac{1}{4 - \cfrac{1}{1 + \cfrac{1}{6 - \cfrac{1}{1 + \cfrac{1}{8 - \cfrac{1}{1 + \cfrac{1}{10 - ...}}}}}}}}}.$$

Note that the denominators, after a slightly shaky beginning, alternate between even numbers followed by a minus-sign and 1s followed by a plus-sign. This elegant pattern continues forever. The nearest-integer continued fraction for INT(e) is made from this one by *bumping down any denominator followed by a minus-sign, then converting the minus-sign into a plus-sign,* and symmetrically, *bumping up any denominator followed by a plus-sign, then converting the plus-sign into a minus-sign.* That same simple algorithm will yield INT(x) for any real number x. That's the whole story. Here, then, is how the nearest-integer continued fraction for INT(e) looks:

$$\mathrm{INT}(e) = 2 + \cfrac{1}{2 - \cfrac{1}{2 + \cfrac{1}{3 + \cfrac{1}{2 - \cfrac{1}{5 + \cfrac{1}{2 - \cfrac{1}{7 + \cfrac{1}{2 - \cfrac{1}{9 + \ldots}}}}}}}}}.$$

While I'm on the subject of INT, I should point out that INT shares a most salient visual feature with Gplot: namely, those delicate, spindly "legs" that rapidly taper off as they approach the corners of the graph. That is an unmistakable sign that something deep links these two graphs. (Incidentally, I expressly chose the word "legs", because those frail, fine-grained, diagonal needles have forever reminded me of the spindly elephant-legs in Salvador Dali's evocative painting *The Temptation of Saint Sebastian*.) If you know INT well, you can't help but recognize its telltale presence in Gplot the instant those surrealistic "legs" hit your retina. But if you don't know INT... well, that key connection just won't get made. Too bad! As I said in my prologue, I just had the luck of being the right person in the right place at the right time.

And speaking of luck, I have only recently understood how lucky was my instinctive choice, back in Regensburg, to use color-coding in hand-plotting Gplot (a decision that came straight out of my prolonged study of η-sequences and INT, many years earlier). Specifically, my color-coding took into account the nearest-integer continued-fraction representation of each rational number whose spectrum I was calculating. For any number whose continued fraction terminated after only *one* denominator (such as 1/5 or 4/5, which equals $1 - \frac{1}{5}$), I colored its bands black. For any value of ϕ whose continued fraction had *two* denominators, I colored the bands using a second color. For a value whose continued fraction had *three* denominators, I colored the bands using a third color. (This kind of color-coding is clearly visible in figures 6 and 7 of the prologue.)

This choice made the recursive nature of Gplot pop out almost instantly to my eye. The reason for this visual popping-out is that the subcells of Gplot that contain distorted copies of the entire graph are automatically highlighted by this kind of coloring. I realized how important this had been for me when I was looking, not long ago, at the graph published in 1969 by Dieter Langbein in *Physical Review* [3] and exhibited in this book in the "Butterfly Gallery". I was wondering why it was that Langbein, who had had so much detail right there in front of his eyes, hadn't beaten me to the punch in spotting the beautiful recursions that constitute the heart and soul of Gplot. Well, one reason is that Langbein didn't know INT, and so he didn't recognize the telltale sign of those spindly legs. A second reason is that in Langbein's graph, there is no color-coding (all bands are black), and so there is nothing to distinguish different kinds of rational numbers (different in the sense of *how many denominators* their nearest-integer continued fractions require). As a consequence, in Langbein's graph, no structure pops out to the eye. That is to say, the very same subbands that so clearly defined the "L" and "R" subcells in my early hand-plotted graph just blend in anonymously with the rest of the subbands in Langbein's plot.

Those, in my opinion, are the two main reasons why I had the good fortune to spot the fundamental recursive nature of Gplot, and why Langbein—as well as the readers of his article—failed to do so. Such is the fickle nature of fate.

6.10 The spectrum belonging to irrational values of ϕ and the "ten-martini problem"

Having brought in continued-fraction representations for the values of ϕ, we can finally broach the subtle but crucial question of the spectra that belong to *irrational* values of ϕ. After all, rational numbers are exceptions along the real line; though they are dense, there are still only countably many of them, while there are *uncountably* many irrationals. To put it another way, if one randomly threw an infinitely fine-tipped dart at a real line scotch-taped to a wall, the probability is vanishingly small that it would land on a rational number. (Of course, it's also vanishingly small that the dart would land on the real line at all, since its width is infinitesimal compared to the height of the wall, but let's not worry about that!) Irrationals are the overwhelming majority of the inhabitants of the real line, and rationals are super-rare exceptions. So of course we want to know what happens when an irrational value of ϕ is chosen.

Let's imagine an irrational number near 1/3, for the sake of concreteness. Just like the golden ratio, this value of ϕ has a nearest-integer continued fraction that never ends. At the very top, however, its first denominator is 3. For that reason, whatever its spectrum is, it will necessarily consist of pieces scattered along the E-axis that, if one just squints a bit, will look very much like the three bands belonging to $\phi = 1/3$. Suppose furthermore that our irrational ϕ is extremely close to 100/301. Then its spectrum will emulate not only the three bands of 1/3, but also, on a much finer-grained level, all of the 100 little tiny bands of 100/301 that cluster around the leftmost band of 1/3, all of the 101 little tiny bands that cluster around 1/3's middle band, and all of the 100 little tiny bands that cluster around 1/3's rightmost band.

But we are of course not done! Since we have posited that this particular value of ϕ is irrational, so that its continued fraction keeps on going down, down, down, it follows that the spectrum for this ϕ will continue to fragment endlessly, as we march down its continued fraction, level by level. Bands (or what look like bands from far away) will be seen to split into subbands, and those subbands will then split into subsubbands, and so forth and so on, with no end in sight. Indeed, in the end (after an infinite number of recursive splittings), there will be no bands left—just an infinite set of scattered *points*—a "Cantor dust", as was discussed in chapter 1.

Incidentally, the construction of the original Cantor set, described in chapter 1, can be profitably generalized. In Cantor's original version, what got removed at each stage was always just the middle third of each remaining interval. However, a more general "sparsification maneuver" could delete not just one but *several* subintervals of each interval (for example, always divide each interval into seven equal subintervals, and then remove the 2nd, 4th, and 6th of them). The end result of this infinitely repeated sparsification maneuver (which we might call "removing three sevenths") will have many properties in common with the original Cantor dust, including that of having measure zero. (This technical term can be taken roughly as saying that the set of isolated points in question has "no weight", as compared with an interval, which, no matter how short, always has positive weight.)

One can also loosen up on the idea that at all stages, the sparsification has to be done in just the same way. Thus, for example, why not remove the *middle third* on the first step, *four ninths* on the second step, *three sevenths* on the third step, *one third* on the fourth step, and so forth? Just as in the construction of the original Cantor set, each new act of sparsification will create an ever-larger number of ever-tinier bands to sparsify further on the next round.

This kind of recursive sparsification process, varying from level to level, would be far closer to what happens in the case of Gplot. There, the basic idea is that for all n, the nth denominator in the nearest-integer continued fraction for ϕ dictates how many subintervals to delete from each band during the nth stage of the process. (Actually, for the sake of precision, it should be pointed out that the sparsification operation performed on the central band at $E = 0$ obeys a slightly different rule, but this is the basic idea.) This infinite sequence of sparsification operations gives rise, for each irrational value of ϕ, to a completely unique Cantor dust dictated by the successive denominators in ϕ's continued fraction.

Figure 6.7 (left panel) illustrates this idea, focusing on the special case where ϕ equals the golden mean. This case, because of the periodicity of ϕ's continued fraction, gives rise not just to a "random" Cantor set, but to a self-similar Cantor set, since at every single stage, essentially the same removal process is effected.

This bizarre but wonderful phenomenon—that for any irrational value of ϕ, the spectrum of Harper's equation is an uncountable set of points constituting a Cantor set of measure zero—was quite a surprise to me, but I soon convinced myself of its truth. Indeed, I devoted a full chapter of my doctoral thesis [4] to proving it as a theorem. Incidentally, the mere fact that a set of points inside an interval is the end result of infinitely many successive sparsification operations does not by itself guarantee measure zero for the final set. To show that measure zero is always the end result for the spectrum of any irrational flux-value ϕ thus required careful analysis, on my part, of the nature of the infinitely many sparsification steps along the way. Luckily I wound up finding a way to show that these steps had the proper relationship to each other to guarantee measure zero.

When I say that I proved this measure-zero Cantor-set theorem, I should more precisely explain that I rigorously proved that it was a necessary consequence of three key postulates that I formulated as a result of carefully observing the nesting behavior of Gplot. From a visual inspection of Gplot, it's abundantly clear that my three postulates are true, but try though I might, I was unable to prove those postulates formally.

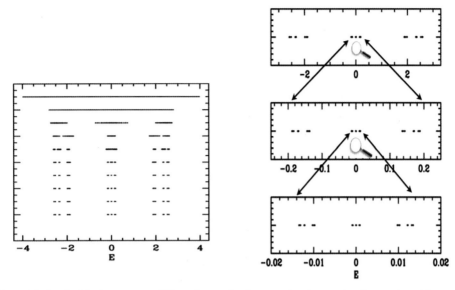

Figure 6.7. On the left, the spectrum of Harper's equation for ten rational approximations to the golden mean. From top to bottom, these rationals are: $\phi = 1, \frac{1}{2}, \frac{2}{3}, \frac{3}{5}, \frac{5}{8}, \frac{8}{13}, \frac{13}{21}, \frac{21}{34}, \frac{34}{55}, \frac{55}{89}$. On the right, we zoom in twice on the region of the E-axis near $E = 0$, allowing us to see the self-similarity of the spectrum very clearly.

My conditional demonstration of the Cantor-set conjecture could thus be likened to a construction firm building the *upper* 20 floors of a planned 40-floor skyscraper without having first built the *lower* 20 floors. For building actual skyscrapers, this may not be the most efficacious method, although it makes for an amusing image. In physics, however, this kind of building process is not atypical. Theoretical physicists are constantly building ornate edifices on shaky if not totally unproven foundations; after a while, they publish their mental constructions, hoping they will withstand the earthquakes of experimental tests.

Although, to my chagrin, I hadn't proven the Cantor-set idea in a full sense, I knew it was true, and it was certainly a beautiful and significant fact, and so, shortly after writing it up in my thesis in 1975, I published it in my sole article about the butterfly in *Physical Review* [5], and a few years later, and somewhat more visibly, I stated it in my book *Gödel, Escher, Bach* [6].

In 1981, the fact—or rather, the conjecture!—that irrational numbers have Cantor-set spectra was vividly brought to mathematicians' attention, when mathematician Mark Kac, in a talk at the annual meeting of the American Mathematical Society, described Harper's equation and its very unusual spectrum, and offered ten martinis to anyone who could prove that my hypothesis was true. However, Kac apparently made no reference to my having originated the claim. Perhaps he did not realize it was due to me. In any case, Kac's amusing (but very serious) challenge was further circulated by his mathematics colleague Barry Simon under the catchy moniker of "the ten-martini problem", and as such, it aroused considerable interest and excitement in the mathematical community. A good number of mathematicians around the world, not only intellectually stimulated but also hoping to gain ten martinis at the expense of their distinguished colleague, jumped into the fray.

It took well over thirty years, but finally in 2009, the so-called ten-martini problem was solved by mathematicians Artur Ávila and Svetlana Jitomirskaya [7]. Unfortunately, Mark Kac was no longer with us when the Ávila–Jitomirskaya paper was published, so he never tasted the joy of lavishing on its two authors the promised ten martinis (plus, without doubt, a few on himself).

6.11 In which continuity (of a sort) is finally established

As I stated in the prologue, what most intrigued me, when Gregory Wannier first described the crystal-in-magnetic-field problem to me, was the weird-sounding proposition that the *rationality* or *irrationality* of a physical parameter—namely, the magnetic-field strength ϕ—could have some actual, measurable effect in the world. That idea made no sense to me, since it implied that all of the digits in the infinite decimal expansion of ϕ not only mattered, but mattered equally. Actually, not even equally, for in fact this idea made the furthest-away (and thus least significant) decimals count the most, since it is only by examining the pattern of digits moving out to infinity that one can tell whether a number is rational or irrational!

This is a crucial idea, so let me spell it out a bit more. Rational numbers have periodic decimal expansions while irrational numbers do not—and by "periodic", I actually mean *eventually periodic*. Consider, for instance, the following real number:

3.14159265358979323846264000000003131313131313131313131313131313131 ...

whose decimal expansion starts out rather bumpily, but eventually winds up settling down into an infinitely boring repetition of just the digit-pair "31". This number will be rational, as long as those 31's keep on going forever.

But how can one know whether a given sequence of digits is eventually periodic or not? It may *look* periodic, as the pattern just exhibited does, but what guarantees that it actually *is* periodic? In the case above, after all, one might well have jumped to the conclusion that the decimal expansion was going periodic when one hit the sequence of eight 0's in a row. But that pattern, though seductive, turned out to be short-lived, and hence deceptive.

I told you that this pattern stays with the 31's forever, but in order for me to be able to guarantee you of that, I had to know the entire expansion, all the way to the bitter end—infinitely many digits. And that's the rub. One has to know *all* of a number's digits, all the way out to infinity, in order to be sure whether those digits eventually do or do not go into a cyclic pattern. And most ironically, what happens at the *outset* (that is, the initial set of digits, even the initial thousand digits) is completely irrelevant to this question!

Therefore, if some physical phenomenon actually depended on ϕ's rationality or irrationality, then ϕ's *magnitude* wouldn't matter at all—only what happened as one looked at the infinitely far-out tail of its decimal expansion. This zany-sounding idea is reminiscent of astrology, which claims that our innermost nature doesn't have to do with local causes, such as the DNA inside every one of our cells, but instead with phenomena involving stars that are hundreds, thousands, or millions of light-years away. Things inconceivably far away and inconceivably remote in time are what determine the course of our lives here on Earth, while things right here and right now don't play a role? Such an idea is of course sheer nonsense. Just as nothing could be more antiscientific than astrology, nothing could be more nonphysical than the claim that some physical phenomenon depends on the rationality or irrationality of some physical parameter. Such a claim makes no sense whatsoever.

Another profound aspect of the nonphysicality of the claim that ϕ's rationality or irrationality matters is that it simply makes no sense to think that all of ϕ's digits are *well-defined*. No physical variable in the universe has an infinitely precise, infinitely sharp value. Do you prefer rational or irrational temperatures? Do you prefer driving at rational or irrational speeds? Can you, in fact, drive down the highway at 60.000000000... miles per hour, where all the digits after the decimal point are "0", forever and ever? Is one's speed ever defined to infinite precision? Can a physicist truly believe that it makes perfect sense to ask for, say, the trillionth digit of the number representing the strength of a magnetic field in a laboratory experiment? If you think this makes sense, you should probably go sign up for a doctorate in astrology (and if, by chance, you should happen to make some *astronomical* discoveries along the way, then maybe you can include them in an appendix to your thesis—on page 137, say).

Through this set of mental explorations of the absurd, we are actually getting very close to the resolution of the problem, because they highlight a major distinction between mathematics and physics, a distinction that became very clear to me as I neared the end of my thesis. My hope was to take the strange graph that I had discovered, and to show that its recursive nature, rather than being paradoxical, implied that the system's observable physical behavior was entirely independent of the rationality or irrationality of ϕ. It was undeniable, of course, that to a mathematician, rational values and irrational values of ϕ gave rise to wildly different spectra. After all, for any rational value, the spectrum consisted of a set of line segments whose total length was a positive number, while for any irrational value, the spectrum consisted of a Cantor dust consisting of infinitely many isolated points whose total "length" (or more technically, whose *Lebesgue measure*) was exactly zero (and that means 0.00000000..., with infinitely many zeros).

Thus if one moved smoothly along the ϕ-axis and graphed the Lebesgue measure of the spectrum of ϕ—let's call it $L(\phi)$—one would create a graph that was almost everywhere zero (at irrational values), and yet, for a dense subset of the ϕ-axis (the rational values), $L(\phi)$ would have positive values. Through careful topological reasoning, I was able to prove that $L(\phi)$ is a function with the peculiar property that it is *continuous at all irrational values and discontinuous at all rational values of ϕ*. Mathematicians rightfully love this kind of crazy-seeming phenomenon, and there's nothing wrong with that. It is a beautiful, paradox-grazing idea, and yet in the end it makes perfect sense. But it has nothing whatsoever to do with the physical world.

Physicists, too, may admire the pristine and exquisite reasoning that establishes all the wonderfully counterintuitive properties of the spectrum of Harper's equation, but when all is said and done, what they want to get out of the mathematics is *physically meaningful ideas*. And this is no less worthy a goal than the goals of mathematicians. Since I had once thought of myself as a budding mathematician and had even gone to graduate school in math for a couple of years, I could see things from either side of the fence. I loved the crazily counterintuitive pure math, but I also wanted to use that math to reach a physically realistic conclusion. And so I began with the undeniable fact that no physical variable is ever infinitely sharply defined. What that means is that there is necessarily an uncertainty $\Delta\phi$ in the value of ϕ. It doesn't matter how small this uncertainty is, but what is crucial is that no matter what the situation is, it is always nonzero. No magnetic field is precisely defined for infinitely many decimals.

And so, what happens if one looks at the spectrum not just for one sharp value of ϕ, but at those of a set of closely surrounding values of ϕ—namely, all those that lie in the interval of size $\Delta\phi$ centered on ϕ? Suppose one were simply to *take the union of* the spectra of all those neighboring values of ϕ—what would that give? This amounts to "jiggling" the graph a little bit back and forth along the ϕ-axis, so that it blurs it a bit. Such jiggling, no matter how mild, creates what I called a *smeared version* of Gplot. In figure 8 of the prologue are displayed the smeared graphs belonging to two different values of $\Delta\phi$ (roughly 1/35 and 1/100). Although to some people these graphs might look a little bit scary, like eerie monsters, they are nonetheless completely "normal" in the sense of being physically meaningful, physically realistic spectra.

Once again using careful topological reasoning (a skill that I'd picked up to some extent while briefly wearing my mathematics graduate-student hat), I studied the effect of jiggling the graph, and I discovered some theorems about smeared versions of Gplot. To my delight, I was able to demonstrate that in a smeared graph, for all values of ϕ, whether rational or irrational, there are at most $\frac{1}{\Delta\phi} + 1$ bands. However small $\Delta\phi$ is, this is a finite number. In other words, the infinitude of the Cantor sets disappears totally when a graph is smeared. Even the very large number of bands for *rational* values with high denominators goes away. The number of bands in a smeared graph is uniformly bounded, for all ϕ. In short, jiggling the graph yields a completely ordinary, "vanilla" spectrum, which, as one moves along the ϕ-axis, varies *continuously* in every way that one could hope for. There are no more sets of measure zero, no more "pathological" behavior; no matter how small $\Delta\phi$ is, the Lebesgue measure of the blurred spectrum at every single value of ϕ will be positive and will vary continuously with ϕ. That is totally reasonable physical behavior.

Here is how I put it in my 1976 *Physical Review* article [5]:

As $\Delta\phi$ approaches zero, the fine structure of the graph is bit by bit recovered; the infinitely fine-grained detail never returns (for positive $\Delta\phi$), but more and more of it is revealed by decreasing the uncertainty $\Delta\phi$. Of course, at the unphysical value $\Delta\phi = 0$, the entire graph returns.

In some ways, this result was the intellectual climax of my doctoral work. I had first found a wildly counterintuitive, nonphysical-seeming spectrum, and then I had shown that despite its craziness, it was nonetheless completely reconcilable with ordinary physics.

6.12 Infinitely recursively scalloped wave functions: Cherries on the doctoral sundae

In my thesis, after exploring the eigenvalues belonging to Harper's equation, I turned briefly to the wave functions that corresponded to them. Given any value of the magnetic flux ϕ, if you pick a particular eigen-energy E, it will determine a one-dimensional wave function ψ_n defined for integer values of n. This wave function at first seems to be defined only on discrete lattice points, rather than along the entire continuum of which the lattice points are only a very sparse subset. This unfortunately makes no sense, since to be physically meaningful, a wave function must be defined at every point in space (whether it is one-dimensional, two-dimensional, or three-dimensional). This sparseness would therefore seem to be a major flaw marring the theory.

However, on examining the mathematics of so-called "magnetic translation operators", I noticed that the existence of the flux quantum *hc/e* provides a way to fill in the wave function along the whole line. The idea, in a very rough nutshell, is that there is a way to combine the flux quantum with the lattice constant a to define a *second* natural length unit (other than a itself) that has the special property of being the one-dimensional wave function's (spatial) period. Exploiting this periodicity allows one to "fold back" the wave function on top of itself over and over again, each time filling in more points along the continuum between lattice sites, where at first there seemed to be no wave function at all. This way of filling in the curve stage by stage can be done only a *finite* number of times for any rational value of ϕ, thus adding points here and there but still leaving gaps almost everywhere—but if ϕ is irrational (and remember that irrationality is by far the most common case), then this folding-back-and-filling-in process can be done *infinitely* many times, which will convert the initially very sparsely defined wave function into a fully defined, completely continuous solution to the Schrödinger equation.

I will not enter into the details here, but this new insight allowed me to calculate and plot out the one-dimensional wave functions for certain rational values of ϕ and certain eigenvalues E. In particular, I plotted the wave functions belonging to three rationals whose continued fractions were closely related:

$$\phi_1 = \frac{1}{5},$$

$$\phi_2 = \frac{2}{11} = \frac{1}{5 + \frac{1}{2}},$$

$$\phi_3 = \frac{17}{93} = \cfrac{1}{5 + \cfrac{1}{2 + \cfrac{1}{8}}}.$$

My idea was that ϕ_1, ϕ_2, ϕ_3 (and further rational values of ϕ) are all approaching a limiting irrational value ϕ_∞, whose nearest-integer continued fraction would start out as theirs do, but would never come to an end. Figure 6.8 shows the three wave functions plotted together. The crudest of them, consisting of only 6 ($=5+1$) points, belongs to ϕ_1, and looks like a chain dangling between two posts. The next one, belonging to ϕ_2, is finer-grained, consisting of 12 ($=11+1$) points, and looks like several scallops linked together to form a chain. The final one, belonging to ϕ_3, is yet more detailed, as it consists of 94 ($=93+1$) points, and it, too, looks like a chain with scallops, but if one looks more closely, one notices scallops on the scallops. Aha! You can see what's coming... If there were a ϕ_4 with one more term in its continued fraction, its wave function would be yet finer-grained and triply scalloped, and so forth and so on. In the limit, then, the wave function belonging to the limiting irrational value ϕ_∞ would fit in with all these approximations, but it would be defined at *all* points, and it would feature "infinitely recursive scalloping", or if you wish, cusps densely distributed over the entire axis. Now that would be quite some wave function! As you can imagine, this unexpected discovery was, in a sense, a delicious cherry on the sundae of my doctoral work.

There is, however, something ironic about this cherry on the sundae. We have just seen that the wave function for an irrational value of ϕ (and thus for almost all values of ϕ, since the rationals constitute a set of but measure zero on the real axis) is infinitely recursively scalloped, meaning that it is constantly changing direction, no matter how tiny the scale on which one inspects it. Such a microscopically zigzagging function is "pathological", in the old 19th-century sense, in that it does not have a well-defined slope anywhere. The irony is that Harper's equation is the endpoint of a pathway of mathematical manipulations whose starting point was a single Bloch band with "chocolate" momentum operators inside it instead of wave numbers k_x and k_y (this is the Peierls–Onsager ansatz). This new differential

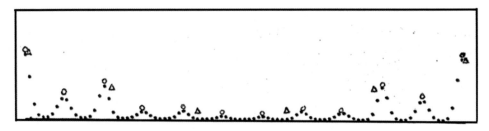

Figure 6.8. The wave functions of three very close rational values of ϕ—namely, $\psi(\frac{1}{5})$ (shown with triangles), $\psi(\frac{2}{11})$ (shown with circles), and $\psi(\frac{17}{93})$ (shown with dots)—superimposed. Each new value has one extra denominator in its nearest-integer continued fraction, and fills in more of the interval than the preceding values did. Scallops are visible at three levels of granularity. As more levels of the continued fraction come into play, finer scallops are involved. In the limit of an irrational value of ϕ, the wave function will be defined at *all* points of the interval, and its shape will be infinitely recursively scalloped.

operator was interpreted as a Hamiltonian, and as such it yielded a version of Schrödinger's equation, which, being a differential equation, assumes that the function satisfying it is *differentiable* (in fact, not just differentiable, but *infinitely* differentiable, as is clear from equations (6.2) and (6.3)). So although we started out with differential operators (which make sense only with a smooth function to act on), we wound up with a wave function that could never be the solution to any differential equation! This bizarre situation grazes paradox in a tantalizing way. Once again, we are seeing the curious results of "magical thinking" in physics—in this, case, the esthetically motivated but nonrigorous heuristic substitution intuitively made by Peierls and Onsager, which led, many years later, to a beautiful set of interrelated discoveries that run deeply against the meaning of the differential operators that gave rise to them. Why it all works out so perfectly is far from fully understood!

6.13 Closing words

It was at this point, in late 1975, that I bowed out of physics. Little did I suspect that from these humble beginnings would flow so many other results in the coming decades. Although I didn't participate in those discoveries, I have watched them from the sidelines with great interest, and it gives me a feeling of pride and privilege to have had the good fortune of playing a role in the launching of this fertile, multifaceted area of research in physics.

Appendix: Supplementary material on Harper's equation

Various butterfly plots shown in other parts of this book were obtained by treating the Harper equation (6.4) as an eigenvalue equation that, given a rational flux-value $\phi = p/q$, involves diagonalizing a $q \times q$ matrix. For $q \geq 3$, this matrix equation can be written as follows:

$$\begin{pmatrix} 2C_1 & 1 & 0 & 0 & \cdots & e^{-ik_x} \\ 1 & 2C_2 & 1 & 0 & 0 & \cdot & 0 \\ 0 & 1 & 2C_3 & 1 & 0 & \cdot & 0 \\ \vdots & \vdots & \vdots & \vdots & \vdots & & \vdots \\ e^{ik_x} & \cdot & & 0 & 0 & 1 & 2C_q \end{pmatrix} \begin{pmatrix} \psi_1 \\ \psi_2 \\ \psi_3 \\ \vdots \\ \psi_q \end{pmatrix} = E \begin{pmatrix} \psi_1 \\ \psi_2 \\ \psi_3 \\ \vdots \\ \psi_q \end{pmatrix},$$

where $C_n = \cos(2\pi n\phi - k_y)$. The lower-left and upper-right corner terms $e^{\pm ik_x}$ in this matrix reflect Bloch's theorem, which assumes periodic boundary conditions—namely, $\psi_n = e^{ik_x}\psi_{n+q}$. In a handful of very simple cases, such as $\phi = 1, \frac{1}{2}, \frac{1}{3}, \frac{1}{4}$, the eigenvalues E can be determined analytically. However, for $q > 4$, the above matrix-eigenvalue problem can only be solved numerically.

This method, involving finding the eigenvalues of a $q \times q$ matrix, is equivalent to the Regensburg group's method, described earlier in this chapter by Douglas Hofstadter and used by him in his explorations, which involves the trace of a product of q successive 2×2 matrices.

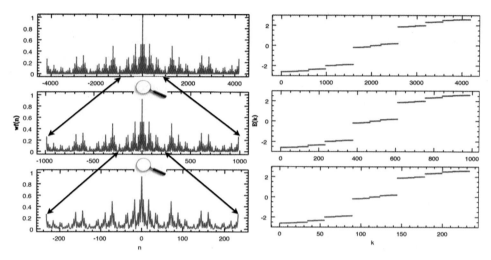

Figure 6.9. The left panel shows the wave functions plotted as a function of lattice sites n for three different lattice sizes N for the $E = 0$ state of the golden-mean flux. From top to bottom: $N = 4181$; $N = 987$; $N = 233$. This figure illustrates the self-similarity of the wave function: it looks just the same when we zoom into the original lattice. Moreover, peaks in the wave function occur at lattice sites whose positions are Fibonacci numbers, which are the denominators of the best rational approximations to the golden mean. For the same three lattice sizes, the panel on the right illustrates self-similarity of the energy spectrum.

In the years since Douglas Hofstadter published his well-known article, the nature of the Harper wave functions and energies belonging to rational flux-values that tend to an irrational limit has been studied. Some of the results are exhibited in figure 6.9, which shows the scale-invariance of the fractal characteristics of the energy and the wave function for the golden-mean flux. We note that the wave function never decays to zero and the heights of the wave-function sub peaks far from the center approach a well defined fraction, a universal number, of the height of the central peak [8, 9].

Some Analytic Results
  I **Analytic expressions for the energy dispersions $E(k_x, k_y)$ for a few simple cases [9]**
  (a) For $\phi = 1$ the energy spectrum consists of a single band with $E = 2(\cos k_x + \cos k_y)$.
  (b) For $\phi = 1/2$, the energies of two bands denoted as E_+ and E_- are
  $$E_\pm = \pm 2\sqrt{\cos^2 k_x + \cos^2 k_y}.$$
  (c) For $\phi = 1/3$, the three bands with energies denoted as E_0, E_1, E_2, where
  $$E_i = 2\sqrt{2} \cos(\theta \pm i\frac{2}{3}\pi).$$ Here $\theta = \frac{1}{3}\arccos[(\cos 3k_x + \cos 3k_y)/2\sqrt{2}]$.
  (d) For $\phi = 1/4$, the energies for four bands, denoted as E_{++}, E_{+-}, E_{-+}, E_{--} are given by $E_{\pm\pm} = \pm\sqrt{4 \pm 2[3 + \frac{1}{2}(\cos 4k_x + \cos 4k_y)]^2}$.

II Brief summary of semiclassical results for energy levels

There have been many theoretical studies of the Bloch electrons in a magnetic field using semiclassical limit of quantum mechanics. These studies express energy levels as a power series expansion in magnetic flux ϕ. For example, semiclassical result for energies for small value of flux ϕ near a rational flux p/q known as the Landau levels is given by,

$$E_n(\phi) = -4 + \phi(2n+1) - \frac{\phi^2}{16}\left[1 + (2n+1)^2\right]$$
$$+ \frac{\phi^3}{64 \times 3}\left[n^3 + (n+1)^3\right] + O(\phi^4). \quad (6.5)$$

The semiclassical analysis can also be used to derive the expression for energy near a rational flux value where $\phi' = \phi - p/q$ is small, known as Wilkinson–Rammal formula. In other words, analytic expression can be obtained near any rational value of flux and this explains Landau-level structure near rational flux values as described in chapter 7.

It turns out that for special cases where two bands touch, forming a conical intersection (known as a Dirac cone) in the (E, k_x, k_y) graph, energy levels (sometimes referred as Dirac levels) differ from the usual Landau level-like behaviour described above and the semiclassical result is given by,

$$E_n = -2\sqrt{2n\phi'}\left(1 - \frac{n\phi'}{2}\right)^{1/2} + O(\phi')^{5/2} \quad (6.6)$$

In other words, in contrast to Landau levels exhibiting linear dependence on ϕ or ϕ', Dirac levels are characterized by a square root dependence in the magnetic flux. For further details on various semiclassical results, we refer readers to the paper by Rammal and Bellissard [10]. The selected bibliography at the end of the book lists some other important papers in the field such as those by Wilkinson, Helffer and Sjöstrand, Rammal and Bellissard.

There exists a great variety of topics that have been discussed in connection with Harper's equation. The selected bibliography at the end of the book lists some of the important papers in this field.

References

[1] Tomonaga S-I 1998 *The Story of Spin* (Chicago: University of Chicago Press)
[2] Azbel' M Ya 1964 Energy spectrum of a conduction electron in a magnetic field *JETP* **19** 634
[3] Langbein D 1969 The tight-binding and the nearly-free-electron approach to lattice electrons in external magnetic fields *Phys. Rev.* **180** 633
[4] Hofstadter D R 1975 The Energy Levels of Bloch Electrons in a Magnetic Field *PhD thesis* University of Oregon
[5] Hofstadter D R 1976 Energy levels and wave functions of Bloch electrons in rational and irrational magnetic fields *Phys. Rev.* B **14** 2239

[6] Hofstadter D R 1979 *Gödel, Escher, Bach: an Eternal Golden Braid* (New York: Basic Books)
[7] Ávila A and Jitomirskaya S 2009 The ten martini problem *Ann. Math.* **170** 303–42
[8] Ketoja K and Satija I I 1994 Renormalization approach to quasi periodic tight binding models *Phys. Lett.* A **194** 64
[9] Wen X G and Zee A 1989 Winding number, family index theorem, and electron hopping in a magnetic field *Nucl. Phys.* B **316** 641
[10] Rammal R and Bellissard J 1990 An algebraic semi-classical approach to Bloch electrons in a magnetic field *J. Physique* **51** 1803

Part III

Topology and the butterfly

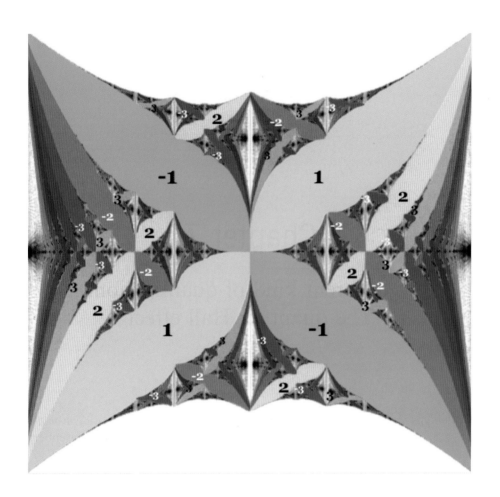

IOP Concise Physics

Butterfly in the Quantum World
The story of the most fascinating quantum fractal
Indubala I Satija

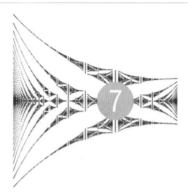

Chapter 7

A different kind of quantization: The quantum Hall effect

Not only is the universe stranger than we think, it is stranger than we can think.
—Werner Heisenberg

The quantum Hall effect is one of the most remarkable of all condensed-matter phenomena, quite unanticipated by the physics community at the time of its discovery in 1980 [1]. The basic experimental observation is the quantization of resistance, in two-dimensional systems, to an extreme precision, irrespective of the sample's shape and of its degree of purity. This intriguing phenomenon is a manifestation of quantum mechanics on a macroscopic scale, and for that reason, it rivals superconductivity and Bose–Einstein condensation in its fundamental importance.

7.1 What is the Hall effect? Classical and quantum answers

The classical Hall effect was named after Edwin Hall, who, as a 24-year-old physics graduate student at Johns Hopkins University, discovered the effect in 1878. His measurement of this tiny effect is regarded as an experimental *tour de force*, preceding by 18 years the discovery of the electron. The classical Hall effect offered the first real proof that electric currents in metals are carried by moving charged particles.

As is shown in figure 7.1, the basic setup of Hall's experiment consists of a very thin sheet of conducting material, which is subjected to both an electric field $\vec{E}$ and a magnetic field $\vec{B}$. The electric field, lying in the plane of the conductor, makes the charges in the conductor move, setting up an electric current I. The magnetic field, on the other hand, is perpendicular to the conductor, and according to the classical laws of electricity and magnetism, it exerts a so-called *Lorentz force* on these moving charges, pushing them sideways in the plane, perpendicular to the current I. This results in an induced voltage, perpendicular to both I and B, which is known as the *Hall voltage*. The Hall resistance, usually denoted by R_{xy}, is deduced from Ohm's law, $V = IR$, so it equals the ratio of the transverse Hall voltage to the longitudinal current I. The Hall conductance, which is the reciprocal of the Hall resistance, is denoted by σ_{xy}.

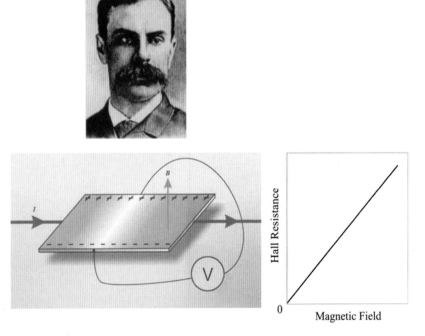

Figure 7.1. The basic setup of Edwin Hall's 1878 experiment. The graph on the right shows that Hall resistance increases linearly with the magnetic field.

Just over a century later, in 1980, a quantum-mechanical generalization of Edwin Hall's classical effect was discovered by German physicist Klaus von Klitzing. What today is called the "integer quantum Hall effect" took the physics world deeply by surprise. Von Klitzing was investigating the transport properties of a certain semiconductor device (namely, a silicon MOSFET, short for "metal-oxide semi-conductor field-effect transistor") at very low temperatures and in high magnetic fields. His experiment revealed two novel features closely related to, but quite distinct from, what Hall had found (which was a purely linear increase in Hall resistivity with the magnetic field). These features were:

- *Plateaus*—that is, step-like structure—in Hall resistance (or conversely, plateaus in Hall conductivity, since conductance is the reciprocal of resistance), as can be seen in figure 7.2.
- Not just plateaus, but *quantization* of resistivity, meaning that the Hall resistivity assumes *periodically spaced* values—namely, integer multiples of a new fundamental constant $\sigma_H = \frac{e^2}{h}$ (where e is the charge on the electron and h, as always, is Planck's constant).

This very surprising discovery earned von Klitzing the Nobel Prize in Physics in 1985. Appendix A contains some excerpts from the announcement of the Nobel Prize [2].

The quantum Hall effect arises in two-dimensional electronic systems, commonly known as *two-dimensional electron gases*, which are immersed in a strong magnetic field. There are a variety of techniques to construct two-dimensional electron gases. Von Klitzing's 1980 discovery relied on the existence of a two-dimensional electron

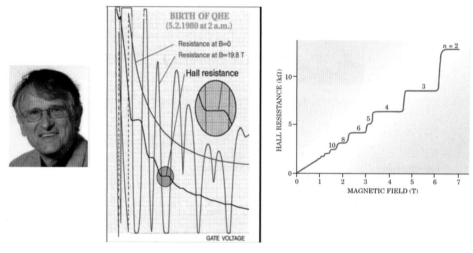

Figure 7.2. On the left, German physicist Klaus Klitzing (Copyright © 2005, Birkhäuser Verlag, Basel. DOI 10.1007/3-7643-7393-8.) and the discovery of the quantum Hall effect. On the right, the *integer* quantum Hall effect, characterized by quantized plateaus of resistance. More specifically, the resistivity of the material assumes only values that are integer multiples of the fundamental physical constant $\frac{e^2}{h}$.

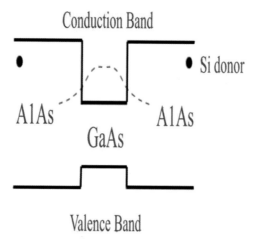

Figure 7.3. Schematic illustration of a rectangularly shaped quantum well in a semiconductor, which becomes the home for a two-dimensional electron gas. The horizontal axis represents an ordinary coordinate of physical position, while the vertical axis represents energy. (To create the energy barriers that define the well, it is common to use an alloy of gallium-arsenide and aluminum-arsenide, as is shown in this figure, rather than pure aluminum-arsenide.) The two dark circles represent positive silicon ions that have donated electrons to the quantum well. (Of course the gas consists of millions of electrons, not just two.) The wave function of the lowest energy level for an electron caught in the quantum well is indicated by the dashed line.

gas in a semiconducting device. By the middle of the 1960s, such systems could be physically realized, thanks to the great technological progress that followed the invention of the transistor roughly twenty years earlier.

Figure 7.3 schematically shows an example of a two-dimensional electron gas, in which the energy bands in a gallium-arsenide/aluminum-arsenide alloy are used to create a *quantum well*. Electrons from a silicon donor layer fall into the quantum well to create the electron gas.

7.2 A charged particle in a magnetic field: Cyclotron orbits and their quantization

7.2.1 Classical picture

We open our explanation of the quantum Hall effect by giving a purely classical description of the motion of an electron with charge $-e$ in the presence of a uniform magnetic field $\vec{B}$, which we will assume is directed along the z-axis. If the electron has velocity $\vec{v}$, it will experience a Lorentz force $-e\vec{v} \times \vec{B}$ due to the magnetic field, and its motion will be described by Newton's equation $\vec{F} = m\vec{a}$, broken down into x- and y-coordinates, as follows:

$$m\ddot{x} = -eB\dot{y} \tag{7.1}$$

$$m\ddot{y} = +eB\dot{x}. \tag{7.2}$$

The general solution to these coupled differential equations is periodic motion in a circle of arbitrary radius R:

$$\vec{r} = R(\cos(\omega t + \delta), \sin(\omega t + \delta)). \tag{7.3}$$

Here, $\omega = \frac{eB}{m}$ is known as the *classical cyclotron frequency*, and δ is an arbitrary phase associated with the motion. Note that the period of the orbit is independent of the radius. In fact, the radius R and the tangential speed $v = R\omega$ are proportional to each other, with constant of proportionality ω. Thus a fast particle will travel in a large circle and will return to its starting point at exactly the same moment as a slow particle traveling in a small circle. Such motion is said to be *isochronous*, much like that of a harmonic oscillator (e.g. a pendulum with small amplitude), whose period is independent of the amplitude of the oscillation.

7.2.2 Quantum picture

To treat the problem quantum-mechanically, we must solve the corresponding Schrödinger equation, and, as in the case of the hydrogen atom, we find that this yields a quantized set of electron energies:

$$E_n = \hbar\omega\left(n + \frac{1}{2}\right), \quad \omega = \frac{eB}{m}, \tag{7.4}$$

where $n = 1, 2, 3, \ldots$. These quantized energy levels are known as *Landau levels*, and the corresponding wave functions as *Landau states*, after the Russian physicist Lev Landau, who pioneered the quantum-mechanical study of electrons in magnetic fields (see figure 7.4).

The Landau wave functions are products of Gaussian functions (bell-shaped curves) and certain polynomials called Hermite polynomials. Without going into

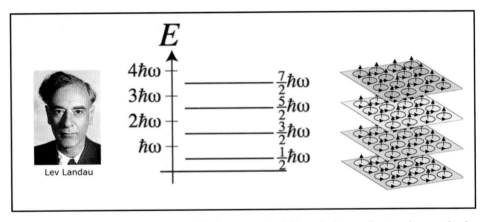

Figure 7.4. The quantization of electron orbits in a magnetic field results in equally spaced energy levels—Landau levels. The spacing of these levels is proportional to the classical cyclotron frequency $\omega = \frac{eB}{m}$. The colorful drawing on the right shows many electrons simultaneously executing cyclotron orbits, where each colored plane in the stack represents a different Landau level.

detail, we will simply state that these wave functions describe waves that spread out over a characteristic distance known as the magnetic length l_B:

$$l_B = \sqrt{\frac{\hbar}{eB}}. \tag{7.5}$$

The existence of both a natural *length* scale l_B and a natural *energy* scale $\hbar\omega$ in this physical situation is a purely quantum phenomenon (as can be seen from the presence of $\hbar$ in the formulas for both of these natural units). In the classical situation, there was no such natural unit of length. Note that the new length scale is nonetheless closely related to the classical cyclotron frequency of an electron in a magnetic field, $\frac{eB}{m}$.

Associated with the natural quantum-mechanical magnetic length l_B is a natural quantum-mechanical unit of *area*:

$$2\pi l_B^2 = \frac{h/e}{B}. \tag{7.6}$$

This formula reveals a natural physical interpretation for the magnetic length, which is that the area that it determines—namely, $2\pi l_B^2$—intercepts exactly one quantum of magnetic flux, $\Phi_0 = \frac{h}{e}$. We can thus write:

$$2\pi l_B^2 = \frac{\Phi_0}{B}. \tag{7.7}$$

Furthermore, it can be shown that each Landau level is highly degenerate, with the degeneracy factor ν being given by the total number of flux quanta penetrating the sample of area A,

$$\nu = \frac{BA}{\Phi_0}, \quad \Phi_0 = \frac{h}{e}. \tag{7.8}$$

The model we have just discussed allows one to understand the quantization of Hall conductivity, as is shown in appendix C. The crux of the matter is that if a Landau level (with ν quantum states) is completely filled, the Hall conductance is quantized. Furthermore, the quantum number associated with Hall conductivity is determined by the number N of filled Landau levels. Thus if the Fermi energy lies in the Nth gap, then the transverse or Hall conductivity is:

$$\sigma_{xy} = N\frac{e^2}{h}. \tag{7.9}$$

Since the Fermi energy resides in a gap between bands (see figure 7.5), quantum Hall systems are insulators, and yet, strangely enough, they exhibit Hall conductivity. How is it possible for a system that, in the bulk, is an insulator, to conduct current? This mystery will be resolved by the intuitive semiclassical arguments given below. Then, for the next two chapters, what will occupy us is the deep topological basis underlying the *quantization* of this strange type of conductivity in the presence of a periodic lattice.

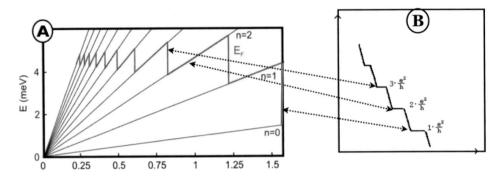

Figure 7.5. On the left side, the so-called "Landau fan" plots Landau levels, labeled by their quantum numbers $n = 0, 1, 2\ldots$, as a function of magnetic field strength. The zigzagging red line shows how the Fermi energy varies with field strength for a fixed electron density. The staircase on the right side shows the quantum Hall states whose quantum numbers $N = 1, 2, 3, \ldots$ correspond to the Fermi energy lying in the first, second, third (etc) gap between Landau levels.

7.2.3 Semiclassical picture

The circularity of the Landau orbits followed by electrons in a two-dimensional electron gas immersed in a magnetic field shows why a sample with such a gas will be an insulator: there is no net flow in any direction, since a circle is a closed loop. However, as is explained below, this insulator, although it does not conduct current in its interior, does conduct current along its *edges* (although only in one direction along each edge). This effect takes place for purely *topological* reasons, independent of the details of the sample's geometry.

In any high-precision measurements, such as those involving the quantum Hall effect, a natural question arises about the consequences of the finite size of the sample—specifically, effects that take place at the sample's boundaries. To describe such effects, we focus on electrons confined inside a slab of finite width. Near the edges of such a sample, the confining potential well produces an upward bending of the Landau levels (see figure 7.6). Wherever a bent Landau level intercepts the Fermi energy, a one-dimensional "edge channel", or "skipping orbit", is formed. Classically, this can be envisioned as shown in figure 7.7—namely, as an electron bouncing along the edge of the sample in a succession of semicircles.

Soon after the discovery of the quantum Hall effect, the importance of these edge channels in the transport properties of the two-dimensional gas was recognized, and several edge-related theories were then developed, based on different approaches, and there is now ample experimental evidence that the current in a Hall bar is indeed flowing very close to the sample's edges.

The edge of a quantum Hall system is a one-way street for electrons.

As is shown in figure 7.7, although electrons in the bulk of the sample cannot move, the electrons along the sample's edge can and do move. An electron near the edge of the sample doesn't have enough room to complete its classical circular orbit. Instead, it will hit the boundary of the sample, be reflected by it, will start another

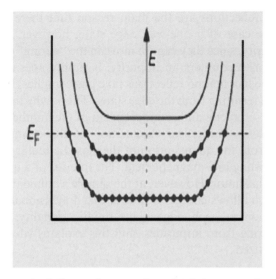

Figure 7.6. Schematic energy-level diagram for a two-dimensional electron gas in a magnetic field with an infinite confining potential at the edges of the sample, as was shown in figure 7.3. The normally flat Landau levels (see figure 7.4) bend upwards as they approach the walls of the potential well. States below the Fermi energy E_F are occupied (solid blue circles). The edge channels are located where the Fermi energy cuts across the Landau levels.

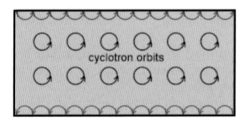
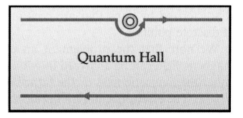

Figure 7.7. In the middle of the left-hand panel are shown cyclotron orbits inside an electron inside a two-dimensional electron gas in a thin sample, as well as two bouncing trajectories (red and blue) of electrons at the sample's two edges. Such classical trajectories, called "skipping orbits", consist of a sequence of successive semicircles. The skipping orbits on both edges of the sample carry current; however, since the two edges are located on opposite sides of the sample, the currents they carry flow in opposite directions. By contrast, the circular orbits *inside* the sample carry no current. In the right-hand panel we see that the edge states not only exhibit "one-way traffic flow" but are also unaffected by the presence of impurities—the electrons simply go around any impurity, never getting reflected backward.

circle and will hit the boundary again, and so forth. Since the electron is continually moving in one direction, we can think of the sample's edge as a one-dimensional "wire". However, this wire is very different from an ordinary one-dimensional wire. In this wire, the electron can move only in *one direction*, whereas in a normal one-dimensional wire, electrons can move in either direction.

This one-way current is very important when impurities are taken into consideration. In an ordinary one-dimensional wire, electrons will be reflected back by an impurity, and such reflections are the main reason that there is resistance at low temperatures. In the case of a one-way edge state, however, electrons cannot be reflected by an impurity, since they cannot move in the "wrong" direction. Therefore, whenever an electron encounters an impurity, it simply goes around it and keeps moving forward. The fact that no reflections take place implies that impurities do not disrupt the transport resulting from the edge states. This is why the quantization of the Hall conductivity is so precise and so independent of the number of impurities.

To summarize this semiclassical explanation of the quantum Hall effect, an energy gap results from the quantization of the closed circular orbits that electrons follow in a sample sitting in a magnetic field. The interior of a quantum Hall sample is thus inert, like an insulator. However, at the sample's boundary, a different type of motion occurs, which allows charge to flow in one-dimensional edge states, though only in one direction at each edge. This unidirectionality makes those states insensitive to scattering from impurities, and this explains why the Hall resistance is so precisely quantized.

7.3 Landau levels in the Hofstadter butterfly

As has been stated above, the quantum Hall effect is tied to the gap structures—the regions of forbidden energy—in the energy spectrum of the electrons in a magnetic field. The coming chapters will explore the quantum Hall effect associated with the butterfly spectrum, where the electrons are subjected both to a magnetic field and to a periodic potential of the crystal lattice. The spectrum of such a system is characterized by gaps that are continuous functions of the magnetic field, except at discrete points.

We note that the problem of an electron moving in a lattice immersed in a magnetic field can be analyzed in two entirely complementary ways, by considering either the magnetic field or the lattice as a *perturbation*:

(1) The electron's motion in the lattice can be thought of as being perturbed by the turning-on of a magnetic field. This is called the *weak-field limit*, meaning that one starts with Bloch states and treats the magnetic field as a perturbation. Harper's equation was originally derived from this point of view, where a single Bloch band in a tight-binding model is perturbed by a weak magnetic field, as was described by Douglas Hofstadter in the previous chapter.

(2) Alternatively, one can think of the electron's motion in a homogeneous magnetic field as being perturbed by the turning-on of a very weak periodic potential (due to a crystal lattice). This is called the *strong-field limit*. Here one begins with the Landau-level scenario and treats the periodic potential of the lattice as a perturbation.

A wonderful and deep surprise is that although one obtains exactly the same equation in both of these cases, in the second case, the magnetic flux per unit cell ϕ is replaced by its reciprocal $1/\phi$ (how this happens is explained in [3]). So far in our discussion, we have adopted the first viewpoint, in which a single Bloch band

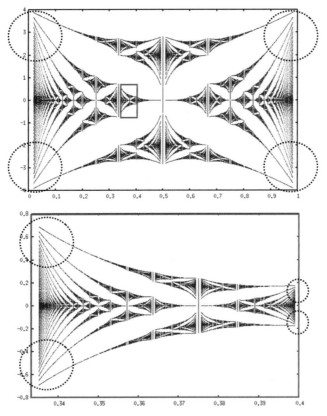

Figure 7.8. The four dotted circles in the upper panel reveal that in the Hofstadter butterfly there are regions in which one can see an almost Landau-level-like spectrum. (Consider a very small flux-value such as $\phi = 0.03$, and look at the very narrow energy levels that are roughly evenly distributed at that flux-value.) In the lower panel we zoom into the area defined by the upper panel's red box. Once again, at the four corners of the sub-butterfly, we again see an almost Landau-level-like spectrum. These energy levels, which are *almost* equally spaced, can be spotted not only at the edges of the full butterfly, but also at the edges of all of its miniature replicas, which occur at magnetic flux-values of all sizes, whether small or large.

subjected to magnetic flux p/q is split into q subbands. By contrast, in the strong-field case, each Landau level carrying one quantum unit of Hall current is split into p subbands (since p is the denominator of $q/p = 1/\phi$).

Certain regions of the butterfly spectrum strongly resemble a spectrum consisting solely of pure Landau levels (see figure 7.4). Such a spectrum arises when there is a magnetic field but no lattice. One would expect this kind of spectrum to continue to hold if one were to turn on an extremely weak lattice as a perturbing potential, and this does indeed happen. Interestingly, however, this to-be-expected effect happens not only for a small magnetic field (where the magnetic length is large and thus the lattice is nearly irrelevant), but also at the boundaries of all miniature butterflies, irrespective of the value of the magnetic flux. This can be seen in the lower panel of figure 7.8. In short, then, the presence of Landau-like levels densely distributed throughout the butterfly, and on all scales, is a consequence of the butterfly's recursive structure. This result can be understood using semiclassical analysis, as is shown in the appendix to chapter 6.

However, as we will show later in the book (see figure 10.1 in chapter 10), in these Landau-level-like regimes of the butterfly spectrum, the Hall-conductivity quantum numbers associated with two neighboring gaps differ by a fixed integer that need not be equal to unity, as is the case for the Landau-level scenario where the Hall conductance is given by equation (7.9). This integer jump is equal to min $[q_L, q_R]$, where q_L and q_R denote, respectively, the denominators of the rational flux-values that determine the left and the right boundaries of a butterfly.

7.4 Topological insulators

It is a marvelous fact that the quantum theory of two-dimensional gases of non-interacting electrons has many hidden treasures waiting to be discovered. Among the recently unearthed treasures are the so-called "topological states of matter", which constitute an emerging frontier of research in condensed-matter physics, involving new and exotic states of matter and new kinds of phenomena in single-particle band theory. Quantum Hall states are the first and simplest examples of *topological insulators* [4].

Topological insulators are a broad class of unconventional materials that insulate in their interior but conduct current along their edges. The edge transport is said to be "topologically protected", meaning that it depends only on the existence of an edge (a topological feature), but not on the exact geometry of the edge. Quantum Hall systems are distinct from all other known states of matter. Close cousins of quantum Hall states are what are known as *quantum spin Hall states*, where instead of one-way edge transport, there is two-way transport, due to the two opposite spin directions, as is shown in figure 7.9.

Topological insulators provide a topological basis for classifying states of matter, with ordinary insulators being topologically trivial. These recent experimental discoveries and theoretical insights (which we will examine more closely in the coming two chapters) have carried the frontiers of condensed-matter physics down new and exciting pathways, which seem to strongly confirm Albert Einstein's faith in the deep rightness of using a geometrical approach to frame and to understand nature's most fundamental laws.

Remarks

There are special conditions under which the Hall conductance exhibits plateaus at certain *fractional* values. This effect—the *fractional* quantum Hall effect—cannot be explained using the framework of a non-interacting two-dimensional electron gas, since electron–electron interactions are key in producing the fractional plateaus. The topic is therefore beyond the scope of this book.

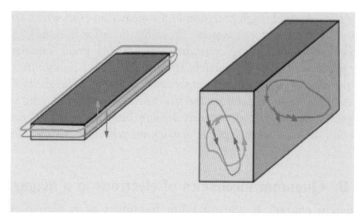

Figure 7.9. Schematic illustration of edge states, or surface states, in quantum spin Hall topological insulators, where electrons with spin up propagate in one direction while electrons with spin down propagate in the opposite direction.

Appendix A: Excerpts from the 1985 Nobel Prize press release [2]

Summary

When an electric current passes through a metal strip there is normally no difference in potential across the strip, if measured perpendicularly to the current. If, however, a magnetic field is applied perpendicularly to the plane of the strip, the electrons are deflected towards one edge and a potential difference is created across the strip. This phenomenon, termed the Hall effect, was discovered more than a hundred years ago by the American physicist E H Hall. In common metals and semiconductors, the effect has now been thoroughly studied and is well understood.

Entirely new phenomena appear when the Hall effect is studied in two-dimensional electron systems, in which the electrons are forced to move in an extremely thin surface layer between, for example, a metal and a semiconductor. Two-dimensional systems do not occur naturally, but, using advanced technology and production techniques developed within semiconductor electronics, it has become possible to produce them.

For the last ten years there has been reason to suspect that, in two-dimensional systems, what is called Hall conductivity does not vary evenly, but changes "step-wise" when the applied magnetic field is changed. The steps should appear at conductivity values representing an integral number multiplied by a natural constant of fundamental physical importance. The conductivity is then said to be quantized. It was not expected, however, that the quantization rule would apply with a high accuracy. It therefore came as a great surprise when in the spring of 1980 von Klitzing showed experimentally that the Hall conductivity exhibits step-like plateaus which follow this rule with exceptionally high accuracy, deviating from an integral number by less than 0.0000001.

Von Klitzing has, through his experiment, shown that the quantized Hall effect has fundamental implications for physics. His discovery has opened up a new research field of great importance and relevance.

Because of the extremely high precision in the quantized Hall effect, it may be used as a standard of electrical resistance. Secondly, it affords a new possibility of measuring the earlier-mentioned constant, which is of great importance in, for example, the fields of atomic and particle physics. These two possibilities in measurement technique are of the greatest importance, and have been studied in many laboratories all over the world during the five years since von Klitzing's experiment. Of equally great interest is that we are dealing here with a new phenomenon in quantum physics, and one whose characteristics are still only partially understood.

Appendix B: Quantum mechanics of electrons in a magnetic field

As was stated in chapter 6, the quantum treatment of a planar electron in a transverse magnetic field $\vec{B}$ comes from replacing the x- and y-components of the momentum vector $\vec{p} = (p_x, p_y)$ (these are "c-numbers", in Dirac's terminology) by a pair of "q-number" operators, as follows:

$$p_x \to -i\hbar \frac{\partial}{\partial x}$$

$$p_y \to -i\hbar \frac{\partial}{\partial y} - eBx.$$

(Here, as in chapter 6, we are using the magnetic vector potential $\vec{A} = B(0, x, 0)$, called the "Landau gauge".) The resulting Schrödinger equation is:

$$\frac{1}{2m}\left[\left(-i\hbar\frac{\partial}{\partial x}\right)^2 + \left(-i\hbar\frac{\partial}{\partial y} - eBx\right)^2\right]\psi(x, y) = E\psi(x, y). \tag{B.1}$$

Since there is no explicit dependence on y anywhere in the above equation, we look for solutions of the form $\psi(x, y) = f(x)e^{ik_y y}$. This leads to the following equation for $f(x)$:

$$-\frac{\hbar^2}{2m}\frac{d^2 f}{dx^2} + \frac{1}{2m}(eBx - \hbar k_y)^2 = Ef(x). \tag{B.2}$$

This equation is identical to the Schrödinger equation for a one-dimensional harmonic oscillator, whose eigenfunctions and eigenvalues are:

$$\psi(x, y) = \phi_n\left(x - \frac{\hbar k_y}{eB}\right)e^{ik_y y}; \quad E_n = \hbar\omega_c(n + 1/2), \tag{B.3}$$

where ω_c equals the cyclotron frequency $\frac{eB}{m}$, and where the functions ϕ_n are harmonic-oscillator wave functions labeled by the same quantum number n that labels the discrete energy levels—that is, the Landau levels. Note that the eigenvalues of this system, unlike its eigenfunctions, are independent of k_y. This results in degenerate levels whose degree of degeneracy is equal to $\nu = BA/\phi_0$ (A being the area of the sample), as was mentioned earlier in this chapter.

Appendix C: Quantization of the Hall conductivity

We start with the expression for σ_{xy}, the classical Hall conductance [1]. Here, Q is the total electrical charge of the system, and A is its area,

$$\sigma_{xy} = \frac{Q}{BA}. \tag{C.1}$$

To obtain a *quantum* expression for Hall conductivity [5], we consider a system with N filled Landau levels. (This means there is no longitudinal current, so the *longitudinal* conductance σ_{xx} is equal to zero.) As each Landau level is ν-fold degenerate, $Q = N\nu e = NBAe^2/h$. (Here we have used equation (7.8), which gives $\nu = BA/\Phi_0 = BAe/h$.) This gives us the desired result:

$$\sigma_{xy} = N\frac{e^2}{h}. \tag{C.2}$$

The validity of this expression has been proven more rigorously, even in the presence of impurities. However, although this explains quantization, it does not give a topological basis for it. That will be done in the following chapters.

References

[1] von Klitzing K 2004 25 years of quantum Hall effect, a personal view on the discovery, physics and applications of this quantum effect *Séminaire Poincaré* **3** 1 http://www.bourbaphy.fr/klitzing.pdf
[2] http://www.nobelprize.org/nobel_prizes/physics/laureates/1985/press.html
[3] Langbein D 1969 *Phys. Rev.* **180** 633
[4] Kane C L and Mele E J 2006 *Science* **314** 1692
  Qi X-L and Zhang S-C 2010 *Phys. Today* January 33–8
[5] Fradkin E Topology and quantum Hall effect *Field Theories of Condensed Matter Systems* (Reading, MA: Addison-Wesley) ch 9

IOP Concise Physics

Butterfly in the Quantum World
The story of the most fascinating quantum fractal
Indubala I Satija

Chapter 8

Topology and topological invariants: Preamble to the topological aspects of the quantum Hall effect

Physics concerns not just the nature of things, but the interconnectedness of the natures of all things.
—Charles Frank, 1976.

As we saw in chapter 7, the integer quantum Hall effect involves the astonishingly precise quantization of Hall conductivity. More specifically, this effect is all about the unexpected discovery that Hall conductivity comes in integer multiples of the fundamental quantity $\sigma_H = \frac{e^2}{h}$, where e is the charge on the electron and h is Planck's constant. The values of laboratory measurements of Hall conductivity turn out to be integer multiples of e^2/h to nearly one part in a billion.

Such extreme precision is absolutely unprecedented in condensed-matter physics. In contrast to atomic and particle physics (especially quantum electrodynamics), in

condensed-matter physics, the difference between theoretical predictions and experimental findings can be quite large—typically in the range of several percent. The error bars here are usually much larger than in atomic and particle physics for the following reasons:
- As the great Wolfgang Pauli once said, "Solid-state physics is the physics of dirt." And what Pauli said was not altogether silly, although it was certainly haughty! Indeed, solids, particularly quantum Hall systems, are typically dirty. They may be filled with impurities, including crystal defects. Such impurities make the data noisy.
- Condensed-matter theories are not as quantitatively accurate because of the complexity of macroscopic physical systems, compared to systems on the size of atoms or smaller, where the phenomena involve very few entities interacting inside a pure vacuum. Phenomena inside solids are very different from that. They have far less symmetry than phenomena taking place in a vacuum.
- A crystal, unlike empty space, is not isotropic. Unlike empty space, a crystal establishes a preferred frame of reference that defines what is moving and what is still. And unlike empty space, a crystal can "enjoy" various states of disorder.
- Also unlike empty space, a crystal has a finite temperature, which gives to all phenomena a statistical blur.
- Once again, unlike empty space, a crystal is made up of a vast number of pieces, which constitute interacting degrees of freedom.
- Last but not least, the fine-structure constant of quantum electrodynamics, $\alpha = \frac{e^2}{\hbar c} \approx \frac{1}{137}$, is a very small expansion parameter whose smallness allows perturbation theory to make accurate predictions in atomic physics. Although perturbative theories such as the BCS theory of superconductivity and Fermi liquid theory are extremely successful and deep theories, the quantitative accuracy of their predictions does not come close to the accuracy one can achieve in atomic and particle physics.

8.1 A puzzle: The precision and the quantization of Hall conductivity

Why is Hall conductivity so precise, and why is it quantized?

The answer to these questions lies in the fact that the integer quantum Hall effect is a very special quantum effect. The effect is a consequence of a *topological state of matter*. In other words, the quantization of Hall conductivity has its roots in topology, a fairly recent branch of mathematics in which an orange and a potato are "the same"—that is, they are topologically indistinguishable from each other (and also from a baseball bat, a soup bowl, and a curving strand of spaghetti). By contrast, a doughnut and a coffee cup (that is, a cup with a handle) are topologically distinct from all those entities, but they are topologically indistinguishable from each other (and also from a rubber band, a funnel, and a piece of macaroni, since all these diverse objects share the feature of having one hole).

The quantization of Hall conductivity, computed in a perfectly non-interacting model, remains pristine despite the inevitable presence of interactions and impurities, and this marvelous fact is a consequence of topological reasons. Indeed, quantum Hall systems are *topological insulators*, as was briefly mentioned in chapter 7. As was hinted in that chapter and as will be explained below, the quantum Hall effect involves a different kind of quantization from that of such familiar observables as energy and angular momentum. It does not depend directly on quantum coherence of the sort that is responsible for the quantized orbits inside an atom. To quote Allan MacDonald, a physicist who has made important contributions to the field, "Quantum coherence plays a supporting role rather than a starring role in this drama." Two-dimensional electron gases exhibiting this type of quantization are cold enough so that quantum coherence holds, and such systems can be characterized by wave functions that evolve according to the Schrödinger equation. The integer quantum numbers of the two-dimensional electron gas are hidden in the wave function.

The quantization of Hall conductivity is due to topological invariants known as *Chern numbers*, named after Shiing-shen Chern (1911–2004), a Chinese-born American mathematician who was a pioneer in differential geometry. The quantization is rooted in the geometry and topology of an abstract space underlying the quantum system. To introduce readers to this exotic type of behavior, we will begin this chapter by discussing two key concepts, both of which can be introduced at a classical level. Indeed, as was once stated by Roman Wladimir Jackiw, a theoretical physicist who championed the subject of quantum anholonomy: *the quantum-mechanical revolution has not erased our reliance on the earlier classical physics.* Here, then, are those two key concepts:

(1) *Topological invariants*: Using simple geometrical shapes, we will show how the intimate relationship between geometry and topology gives rise to integer topological invariants.

(2) *Anholonomy*: This concept applies to a physical system that fails to return to its original state after executing a closed pathway (and thus seeming to "return home") in a curved space. The fascinating Foucault pendulum, which does not return to its starting state after a full rotation of the Earth, illustrates this important phenomenon. Sharing a common mathematical theme with anholonomy is the more abstract notion of a *Berry phase*, in which a particle's initial wave function, as a result of the particle's undergoing a cycle, acquires a phase factor of purely geometric origin.

8.2 Topological invariants

Topology originated from a branch of geometry that describes manifolds (a fancy term for "shapes") in two-dimensional, three-dimensional, or higher-dimensional spaces. (Included in the set of such spaces are very abstract spaces that go well beyond the geometrical notions that we are used to; for example, they include Hilbert spaces, a notion used in quantum physics, which have infinitely many dimensions.) If one manifold can be continuously deformed into another one, we say the two have the same topology; otherwise, they are topologically different. Examples: the surface of a sphere and the surface of a cube are topologically equivalent, while the surface of a sphere and the surface of doughnut (a torus) are topologically different, because a torus has a hole while a sphere has none.

To distinguish different manifolds, mathematicians developed *topological indices*, which are topologically invariant quantities that take on integer values. For two objects having the same topology, the index takes the same value. Otherwise, the values are different.

8.2.1 Platonic solids

To illustrate topological invariants, we will use the example of convex polyhedra in three-dimensional space—the *Platonic solids* [1], named after the ancient Greek philosopher Plato who, in his dialog the *Timaeus*, theorized that everything in the universe was composed of exceedingly small objects shaped like these regular solids. In three-dimensional Euclidean geometry, a Platonic solid is a convex polyhedron all of whose faces are mutually congruent regular polygons, and at whose every vertex the same number of faces meet. There are five distinct objects in 3-space that meet those criteria (see table 8.1), and each is named after its number of faces—the *tetrahedron* (four faces), the *hexahedron* (six faces), the *octahedron* (eight faces), the *dodecahedron* (twelve faces), and the *icosahedron* (twenty faces). The results described below apply to any convex polyhedron, and more generally to any polyhedron whose boundary is topologically equivalent to a sphere and whose faces are topologically equivalent to disks.

Table 8.1. For all five Platonic solids, the Euler index $V - E + F$ (number of *vertices* minus number of *edges* plus number of *faces*) equals 2. This number is a topological invariant, because its value remains constant no matter how the object is distorted (as long as it is not broken in any way).

Name	Image	Vertices V	Edges E	Faces F	Euler characteristic: $V - E + F$
Tetrahedron		4	6	4	2
Hexahedron or cube		8	12	6	2
Octahedron		6	12	8	2
Dodecahedron		20	30	12	2
Lcosahedron		12	30	20	2

Let the number of faces of such a polyhedron be F, let its number of edges be E, and let its number of vertices be V. We then define a new quantity χ, known as the *Euler index*, named after the 18th-century Swiss mathematical genius Leonhard Euler, who, among his almost endless achievements, was responsible for much of the early work in topology:

$$\chi = F - E + V. \tag{8.1}$$

Surprisingly, no matter which polyhedron one chooses, the number obtained when one computes χ is always equal to 2. The cube, for example (referred to earlier as a "hexahedron"), has 6 faces, 12 edges and 8 vertices, and, as was claimed above, $F - E + V = 6 - 12 + 8 = 2$. This easy-to-compute integer χ is one of the simplest examples of a topological invariant, capturing a deep topological aspect of an object's shape or structure, regardless of the way the object is bent.

A great discovery of the last few decades is that a topological invariant also rules the dynamics of the electrons in the quantum Hall effect. Just as the Euler index χ is a topological invariant belonging to a simple geometric entity, the integer that underlies Hall conductivity is a more abstract topological invariant called a "Chern number". Just as Euler's χ does not depend on shape, the Chern number does not depend on the shape or other details of the experiment.

8.2.2 Two-dimensional surfaces

To gain a feel for some topological invariants applying to objects other than polyhedra, we turn next to a topological index defined for two-dimensional closed surfaces. This index, like most topological invariants in physics, arises as the integral of a geometric quantity. It is defined as follows:

$$\chi = \frac{1}{2\pi} \int_S \kappa \, ds. \tag{8.2}$$

Here κ is the *local curvature* of the surface—a concept that we shall now define.

First of all, let us consider curves in a Euclidean plane. If the curve is a circle (the simplest case), then the reciprocal of its radius, $\kappa = 1/R$, is defined to be its local curvature (and the curvature is the same at every point of the circle). Clearly, the smaller the circle, the greater the curvature, and vice versa. And a straight line, being a circle of infinite radius, has zero curvature, which makes perfect sense, since it is not curved at all.

For curves that are not circular, the curvature will vary from point to point. The local curvature will of course be greater at points where the curve turns more sharply, and lower where it approaches straightness. To obtain the precise numerical value of the curvature, one finds the circle that best fits the curve at the point in question, and the inverse radius of that circle is defined to be the local curvature at that point.

This notion of local curvature can be extended from two-dimensional curves to three-dimensional manifolds (surfaces in 3-space), such as a tin can, a fruit bowl, an egg, someone's cheek, and so forth. The most naïve approach to defining the local curvature of such an object (that is, the idea based on the simplest possible analogy to the two-dimensional case) would be, given a point on the object's surface, to try to find "the best-fitting circle" at that point. Unfortunately, however, this naïve approach doesn't work, because in general, many different circles of different radii will fit snugly against the object at the chosen point, since at each point on the object's surface, there are infinitely many different directions that one could choose.

To be very concrete, consider a straight section of pipe with a circular cross-section. An ordinary ruler (which has zero curvature) can be placed on it lengthwise, and the two objects will match exactly. This suggests that at every point, the piece of pipe, like the ruler, has *zero* local curvature. On the other hand, a circular washer of the proper size will also fit snugly inside the pipe. This fact suggests, contrariwise, that at every point, the pipe has the same local curvature as the washer's circumference (which is *not* zero). This discrepancy shows why generalizing the notion of curvature to manifolds in three-dimensional space is tricky. Nonetheless, there is an elegant way to make the generalization.

Imagine a smooth fruit bowl longer than it is wide. Now imagine standing a thin napkin ring (balanced upright on its circumference) on the bowl's lowest point and twiddling an imaginary knob that increases or decreases the ring's radius. The goal of such twiddling is to make the ring match the bowl's shape as exactly as possible at that spot. If you orient the ring so it is aligned with the bowl's *long* axis (say, east–west) and then twiddle its radius until you find the best fit, you will wind up with a relatively *large* radius (i.e., low curvature), whereas if you align the napkin ring with the bowl's *short* axis (north–south) and again twiddle in search for the best fit, you'll wind up with a ring with a relatively *small* radius (i.e., high curvature). And if you orient the napkin ring in some intermediate direction (northwest–southeast, say), you'll wind up with an intermediate-size radius for the best-fitting one. In fact, there will be a continuum of different radii, depending on the angle at which you orient the napkin ring.

The key idea for defining the Gaussian curvature of the bowl at its bottom (or at any other point) starts with the fact that among all these differently oriented circles, there is one whose curvature κ_{max} is *greatest* and one whose curvature κ_{min} is *least*. These two circles' curvatures are known as the manifold's *principal curvatures* at that point. (By the way, some 250 or so years ago, Euler proved that for any point of any manifold, the extremal circles are always oriented at 90 degrees to each other.) We now define κ, the *local Gaussian curvature* of the bowl at this point (or at any given point), as the *product* of the two extremal curvatures: $\kappa = \kappa_{max}\kappa_{min}$. For the pipe, since the *minimal* curvature is 0, it doesn't matter what the maximal curvature is, since their product will always be 0. Thus the local curvature at every point of the pipe equals 0. On the other hand, for the fruit bowl, the two perpendicular napkin rings standing upright on the bowl's bottom both have non-zero curvature, so their product will be non-zero.

The Gaussian curvature of a manifold at any specific point turns out to be an *intrinsic* measure of curvature at that point, which means that its value could, in principle, be calculated by a being that inhabits the surface itself, without using circles (or napkin rings) that stick upwards or outwards into an extra spatial dimension; one does not have to look, at or even be aware of, the space in which the object is embedded. Thus an ant living on a basketball could in principle calculate the basketball's curvature without ever knowing about three-dimensional space.

For a sphere of radius R, $\kappa_{max} = \kappa_{min} = \frac{1}{R}$, so $\kappa = \frac{1}{R^2}$. Consider next a horse's saddle, and pick some point near its middle. At that spatial point, if you move backwards or forwards (i.e., headwards or tailwards, if the saddle is sitting on a horse), the saddle curves *upwards*, while if you move sideways, the saddle curves *downwards*. For this reason, we assign the two curvatures opposite signs, so that $\kappa_{max} > 0$ and $\kappa_{min} < 0$. Thus the Gaussian curvature at that point, $\kappa_{max}\kappa_{min}$, will be negative.

Finally we can return to the integral exhibited earlier, which integrates the Gaussian curvature over the entire surface of the manifold in question. The value obtained is called the manifold's *total* Gaussian curvature. And it turns out, rather miraculously, that for any two-dimensional closed manifold, the total Gaussian curvature is always an integer. This is a consequence of a beautiful theorem, the *Gauss–Bonnet theorem* in differential geometry. It is named after Carl Friedrich Gauss, who was aware of a version of the theorem but never published it, and Pierre Ossian Bonnet, who published a special case of it in 1848.

8.2.3 The Gauss–Bonnet theorem

$$\chi = \frac{1}{2\pi} \int_S \kappa \, ds = 2(1-g). \tag{8.3}$$

In this equation, g is the *genus* of the object in question—that is, the number of handles it possesses (always an integer, of course). Both g and the number χ are topological invariants. For a sphere, which clearly has no handle, $g = 0$ and hence $\chi = 2$. For a torus, $g = 1$ and thus $\chi = 0$. A coffee mug has one handle, and so, topologically speaking, it is a torus; that is to say, *coffee mug = doughnut*(at least as viewed by a topologist). For a delightful discussion of this theorem, we highly recommend the article by Daniel Henry Gottlieb [2].

The Gauss–Bonnet theorem states that the integral of the curvature over the whole surface is "quantized" (restricted to integer values, in this case), and is a topological invariant. This theorem, relating geometry to topology, was generalized by Shiing-Shen Chern to quantum systems, in order to describe the geometry of wave functions, which are *complex*-valued functions. In the generalized Gauss–Bonnet–Chern theorem, although the lefthand side is an integer, it is not always an *even* integer, and thus g cannot be interpreted as the number of handles, as is explained later in this chapter.

The key question for us is how to define curvature in quantum systems, characterized by wave functions in a Hilbert space. To answer this question, we now introduce *anholonomy*, a phenomenon that is rooted in curved space and that therefore relates directly to the notion of curvature for classical as well as for quantum systems.

8.3 Anholonomy: Parallel transport and the Foucault pendulum

At any given moment, a physical system's state is characterized by a set of numerical parameters, and in classical physics, most systems studied have the property that when they are modified and then brought back to their original states, *all* of their parameters have resumed their original values. In other words, the excursion to other states and back leaves no lasting trace. When this is the case, the system is said to be *holonomous*. By contrast, an *anholonomous* system is one in which at least one parameter describing the system does *not* return to its original value, so in a sense, the system has a kind of "memory" of having traveled to other "places" and then having come back "home".

Parallel transport

As a first illustration of anholonomy, we consider a vector moving on a surface. We will stipulate that the vector moves such that it remains parallel (i.e., tangent) to the surface, and also that it keeps a fixed angle with respect to the pathway along which it is moving. This kind of motion of a vector is known as *parallel transport*. If the vector undergoes a series of parallel transports and eventually comes back to where it started, and if the surface in which it is traveling is flat (as in the left half of figure 8.1), then the vector will return to its initial state unchanged.

However, if the surface is *curved*—for example, the surface of a sphere (the right half of figure 8.1)—then the vector will be found to point in a different direction after traveling around a closed loop, even though it never rotated. The change in direction will, in fact, be proportional to the integral of the Gaussian curvature of the surface contained within the loop, which in turn is proportional to the solid angle enclosed

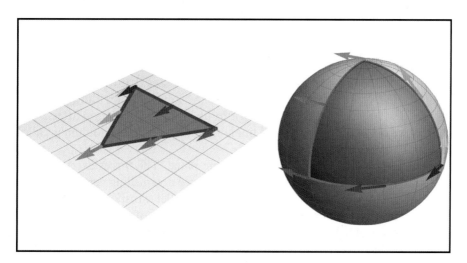

Figure 8.1. The parallel transport of a vector in a flat space and in a curved space. Reproduced with permission of MPQ&LMU Munich.

by the loop. The angular difference between the vector's direction before and after such a loop will depend only on the chosen loop and the nature of the surface, but not on anything else, such as the speed of the vector's movement. This is why the effect is called "geometric". This phenomenon has in fact been appropriated, in modern geometry, to define local curvature as follows:

> *The local curvature at a specific point of a given surface is defined as the angular shift of a vector after it has been parallel-transported around an infinitesimal loop located at that point, divided by the area of the loop.*

This definition is immediately generalizable to quantum states in parameter space, and thus it provides a way to define the curvature of a Hilbert space. However, before venturing into the deep waters of *quantum* anholonomy, we will look carefully at a case of anholonomy in classical physics—namely, the Foucault pendulum.

The Foucault pendulum: an example of classical anholonomy

An iconic example of anholonomy and parallel transport in a classical system is the Foucault pendulum. In 1851, the French physicist Léon Foucault introduced a simple device that he designed in order to demonstrate very concretely the Earth's rotation. This device turns out to be a canonical example of anholonomy, in which a variable of a physical system does not return to its initial state after the system has made a "round trip".

Foucault pendulums are often shown in museums, where they swing back and forth above a stationary horizontal circle on the floor; the circle's center is directly below the balance point of the pendulum. What one observes is, firstly, that such a pendulum's direction of swing changes noticeably in the course of a few minutes, but also—and more surprisingly—that after 24 hours have elapsed (at which point one would naturally think that the system would have "come back home again"), it is not swinging in the same plane as it was swinging in at the outset (figure 8.2).

That is, as the Earth rotates through an angle of 2π radians (360 degrees), the system's cyclic pathway results in the plane of oscillation of the pendulum rotating through a *smaller* angle. An observer on Earth witnesses that the orientation of the pendulum—that is, of its plane of oscillation—slowly rotates during the course of the day, and in general does not return to its original orientation after 24 hours. The difference between the initial and final orientations of the pendulum is called the "phase shift", and is our main focus here.

The pendulum's trajectory starts out in one plane. The forces exerted on it (the Earth's gravity and the tension in the wire) produce a vanishing vertical torque, and so the plane of the pendulum's swing undergoes parallel transport as the Earth turns, carrying the pendulum along a circular pathway (a line of latitude).

The precession of a Foucault pendulum can be calculated in classical mechanics using the Coriolis force. However, in addition to this traditional type of explanation, the Foucault precession has a purely geometrical explanation, which is not only simpler but also more elegant. Let us look into this.

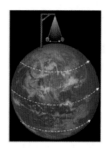
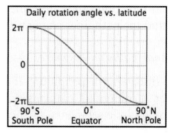

Figure 8.2. Image from November 2, 1902, celebrating the 50th anniversary of the original demonstration by Léon Foucault of the rotation of the Earth under the grand dome of the Panthéon in Paris in 1851. After just one hour, the public could see that the plane of oscillation of the pendulum had already rotated by 11 degrees. As is shown in the graph on the right, there is no rotation of the pendulum's direction if it is swinging at the equator, and one full turn (360 degrees) if it is swinging at the north or the south pole; at all other latitudes, the amount of rotation varies as a continuous function of the latitude.

8.4 Geometrization of the Foucault pendulum

Before we discuss the "geometrization" of the Foucault pendulum, we note the following.

- *Pendulum moving on a Euclidean plane.*

  A pendulum that is slowly carried along any pathway in a Euclidean plane always maintains a fixed direction in space. This is because gravity acts downward (perpendicular to the plane), so it cannot affect the orientation of the pendulum.

- *Straight lines on a sphere are great circles.*

  To understand which paths on a sphere should be called "straight", we resort to Newton's first law, and define a straight line to be a path that a particle follows in the absence of external forces. In this case, the particles in question are moving on the surface of a sphere. On the Earth's surface, the *equator* is an example of such a "straight line". It divides the Earth into two mirror-symmetric halves. In the absence of external forces, any particle set in motion along the equator cannot distinguish the two hemispheres, and thus has to stay on the equator. (Of course the equator is not a straight line in 3-space. If we were to allow the particle to leave the surface of the Earth, it would travel along a different trajectory.) If the sphere is rotated, the equator changes its orientation, but it remains a "straight line". All such "rotated equators" on a sphere are called *great circles* or *geodesics*. Alternatively, we may think of a great circle as the intersection of the sphere with a plane that passes through the sphere's center.

When a pendulum is carried around a great circle, the angle between the pendulum's swing and the great circle never changes. If, however, the pendulum is taken along a path that is not a great circle—for example, a path of fixed latitude other than the equator—then the angle between the pendulum's swing and the path does change.

Triangular paths on a sphere

Before discussing a pendulum that moves along a circle at a fixed latitude, let us consider a triangular pathway C on the sphere—that is, a pathway consisting of three great-circle segments, as shown in figure 8.3. Let θ_1, θ_2, and θ_3 be the angles at the vertices of the triangle. We want to see how the pendulum's plane of oscillation changes as it follows the closed triangular pathway.

There is no change while the pendulum moves along any one segment. There is a change, however, each time that the pendulum moves from one segment to the next. At those moments, the angle between the plane of oscillation and the path is changed by the angle between the two segments at the point where they meet. Therefore, the total change of the pendulum's plane of oscillation—the pendulum's phase shift $\alpha(C)$—is the sum of these three discrete changes, minus π. (The π comes from the fact that in a Euclidean plane, where the phase shift is zero, the sum of the angles of a triangle equals π.) Thus the pendulum's total phase shift is given by the following formula:

$$\alpha(C) = \theta_1 + \theta_2 + \theta_3 - \pi. \tag{8.4}$$

If we combine this with the definition given above for local curvature (namely, the angular shift of a vector that has been parallel-transported around a loop, divided by area of the loop), we get a formula relating the phase shift for a triangular pathway C on a sphere of radius R to the area A of the triangle—namely, $\alpha(C) = A/R^2$.

The quantity $\alpha(C)$ is the solid angle subtended at the center of the sphere by the triangle defined by the path C. The net phase shift is a continuous function of the path, and so is the enclosed area. Therefore, the above equation, $\alpha(C) = A/r^2$, also holds in the general case, which proves the Gauss–Bonnet theorem for all closed paths C of area A on the sphere. In other words, equation (8.4) is the Gauss–Bonnet formula for a geodesic triangle, $\int \kappa \, dS = \theta_1 + \theta_2 + \theta_3 - \pi$, where the integration is carried out over the surface of the triangle.

Foucault pendulum trajectories: paths of constant latitude on the sphere

Consider now a pendulum moving around a circle C at a fixed latitude θ_0. (In the case of a Foucault pendulum, of course, instead of someone moving the pendulum

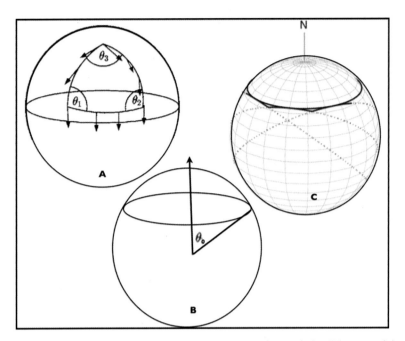

Figure 8.3. (A) A spherical triangle, whose three sides are segments of great circles. When a pendulum (whose oscillation direction is represented by the arrows) is taken along such a triangular path, the angle of the pendulum's swing with respect to each great-circle segment remains constant. Thus, only the angles *between* two segments contribute to the anholonomy—that is, to the overall angular shift of the pendulum. (B) The path of the Foucault pendulum at latitude θ_0, over the course of one day. (C) A circle of fixed latitude approximated by segments of great circles tangent to it. Adapted from [5] and [6] with the permission of the American Association of Physics Teachers.

along the fixed-latitude circle C, it is simply carried along such a pathway by the Earth's rotation.) Although C is not a great circle, we can approximate it by a spherical polygon—that is, the union of a large number of short segments of great circles, where each such segment runs along the path for a short distance (see figure 8.3(C)). The pendulum keeps a fixed angle with respect to each of the three geodesic segments. The net phase shift of a Foucault pendulum is thus equal to the sum of the three vertex angles of the spherical polygon, which, by Gauss–Bonnet, equals the solid angle subtended at the center of the Earth by the cyclic path.

Strictly speaking, the surface whose intrinsic curvature gives rise to the phase shift of a Foucault pendulum is not the Earth's *physical* surface, but a more *abstract* surface—namely, the surface of *constant gravitational field* along which the pendulum moves. (By this is meant any surface on which the field's magnitude is constant, but not the field's direction.) Conveniently, however, when we are talking about the Earth's gravitational field (the field due to a point having mass), all such abstract surfaces are all perfectly spherical, and they all have their center at the Earth's center, and therefore the surface giving rise to the Foucault precession coincides with the Earth's (nearly perfectly) spherical shape. This makes it very natural to phrase the precession result in terms of the Earth's physical curvature.

In summary, unlike a cyclic pathway followed on a *flat* surface, which makes a system always return to its initial state (a holonomic process), a cyclic pathway on a *curved* surface results in a mismatch between the system's initial and final states.

> *By describing the classical Foucault pendulum as an oscillator transported around a cyclic pathway on a surface of a sphere, we "geometrize" it.*

This important insight is summarized in the following way by Alfred Shapere and Frank Wilczek in their book *Geometric Phases in Physics* [3]:

> *How does the pendulum precess when it is taken around a general path C? For transport along the equator, the pendulum will not precess. [...] Now if C is made up of geodesic segments, the precession will all come from the angles where the segments of the geodesics meet; the total precession is equal to the net deficit angle, which in turn equals the solid angle enclosed by C (modulo 2π). Finally, we can approximate any loop by a sequence of geodesic segments, so the most general result (on or off the surface of the sphere) is that the net precession is equal to the enclosed solid angle. This result may seem rather esoteric, but its generality and geometric nature suggest its depth.* **In fact, the mathematics describing it is essentially identical to that describing the motion of a charged particle in the field of a magnetic monopole.**

8.5 Berry magnetism—effective vector potential and monopoles

The boldface sentence closing the above quote applies, remarkably enough, not only to the precession of a Foucault pendulum but also to the Berry phase—the quantum form of anholonomy that arises in a wide variety of physical phenomena. Below we highlight some of the key aspects of this universal picture, with additional details given in the appendix. The discussion is applicable to both classical and quantum systems, and it forms a preamble to our revelation, in the following chapter, of the secrets of the topological quantization of the quantum Hall effect.

Rather surprisingly, it turns out that fictitious magnetic monopoles—the exotic hypothetical particles pictured and briefly described in figure 8.4—are hidden in the mathematics of anholonomy. The key variable in an anholonomic system is γ, the shift after the system goes around some cyclic pathway and returns home. For the Foucault pendulum, for example, γ is the total twist angle, or the precession angle, of the pendulum after a 24 h period has elapsed.

Michael Berry's central result, as described in [3] (see the appendix for details), states that the anholonomic shift γ can be written in two different ways—either as a *surface* integral (the left side of the equation below) or as a *line* (contour) integral (the right side):

$$\gamma = i \int_S \left[\frac{d\mathbf{n}^*}{dx_1} \frac{d\mathbf{n}}{dx_2} - \frac{d\mathbf{n}^*}{dx_2} \frac{d\mathbf{n}}{dx_1} \right] dx_1 dx_2 = i \oint \mathbf{n}^* \cdot \frac{d\mathbf{n}}{dx} dx. \qquad (8.5)$$

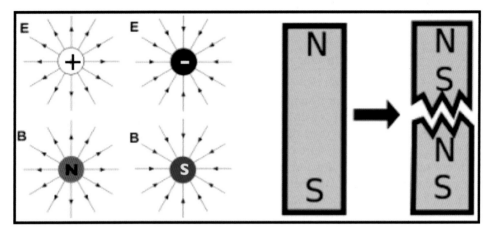

Figure 8.4. A magnetic monopole—an isolated magnet having only one magnetic pole, either South or North—is a hypothetical elementary particle suggested in 1931 by British physicist Paul Adrien Maurice Dirac, in analogy to a point electric charge (an electric monopole, either positive or negative). Magnetism in the real physical world (e.g. in bar magnets and electromagnets) does not arise from magnetic monopoles. In fact, no magnetic monopole has ever been experimentally observed, nor is there any experimental evidence suggesting that magnetic monopoles exist somewhere in our universe. Nonetheless, the grand unified and superstring theories of particle physics predict their existence. Moreover, some condensed-matter systems contain *effective* magnetic monopoles as quasiparticles, which means that such systems exhibit certain behaviors that are *mathematically analogous* to the presence of magnetic monopoles.

Here, **n** is a complex unit vector whose phase encodes the twist or the precession of the pendulum as described in the appendix, while the variables x_1 and x_2 coordinatize the surface that the pendulum is moving on. To bring this equation closer to the comfort level of nonspecialists, we note that on a spherical surface, the integrand in the surface integral above is just the solid angle subtended, at the center of the sphere, by the system's cyclical pathway.

The sole reason for introducing the complex vector **n** here—an ingenious trick—is to bring out the hidden analogy that such a system has with quantum wave functions, which are also complex-valued. Indeed, in the next chapter, we will see that the identical formula, but with **n** replaced by the wave function, gives the Berry phase. We also note in passing that the two formulas given above for γ, one calculating it as a surface integral and the other as a line integral, are equal, thanks to Stokes' theorem in vector calculus, which relates a surface integral to a line integral.

Let us take a sphere (such as the Earth) as our surface. In that case, one can show that γ is the solid angle subtended by the surface S at the sphere's center. If we integrate over the whole sphere, we obtain the total curvature of the sphere, which is 4π. Dividing the total curvature by 2π gives us the topological invariant χ, which is equal to 2, for a sphere. In other words, we can relate equation (8.5), which gives the twist angle of the pendulum, to equation (8.3)—the Gauss–Bonnet theorem—as follows:

$$\frac{\gamma}{2\pi} = \frac{i}{2\pi} \int_S \left[\frac{d\mathbf{n}^*}{dx_1} \frac{d\mathbf{n}}{dx_2} - \frac{d\mathbf{n}^*}{dx_2} \frac{d\mathbf{n}}{dx_1} \right] dx_1 \, dx_2 \equiv \chi = \frac{1}{2\pi} \int_S \kappa \, ds. \quad (8.6)$$

This equation reveals the profound fact that *anholonomy can be thought of as a kind of curvature*. We note that χ will be an integer—a topological quantum number—provided that we integrate over the entire surface S.

To give readers a head start on seeing how all this relates to the topological quantum numbers that characterize the different levels of Hall conductivity, or to the topological quantum numbers that decorate every nook and cranny of the Hofstadter butterfly, we point out that if the vector $\mathbf{n}$ plays the role of the wave function, then γ plays the role of the Berry phase and χ plays that of the topological quantum number. In the Berry-phase literature, the quantity κ, which emerges out of the analogies in equation (8.6) like a rabbit popping out of a hat, is known as the "Berry curvature".

We hope readers will experience an "aha" moment upon seeing how cleverly nature exploits one single idea but realizes it differently in many highly diverse contexts. Michael Berry himself put it as follows: "A circuit tracing a closed path in an abstract space can explain both the curious shift in the wave function of a particle and the apparent rotation of a pendulum's plane of oscillation." In other words, although the classical and the quantum systems that exhibit anholonomy involve very different physics, they have topological aspects that are captured in one and the same mathematical language [4].

Monopoles: where do they come from?

From the above formula for anholonomy (whether classical or quantum) emerges an elegant theoretical description that deepens our understanding of these phenomena.

Physicists have a deep drive to seek simple ways to picture anything new they encounter. They pursue this goal with great alacrity, driven by their natural faith that all physical phenomena, when viewed from the right perspective, are connected in some way. Their journey usually begins with questions like "Where does this phenomenon belong within the framework of familiar equations or familiar laws of physics?" or "To what familiar physical phenomenon might this new phenomenon be analogous?" A sense of simplicity and elegance plays a central role in reincarnating, in exotic new contexts, mathematical ideas that originally sprang up in completely different contexts. To show how this kind of self-questioning helped physicists to discover the proper ideas yielding a deep understanding of the quantum Hall effect, we now focus on analogies linking anholonomy to previously known phenomena.

Since anholonomic processes involve phase shifts, it is intuitively plausible that such processes might have some deep analogical links to Maxwell's equations.

Indeed, readers familiar with particle physics will recall that phases are associated with gauge fields, and Maxwell's equations are perhaps the simplest form of gauge fields known to physicists. In fact, the gauge invariance of Maxwell's equations is the core reason underlying the fundamental law of conservation of charge.

As it turns out, this intuitive guess that there might be an analogical link connecting electromagnetic phenomena to anholonomy is deeply correct. The curvature κ in equation (8.6) can be interpreted as an *effective magnetic field*. This fictitious magnetic field, today commonly called the "Berry curvature", along with its corresponding vector potential (the "Berry connection"), forms the crux of the mathematics of anholonomic processes. This radically novel way of describing anholonomy, often called "Berry magnetism", has many striking features, some of which are *disanalogies* with genuine electromagnetism. For instance, unlike the actual equations of Maxwell, which have no solutions involving magnetic monopoles, Berry magnetism permits the existence of (effective) magnetic monopoles.

To see how the mathematics underlying anholonomic phenomena can be manipulated to give something that has the formal appearance of electromagnetic theory, let us rewrite equation (8.5) in the following way:

$$\gamma = i \int_S \left[\frac{d\mathbf{n}^*}{dx_1} \frac{d\mathbf{n}}{dx_2} - \frac{d\mathbf{n}^*}{dx_2} \frac{d\mathbf{n}}{dx_1} \right] dx_1 \, dx_2 \equiv \int_S \vec{B}_{\text{eff}} \cdot d\vec{S} = \oint \vec{A}_{\text{eff}} \cdot d\vec{R} \equiv \int_S \kappa_b dS, \quad (8.7)$$

where $\vec{B}_{\text{eff}} = i[\frac{d\mathbf{n}^*}{dx_1}\frac{d\mathbf{n}}{dx_2} - \frac{d\mathbf{n}^*}{dx_2}\frac{d\mathbf{n}}{dx_1}] = \nabla \times \vec{A}_{\text{eff}}$. These equations suggest that $\vec{B}_{\text{eff}}$ can be viewed as a kind of curvature—the Berry curvature, denoted by κ_b. Recall that Berry curvature is the curvature of some type of abstract space. In the special case of the Foucault pendulum moving on the surface of a spherical Earth, that abstract space coincides with the concrete surface of the Earth.

Readers can easily check that $\vec{A}_{\text{eff}}$ has the following property,

$$\text{If} \quad \mathbf{n} \to e^{i\beta}\mathbf{n}, \quad \text{then} \quad \vec{A}_{\text{eff}} \to \vec{A}_{\text{eff}} - \vec{\nabla}_R \beta. \quad (8.8)$$

Equation (8.8) is strongly reminiscent of the gauge transformations obeyed by a true magnetic vector potential (also sometimes called a "gauge potential" in electrodynamics, as described in chapter 5). If we interpret $\vec{A}_{\text{eff}}$ as a kind of effective vector potential, then $\vec{B}_{\text{eff}}$ will be the "magnetic field" associated with that vector potential.

This analogy gives birth to a variation on the theme of electromagnetism—namely, the above-mentioned Berry magnetism. Incidentally, in the case of a Foucault pendulum moving on the surface of a spherical earth, the fictitious magnetic field $\vec{B}_{\text{eff}}$ equals $\frac{\vec{R}}{R^3}$. This "magnetic field" can be thought of as the field due to a (fictitious) magnetic monopole sitting at the center of the earth (whose surface is assumed to be a sphere). Next we show how all this is related to Maxwell's equations.

Accommodating monopoles into Maxwell's equations

A seeming contradiction arises here, due to the basic theorem of vector calculus that says that the divergence of a curl is always zero. In this case, $\vec{B}_{\text{eff}}$ is defined as the curl of the vector potential $\vec{A}_{\text{eff}}$ (symbolically, $\vec{B}_{\text{eff}} = \nabla \times \vec{A}_{\text{eff}}$), which would imply that the divergence of $\vec{B}_{\text{eff}}$ (that is, $\nabla \cdot \vec{B}_{\text{eff}}$), must equal zero everywhere, but this is equivalent to saying that there are no point sources of magnetism, ergo no monopoles. How, then, can we have $\vec{B}_{\text{eff}} = \nabla \times \vec{A}_{\text{eff}}$, while at the same time having $\nabla \cdot \vec{B}_{\text{eff}}$ not equal to zero?

The resolution of this contradiction comes from realizing that the vector potential induced by anholonomy—that is, $\vec{A}_{\text{eff}}$—must be a *singular* function. In other words, nature can reconcile the existence of monopoles with Maxwell's equations as long as there are vector potentials that "act badly" in certain regions of space.

Fortunately, physicists are already familiar with this type of scenario, thanks to Dirac, who showed, in his 1931 article, that the presence of a monopole amounts to having a vector potential that is singular along a line (usually called a *Dirac string*). Dirac proposed that a magnetic monopole could be envisioned as a semi-infinitely long thin string of magnetic flux (that is, a line that has one end in a finite region of space, but in the other direction goes out forever—like a water hose with a nozzle that one can hold in one's hand, but whose source of water is infinitely far away). The accessible end of the Dirac string, where the magnetic flux spills out, acts like a magnetic point charge (in other words, an isolated North or South pole, with no partner anywhere).

Dirac's ideas, building on pioneering work done in 1918 by the German theoretical physicist Hermann Weyl, led him to a very striking result: namely, that if even one magnetic monopole exists, then all electric charge in the universe must be quantized (that is, the electric charge of any subatomic particle must always be an integral multiple of a fixed, fundamental, minimal amount of charge). Put otherwise, the existence of magnetic monopoles would immediately explain why an electron cannot be sliced in half. Dirac was so thrilled with this revelation that he concluded his famous paper with the characteristically British understatement, "One would be surprised if Nature made no use of it."

Paul Dirac forged a completely new style of doing theoretical physics, one that has been deeply influential ever since. By a long-standing tradition, distinguished visitors to the University of Moscow are invited to write on a blackboard a short statement for posterity. The blackboard quote (reproduced with the permission of Professor Sardanashvily, Moscow State Univerisity) that Dirac left behind after his visit there eloquently sums up his philosophy.

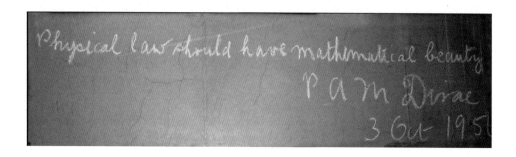

The elegance, inherent beauty, and universality of anholonomic phenomena in classical and quantum physics, including Yang–Mills theory, which revolutionized particle physics in the 1950s and 1960s, constitute an impressive testimony to what Dirac envisioned and believed in. And yet, we still do not really know why Dirac's philosophy seems to apply so accurately to nature. These elusive ideas are closely related to what Eugene Wigner (a Hungarian-American theoretical physicist who received the Nobel Prize in Physics in 1963) called "the unreasonable effectiveness of mathematics in the physical sciences."

In the upcoming chapter, we will continue our exploration of how Berry magnetism relates to quantum anholonomy—namely, we will look at the Berry phase as well as some strange interlopers, such as magnetic monopoles and Dirac strings—with all these ideas linked by the common theme of gauge transformations. Our ultimate aim in that chapter will be to reveal the secrets behind the quantization of Hall conductivity.

8.6 The ESAB effect as an example of anholonomy

Before concluding this chapter, we wish to point out that the ESAB effect, which we discussed in chapter 5, is an example of quantum anholonomy, where what is parallel-transported is not a tangent vector but the phase of the wave function of an electron. This is the phase that plays the key role in the drama of "quantum weirdness".

On the subject of the ESAB effect, John Preskill, an American theoretical physicist and currently the holder of the Richard P Feynman Chair of Theoretical Physics at Caltech, has stated, "This effect is to electrodynamics just as the cone is to Riemannian geometry. As a resident of a cone can infer the existence of the curvature at the tip without ever visiting the tip directly, the electron that propagates in a field-free region can know about the nonvanishing magnetic field inside a perfectly shielded solenoid, even though it never experiences the field directly."

We remind readers that a cone, just like a straight segment of circular pipe, is a manifold that is *flat* (i.e. has zero intrinsic curvature) almost everywhere. The *non-zero* curvature of a cone is completely concentrated at just one single point: its tip. Parallel transport of a vector around a closed path on a cone results in a rotation of the vector (by an angle equal to the so-called "deficit angle" of the flattened cone, which is the angle between the two edges if the cone is sliced open up to its tip, and laid flat on a table) as long as the closed path encircles the cone's tip. Residents of the flat cone can thus detect the far-away non-flatness of their surface without ever visiting the tip, simply by carrying out parallel transport on a looping pathway that encircles it. It is the cone's tip that plays the role of a magnetic monopole in the ESAB effect.

As was explained above, the magnetic field of a monopole can be thought of as a kind of curvature, and this reveals the ESAB effect to be an essentially *geometrical* phenomenon. In an ESAB setup with a particle of charge q (e.g. an electron), the effective magnetic field—the field of the monopole—happens to be equal to $q/\hbar$ times the magnetic field of the solenoid, which is exactly zero everywhere except inside the solenoid—the region where the wave function of the electron vanishes. This makes the region of space that is visitable by the electron not simply connected; and this in turn provides an example of the kind of singularity that necessarily characterizes a Dirac string.

Appendix: Classical parallel transport and magnetic monopoles

Below we summarize Michael Berry's discussion of classical parallel transport on the surface of a sphere, casting it in a form generalizable to quantum mechanics. It shows how anholonomy can be expressed as an effective magnetic flux due to an effective magnetic monopole sitting at the sphere's center.

Let us consider a vector $\hat{e}_1$ that is transported by changing the radius vector $\hat{r}$ under two constraints: firstly, that $\hat{e}_1 \cdot \hat{r} = 0$, and secondly, that the orthogonal triad made up of $\hat{e}_1$, $\hat{r}$, and $\hat{e}_2$, where $\hat{e}_2 = \hat{e}_1 \times \hat{r}$, must not twist. When $\hat{r}$ returns to its original direction after a circuit on the sphere, $\hat{e}_1$ does not return to its original direction. The law of parallel transport is given by:

$$\frac{d\hat{e}_1}{dt} = \omega \times \hat{e}_1, \tag{8.9}$$

where ω is the angular velocity of the triad ($\hat{e}_1$, $\hat{r}$, $\hat{e}_2$).

We now define a complex vector ψ in a plane perpendicular to $\hat{r}$ as follows:

$$\psi = (\hat{e}_1 + i\hat{e}_2)/\sqrt{2}. \tag{8.10}$$

In terms of ψ, the parallel-transport law—that is, equation (8.9)—becomes:

$$\text{Im}\, \psi^* \cdot \frac{d\psi}{dt} = 0, \tag{8.11}$$

where "Im" stands for the imaginary part (of a complex quantity).

To find the anholonomic shift γ of a system, we follow the passage of ($\hat{e}_1$, $\hat{e}_2$) relative to a local basis of unit vectors $\hat{u}$, $\hat{v}$, defined at each point on the sphere. Specifying a local basis is equivalent to specifying the complex unit vector

$$\mathbf{n} = (\hat{u} + i\hat{v})/\sqrt{2}, \tag{8.12}$$

where $\psi(x) = \mathbf{n} e^{-i\gamma(x)}$ and with $\gamma(x)$ being the angle between the transported $\hat{e}$ and the local $\hat{u}$. This gives the anholonomy shift as:

$$\gamma = \oint \frac{d\gamma(t)}{dt} dt = i \int_S \left[\frac{d\mathbf{n}^*}{dx_1} \frac{d\mathbf{n}}{dx_2} - \frac{d\mathbf{n}^*}{dx_2} \frac{d\mathbf{n}}{dx_1} \right] dx_1\, dx_2 = i \oint \mathbf{n}^* \cdot \frac{d\mathbf{n}}{dx} dx. \tag{8.13}$$

The integrand, when expressed in spherical polar coordinates, can be interpreted as the magnetic field of a monopole [3]. The net anholonomy then becomes the magnetic flux that is enclosed by the surface that caps the circuit.

References

[1] http://en.wikipedia.org/wiki/Platonic_solid
[2] Gottlieb D H 1996 All the way with Gauss-Bonnet and the sociology of mathematics *Am. Math. Mon.* **103** 457–69
[3] Berry M V 1989 The Quantum Phase *Five Years After (Geometric Phases in Physics)* ed A Shapere and F Wilczek (Singapore: World Scientific)
[4] Berry M V 1988 *Sci. Am.* **46** December
[5] von Bergmann J and von Bergmann H-C 2007 Foucault pendulum through basic geometry *Am. J. Phys.* **75** 888
[6] Jordan T F and Maps J 2010 Change of the plane of oscillation of a Foucault pendulum from simple pictures *Am. J. Phys.* **78** 1188

IOP Concise Physics

Butterfly in the Quantum World
The story of the most fascinating quantum fractal
Indubala I Satija

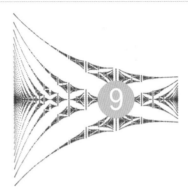

Chapter 9

The Berry phase and the quantum Hall effect

Some questions one poses to oneself, but certain questions pose themselves.
— Henri Poincaré

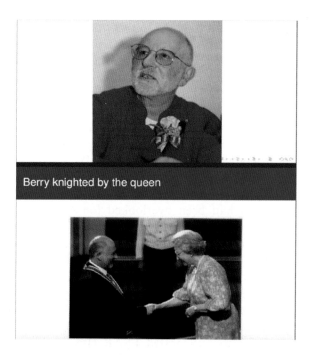

Berry knighted by the queen

9.1 The Berry phase

We now sail into the quantum world, which features anholonomic effects associated with the intrinsic curvatures of certain abstract spaces. Anholonomic processes and the topological invariants connected with them turn out to be an inherent aspect of the quantum world. In fact, the ultra-precise quantization of Hall conductivity is a salient member of this family of phenomena. Moreover, the integer quantum numbers that label the gaps of the Hofstadter butterfly are also subtly encoded in the geometry and topology of a two-dimensional electron gas immersed in a magnetic field.

As was hinted in the previous chapter, quantum anholonomy is strongly analogous to classical anholonomy, which occurs when a vector undergoes parallel transport along a closed pathway on a curved surface. However, quantum anholonomy, rather than involving the change of a visible *angle*, involves a change in the *phase* of a quantum state. Luckily, the phase of a complex number z, although it is more abstract, can still be thought of as a kind of angle—namely, the angle (often called the "argument") made by z, when it is plotted in the complex plane, with the real axis. Thus quantum anholonomy can still be described as a geometrical effect, coming about as the parameters defining the physical system's state vary while the system traces out a cyclic loop in a more abstract type of space.

This remarkable quantum effect was first described clearly by Michael Berry, a mathematical physicist at the University of Bristol, in a paper in 1984 [1], and it became known as the "Berry phase". Subsequently, Berry has been honored with numerous scientific and mathematical awards, as well as being knighted by Queen Elizabeth in 1996. (Images used with the permission of Professor Sir Michael Berry.)

There is an amusing historical footnote to this story. In August of 1983, Michael Berry, shortly after submitting his seminal paper "Quantal phase factors accompanying adiabatic changes" for publication, described his results to Barry Simon, who instantly saw a connection to holonomy and the curvature of Hilbert space. In Simon's opinion, this viewpoint takes at least some of the mystery out of what Berry had called "remarkable and mysterious results". Incidentally, Simon's work [2] on this topic was published in 1983, while Berry's paper, to which Simon's paper was a reaction, appeared in 1984. As Berry put it in one of his writings, "Thanks to a referee's delay and an accident of astronomy, his paper appeared in 1983, mine in 1984."

Berry's discovery, rooted in general principles of quantum physics, was initially quite a surprise to the physics community. It was hard to understand how such a fundamental phenomenon had been overlooked for almost fifty years after the quantum revolution. As the great generality of Berry's phase gradually emerged, so did the appreciation of its beauty and importance, and along came the surprising realization that physicists were already very familiar with several phenomena that are in fact special cases of the Berry phase. Among them were Indian physicist Shivaramakrishnan Pancharatnam's discovery, in 1956, of a geometric phase in the context of optical polarization, as well as the discovery, first made in 1949 by Werner Ehrenberg and Raymond Siday and later made in 1959 by Yakir Aharonov and David Bohm, of the now-celebrated ESAB effect, which was discussed in chapter 5.

In short, Berry's discovery provided a deeper understanding of, and a significant unification of, a set of highly diverse phenomena that had formerly been thought of as unrelated effects. Until Berry's article appeared, the physics community was unaware of the fact that all those phenomena belonged to a single family. This family even includes the celebrated gauge theories, so central to today's particle physics. The story is elegantly told in the book *Geometric Phases in Physics*, edited by Shapere and Wilczek ([3]).

As was noted on the special occasion simultaneously commemorating the 50th anniversary of Aharonov and Bohm's paper and the 25th anniversary of Berry's paper, "Like good wine, the Aharonov–Bohm effect and the Berry phase become finer and more appreciated with time. Even though they appear in textbooks and encyclopedias, the number of researchers thinking and publishing papers on these two effects is increasing annually. The Aharonov–Bohm effect and the Berry phase keep being observed in new systems, and with every day that passes, novel applications are routinely found."

The mathematics that explains the topological nature of the Berry phase is rather technical. Through examples given below, we attempt to convey the essence of what it is about, leaving various technical details to appendices A, B, C and D. Our discussion here will highlight the diversity, universality, and richness of the notion of Berry phase. Despite the complexity of the phenomenon, there is still something very simple and profound about it, a foretaste of which readers already got in chapter 8.

To put it very succinctly, the smoothly varying Berry curvature of an abstract space, thanks to the Gauss–Bonnet theorem for quantum systems (which we will call the "Gauss–Bonnet–Chern theorem"), gives us the Berry phase (an integer), which sets the stage for finding topological invariants. This flow of abstract ideas can be roughly schematized as follows:

anholonomies → *curved abstract spaces* → *Gauss–Bonnet–Chern theorem*
→ *topological invariants* → *topological quantization*

Berry magnetism: effective vector potentials, magnetic fields, and monopoles

Before illustrating the Berry phase in specific quantum systems, we summarize the key points. Our discussion will highlight the analogies that tie quantum anholonomy very tightly to classical anholonomy.

The Berry phase can be described in terms of an effective vector potential and the corresponding effective magnetic field, and in fact to effective magnetic monopoles (whose counterpart does not exist in real electromagnetism). The Berry phase is therefore intimately tied to Dirac strings and to vortex structures in wave functions. Both of these phenomena are related to the singular nature of the vector potential, which plays the key role in determining the existence or nonexistence of monopoles. Furthermore, in the quantum Hall example discussed below, the fact that the effective magnetic monopoles are found only inside spectral gaps (regions of energy that are quantum-mechanically forbidden) reflects the essence of the quantum Hall effect—namely, its origin in edge effects, as described in chapter 5.

- *Berry curvature*

  The mathematical framework underlying the Berry phase is outlined in appendix A. Interestingly, the final expression for quantum-mechanical anholonomy, denoted by γ, is precisely parallel to the expression for classical anholonomy, which was given in equation (8.4). The key point here is that the cyclic "trip" during which the phase of the wave function ψ changes by γ takes place in a two-dimensional *parameter space* denoted by $\vec{R} = (R_x, R_y)$. The final result is expressed either as a surface integral in R_x–R_y space or as a line integral taken over a closed loop in this space. These two formulas, which by Stokes' theorem are equivalent, involve an effective magnetic field or an effective vector potential, respectively. We can see this below:

  $$\gamma = i \int_S \left[\partial_{R_x}\psi^* \partial_{R_y}\psi - \partial_{R_x}\psi \partial_{R_y}\psi^* \right] dR_x \, dR_y = i \oint <\psi|\nabla_{\vec{R}}|\psi> \cdot d\vec{R}$$

  $$\equiv \int_S \vec{B}_{\text{eff}} \cdot dR_x dR_y = \oint \vec{A}_{\text{eff}} \cdot d\vec{R} \equiv \int_S \kappa_b \cdot dR_x dR_y.$$

  Here, the effective magnetic field $\vec{B}_{\text{eff}}$ is equal to $i\, (\partial_{R_x}\psi^* \partial_{R_y}\psi - \partial_{R_x}\psi \partial_{R_y}\psi^*)$, which can be interpreted as a curvature—the *Berry curvature* κ_b, which was discussed in the previous chapter. In this formula for quantum anholonomy, the (complex-valued) wave function ψ plays the role played by the (complex) vector **n** in the previous chapter.

  The reason one can interpret the expression $<\psi|\nabla_{\vec{R}}|\psi>$ as an effective vector potential is that if we systematically change the phase of the wave function ψ (thus carrying out a gauge transformation), that change requires a simultaneous change in $\vec{A}_{\text{eff}}$ in order to keep $\vec{B}_{\text{eff}}$ unchanged, and that set of coordinated changes and constitutes an exact parallel to what happens in electrodynamics, as can be seen below:

$$\text{If } \psi \to e^{i\beta}\psi, \text{ then } \vec{A}_{\text{eff}} \to \vec{A}_{\text{eff}} - \vec{\nabla}_R \beta, \text{ assuring that } \vec{B}_{\text{eff}} \to \vec{B}_{\text{eff}}. \quad (9.1)$$

- *Monopoles: extraterrestrial ghosts*

    Effective magnetic monopoles emerge as the star players in the Berry phase drama. The reconciliation of Berry magnetism with Maxwell's equations, which allow no magnetic monopoles, is a consequence of the fact that the effective vector potential becomes *singular* in certain regions of space (something that never happens in genuine electromagnetism). However, the word "space" might be misleading. We thus wish to emphasize the fact that the effective magnetic monopoles behind the scenes of the Berry phase do not reside in ordinary three-dimensional space, but in a more abstract space.

    In general, the physics community has long been accustomed to physical phenomena that take place in highly abstract spaces, and it enthusiastically embraces very abstract theories, such as that of the Berry phase, which is tied to the curvature of a Hilbert space, and this abstract kind of curvature can in turn be envisioned as resulting from a monopole. In fact, condensed-matter theoreticians feel a special pride when some new type of particle that has never before been seen in the real world makes its debut in a condensed-matter system, even if this "sighting" takes place only in some highly abstract space in the theory, rather than in physical space. The discovery of such a high abstraction that nonetheless accurately describes some aspects of the physical world is a remarkable triumph of the human mind.

    The effective monopoles that we encounter in Berry phases are rather ghostly, existing only in a curious nonphysical space, and signaling strange characteristics of a wave function. For instance, the topological aspect of the Berry phase may appear as a *vortex*—a tiny whirlpool-like structure—in the wave function of the particle, reflecting the fact that it is impossible to choose the phase of the wave function at all points in space in such a way as to make the wave function normalizable. (This technical idea is spelled out in greater detail in appendix C.) Berry monopoles also arise in the equally abstract venue of *strings*—namely, the *Dirac strings* that can be interpreted as effective vector potentials having singularities in certain regions.

    In his beautiful theory of geometric phases, Michael Berry showed that monopoles reside only in special spots of the parameter space where there is a degeneracy—namely, those spots where two eigenvalues coalesce. In the examples below, such a point is a *crossing point* where two different quantum states share the same energy. Without dwelling on how such a degeneracy could "know" about the topology, we simply note that the adiabatic formulation of Berry phases requires the energy eigenvalues to be well separated, and hence one should expect some sort of "catastrophe" to occur at degeneracy points, since they violate this requirement.

- *Topological quantum numbers*

    Topological quantum numbers, such as those associated with quantum Hall conductivity, are related to the "magnetic charge" g of effective magnetic monopoles, which plays the role of a proportionality constant in the equation $\vec{B}_{\text{eff}} = g\frac{\vec{R}}{R^3}$ (an exact parallel to Coulomb's law, in which electric charge acts as a proportionality constant defining the strength of the electric field due to a point charge). In a quantum Hall situation, where the effective monopole shows up as a vortex in the wave function, what determines these quantum numbers is the total vorticity that one encounters while "swimming" in a certain limited region of the two-dimensional landscape known as "reciprocal space", or k-space—specifically, in the Brillouin zone in k-space (or more precisely, in the *magnetic* Brillouin zone, which is very closely related to the ordinary one).

9.2 Examples of Berry phase

We now give two examples of the notion of Berry phase. As we have stated before, they arise as a result of *quantum anholonomy*, in which the phase of a wave function changes as the system follows a cyclic pathway in an abstract space. Here we will also encounter exotic spaces inhabited by strange creatures—monopoles, vortices, and Dirac strings—all closely related to the Berry phase. These examples will set the stage for understanding the quantization of Hall conductivity.

In the examples below, the parameter $\vec{R}$ may be a real magnetic field (denoted by $\vec{B}$), and it is thus very important to distinguish between $\vec{B}$ (the *parameter*) and $\vec{B}_{\text{eff}}$ (the *effective magnetic field*, whose sources are the effective monopoles). Unlike the parameter $\vec{B}$, whose dimensions are those of dimensions of a normal magnetic field, the effective magnetic field $\vec{B}_{\text{eff}}$ (or Berry curvature) has dimensions of the inverse square of a real magnetic field. Therefore, the "magnetic flux" due to $\vec{B}_{\text{eff}}$ passing through a two-dimensional surface in $\vec{B}$-space is dimensionless. Keeping these subtle points in mind helps one not to confuse the Berry magnetic field with the normal one.

Example 1. Spin in a magnetic field

A simple example of the Berry phase is the textbook problem of an electron in the presence of a magnetic field $\vec{B}$ having constant magnitude but changing in direction, as shown in figure 9.1. The electron is fixed in space, so that its only degree of freedom is its spin. In such a situation, the electron's position and momentum play no role in the equations. This type of system is a familiar one in classical physics, and is described by the following Hamiltonian:

$$H = -\frac{1}{2}\vec{B} \cdot \vec{\sigma}, \tag{9.2}$$

where $\frac{\sigma}{2}$ represents the spin of the electron, and $\sigma^2 = 1$. If we introduce a unit vector $\hat{B} = \frac{\vec{B}}{B}$, then the endpoints of $\hat{B}$ trace out a unit sphere called the "Bloch sphere", as

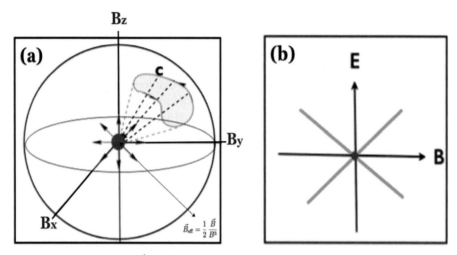

Figure 9.1. An electron with spin $\frac{1}{2}$ moves along a loop in an abstract parameter space. Panel (a) shows the geometry of the Berry phase. The electron's state vector, or wave function, acquires a phase shift as the parameters change adiabatically (i.e., very slowly) along the closed cycle C. In the case shown in the figure, there is a spherically symmetric effective magnetic field $\vec{B}_{\text{eff}}$ due to the effective monopole at the origin, and the Berry phase is proportional to the solid angle subtended at the origin by the cyclic pathway C traced out by the electron's quantum state. The green straight lines in panel (b) show that for any value of the magnetic field there are two possible values of the electron's energy, one positive and one negative; these are due to the electron's two possible spin-states (spin-up and spin-down). Where these two lines cross (at $\vec{B} = 0$), there is a degeneracy in the energy. It can be shown that such degeneracies give rise to Dirac monopoles.

is shown in figure 9.1. The points on the sphere can be parametrized by the pair of angles (θ, ϕ), where:

$$\cos\theta = \hat{B}_z; \qquad e^{i\phi} = \frac{B_x + iB_y}{\sqrt{B_x^2 + B_y^2}}; \qquad \vec{B} = \vec{B}(\theta, \phi).$$

Figure 9.1 shows the key aspects of this problem. (Appendices B and C provide additional details.) The problem of a spin-$\frac{1}{2}$ particle in a magnetic field is mapped to the problem of a magnetic monopole whose magnetic field $\vec{B}_{\text{eff}}$, pointing radially outwards from the origin of the parameter space $\vec{B}$, is given by:

$$\vec{B}_{\text{eff}} = \frac{1}{2}\frac{\vec{B}}{B^3}. \qquad (9.3)$$

The pre-factor of $\frac{1}{2}$ comes from the spin of the particle, which, in multiples of $\hbar$, is equal to $\frac{1}{2}$. The Berry phase due to the spin-$\frac{1}{2}$ particle sweeping out a closed loop on the Bloch sphere (representing a magnetic field of constant magnitude) will be equal to the solid angle subtended at the center of the sphere by the cyclic pathway. This elegant result is precisely analogous to the case of the Foucault pendulum.

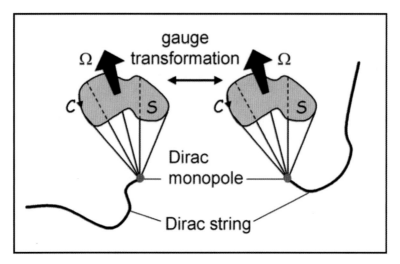

Figure 9.2. Schematic representation of the magnetic flux Ω (where $\Omega = \gamma$) due to a Dirac monopole (the red dot) through the loop C, and also showing the effect of a gauge transformation, which moves the Dirac string. Reproduced with permission of Dr Patrick Bruno, European Synchrotron Radiation Facility.

This quantum anholonomy is also reflected in the coalescing of two energy eigenvalues (as shown in figure 9.1(b)), which gives rise to a degeneracy point where a monopole is located. In other words, unlike the Foucault-pendulum situation, where the classical anholonomy can be interpreted as resulting from a (gravitational) monopole located at the center of the spherical earth, the effective monopole in the spin-$\frac{1}{2}$ situation "lives" in the more abstract parameter space of $\vec{B}$, and in fact it sits at that space's origin.

In appendix C, we show that the monopole singularity is due to the lack of a good global gauge; this gives rise to a vector potential that has a Dirac-string singularity or a vortex in the wave function. Figure 9.2 gives a schematic drawing of a gauge transformation that can move the Dirac string without having any effect on the magnetic flux that determines the anholonomy.

Example 2. A lattice with two bands: a simple model of the quantum Hall effect

The simplest model of the quantum Hall effect is a lattice in a magnetic field whose allowed energies lie in two bands separated by a gap. Such a system is an insulator when one of its bands is filled and the other one is empty. The three panels of figure 9.3 show the essential aspects of such a quantum Hall system: a Brillouin zone having the topology of a torus's surface, edge states residing in the gap (not on the torus's surface), and a wave function that is singular at certain isolated points in the Brillouin zone. (Appendix D provides further details.)

Such a system is described by an effective Hamiltonian of this form:

$$H(k_x, k_y) = -\vec{h}(k_x, k_y) \cdot \vec{\sigma}. \qquad (9.4)$$

This bears a strong resemblance to the Hamiltonian of a spin-$\frac{1}{2}$ particle in a magnetic field, with one exception: the magnetic field $\vec{h}$ depends on the wave vector

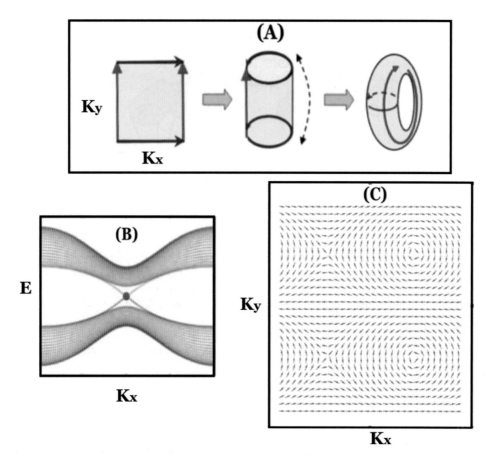

Figure 9.3. This figure summarizes the key aspects of Berry magnetism in a quantum Hall system. Panel (A) shows the Brillouin zone for a two-dimensional square lattice. This zone in reciprocal space (the space of Bloch wave-numbers (k_x, k_y)) is periodic in both of its dimensions, which means that its left and right edges should be identified (i.e., brought together and "glued", as is suggested by the two parallel vertical red arrows), and analogously, so should the top and bottom edges (the parallel horizontal black arrows). Carrying out these two "gluing" operations (shown in the middle) results in a manifold in the form of a torus (see the right side). The torus's surface is thus simply a different way of representing the Brillouin zone. Panel (B) is a graph of energy versus wave number, showing the nature of a two-band structure, for a finite crystal (that is, a crystal with edges). The two bands are separated by a gap. However, in a system with edges, there exist extra states whose energies lie in the gap between the two bands. The energy of such states depends on k_x (in the graph roughly linearly), and at the spot in k_x-space where those energies cross (indicated by the red dot), there is a degeneracy. (Note the resemblance to the right side of figure 9.1.) Any such energy degeneracy is the telltale sign of a magnetic monopole. Since the degeneracy lies in the gap, the monopole is *inside* the torus, not on its surface (thus not in the Brillouin zone). The graph in panel (C) shows, using arrows, the phase of the wave function, plotted in the Brillouin zone. There are four clearly visible vortices where the wave function's phase's direction is ill-defined, meaning also that its amplitude vanishes. These vortices are thus the locations of *singularities* in the wave function.

$\vec{k} = (k_x, k_y)$, which lies on the surface of a *torus*, in contrast to the Bloch *sphere*, which applies to a spin-$\frac{1}{2}$ particle in a constant magnetic field. This alternative Hamiltonian gives rise to a two-band spectrum (the pink regions in figure 9.3(B)), separated by a gap (the white region between the two bands).

Second, we note that the energy spectrum shown in figure 9.3 describes a two-dimensional system of finite size, such as a surface of a cylinder of finite length. Unlike a sample of infinite size, systems having finite size can have eigen-energies that lie inside the gap between the bands. In other words, while the bands (pink) represent all the possible energies of an *infinite* system, the two levels (blue and green) inside the gap correspond to two extra states (*edge modes*) that belong to wave functions localized at the two boundaries of the (non-infinite) sample. The reason that some states can exist inside the inter-band gap is that there are *edges* in this physical setup. As was pointed out by Barry Simon [2], the Hamiltonian for a sample of finite size can be obtained via a smooth interpolation of the Hamiltonian given in equation (9.4), which describes an infinite system. Appendix D explicitly shows this interpolation for the simple two-band model described here.

The third key point is that the system described by the effective Hamiltonian in equation (9.4) must have its magnetic monopole located at $\vec{h}(k_x, k_y) = 0$, just as in the earlier example of the spin-$\frac{1}{2}$ electron, where a monopole was located at the origin of the magnetic field (i.e., at $\vec{B} = 0$). As is stated in appendix D, this cannot occur inside either band, but occurs instead in the gap between them. The crossing-point of the two edge-mode energies (the red dot in figure 9.3(B)) is the degeneracy point, and it thus "houses" the monopole—an incarnation of the quantum anholonomy that is intrinsic to such systems. It is important to note that the monopole resides inside the gap—that is, *not on the surface* of the Brillouin zone—and it holds the secret to the topology of quantum Hall systems.

A graph of the phase of the wave function in k-space, shown in figure 9.3(C), shows a key feature of the system—namely, the absence of any good global gauge [4]. This manifests itself through the presence of *vortices* at certain points in the Brillouin zone. It can be shown that, analogously to the spin-$\frac{1}{2}$ situation on the Bloch sphere (discussed in detail in appendix C), there exists no single global gauge that applies to the wave function at all points in the Brillouin zone. The vortices thus represent singularities in the wave function. At such points, the wave function's phase is undefined, since its amplitude vanishes, which gives rise to a vortex in k-space. In the present case, the vorticity at such points is the anholonomy associated with the wave functions. For further details, we refer readers to the paper by Hatsugai [5].

9.3 Chern numbers in two-dimensional electron gases

The two-band model of the quantum Hall effect just described leads naturally to a generalization of the quantum Hall effect to other systems, such as the Hofstadter butterfly, whose fractal pattern reflects all possible quantum Hall states for non-interacting electrons in a lattice.

As was explained in chapters 5 and 7, the energy spectrum of electrons in a crystal consists of bands. When the Fermi energy lies in a gap between bands, we have a system some of whose bands are fully occupied at zero temperature. Associated with any filled band n, we can define the Berry phase γ_n for the wave function as the integral of an effective magnetic field, B_{eff}, also known as the Berry curvature κ_b, over the entire Brillouin zone, as described in detail in the previous chapter, and also in appendix A below. Using the mathematical framework given in appendix A (see equations (A.8) and (A.9)), one obtains the following expression for the geometric phase γ_n and the corresponding topological invariant σ_n, known as the Chern number of the nth band, and associated with the wave function ψ_n:

$$\sigma_n = \frac{\gamma_n}{2\pi} \equiv \frac{1}{2\pi} \int_{\text{torus}} \kappa_b \, dk_x dk_y = \frac{i}{2\pi} \int_{\text{torus}} \left[\partial_{k_x}\psi_n^* \partial_{k_y}\psi_n - \partial_{k_x}\psi_n \partial_{k_y}\psi_n^* \right] dk_x \, dk_y. \quad (9.5)$$

A comparison of this equation with equation (8.2) reveals the tight analogy between Chern numbers and the Euler index. Specifically, the Chern numbers σ_n in equation (9.5) are quantum analogues of the Euler index χ. The fact that χ can take on only integer values—the "quantization" of χ, so to speak—is due to the fact that one is integrating the local curvature over the entire manifold. Analogously, the quantization of σ_n stems from the fact that we are integrating the Berry curvature over the entire Brillouin zone, taking into account the "filled-band condition" (that is, the fact that all the states in reciprocal space are occupied). This condition holds because the system in question is an insulator.

We can concisely summarize the tight analogy between classical and quantum-mechanical anholonomies in the following list of parallels:

$$\text{local curvature} \longleftrightarrow \text{Berry curvature}$$
$$\text{total curvature} \longleftrightarrow \text{Berry phase}$$
$$\text{Euler index } \chi \longleftrightarrow \text{Chern number } \sigma_n$$
$$\text{monopole in real space} \longleftrightarrow \text{monopole in reciprocal space}$$

Finally, we note that the Chern number σ_n is an integer—a topological invariant belonging to band n. If there are N filled bands, so that the Fermi energy lies in the gap just above the Nth band, then we can define a topological invariant σ associated with the entire system, as follows:

$$\sigma = \sum_{n=1}^{N} \sigma_n \quad (9.6)$$

9.4 Conclusion: The quantization of Hall conductivity

After all of this abstract mathematics and these subtle physical concepts, we finally arrive at the remarkable result that the Chern numbers given by equation (9.5), when they are summed over all the bands, as in equation (9.6), yield the quantum number associated with Hall conductivity. In other words, the

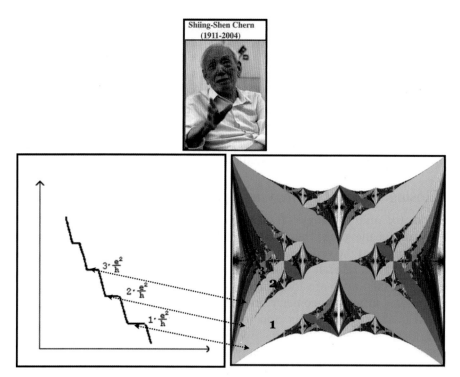

Figure 9.4. Each gap in this colorful butterfly is a distinct topological state labeled by a Chern number, and each Chern number is associated with a particular quantum Hall state, as is shown by the left-pointing arrows. (Photograph: Author: George M Bergman, Source: Archives of the Mathematisches Forschungsinstitut Oberwolfach.)

experimentally accessible quantum Hall conductance σ_{xy} is given by this theoretical expression:

$$\sigma_{xy} = \left[\sum_n i \int_{torus} [\partial_{k_x}\psi_n^* \partial_{k_y}\psi_n - \partial_{k_x}\psi_n \partial_{k_y}\psi_n^*] dk_x\, dk_y\right] \frac{e^2}{2\pi h}$$

$$= \sigma \frac{e^2}{h}.$$

This remarkable result, the climax of the story, as summarized in figure 9.4, due to David Thouless and collaborators, is commonly referred to as the *TKNN formalism* [6]. The model in which they developed these ideas was a two-dimensional electron gas in a lattice in a homogeneous magnetic field perpendicular to the lattice—in other words, the same model as was used by Hofstadter in his calculations of the butterfly spectrum. In this model, one takes into account the interactions of the electrons with the nuclei in the lattice, but one ignores any interactions among the electrons themselves. This model allows a rather simple mathematical treatment, and yet it already contains the essential features that give rise to the quantization and the stability of Hall conductivity.

This result is the conclusion of a fascinating story as summarized in figure 9.4. Thanks to its combination of abstractness and simplicity, it represents a great triumph of theoretical physics. It shows that certain beautiful and simple ideas of geometry and topology, presented in these past two chapters, constitute the heart and soul of one of the most exotic phenomena of all of contemporary physics—the quantum Hall effect. Moreover, the integers labeling the gaps in the infinitely recursive butterfly encode this fact in their own beautiful way. It is hard to believe that all this complexity was already lurking unsuspected in the colorful plot that a graduate student in Regensburg drew by hand in his notebook, some forty years ago, using numbers provided to him by a desktop computer that had roughly the computing power of a hand-held calculator.

9.5 Closing words: Topology and physical phenomena

The geometry and topology of curved spaces are central elements of general relativity and of the gauge theories of contemporary particle physics. This chapter has pointed out that the very same theme also underlies the remarkable phenomenon of the quantization of Hall conductivity.

Quantum Hall states are the simplest examples of a wide class of condensed-matter systems in which geometry and topology play a starring role. Even in the fractional quantum Hall effect and in quantum spin Hall effects, where Chern numbers play no role whatsoever, topology is still the key factor. The fact that geometry and topology can profoundly illuminate a wide variety of subtle and important physical phenomena has, over the past forty years, opened up a fundamental new way of looking at diverse areas of physics.

Approaching the mysteries of physics through geometry, a philosophy in which Albert Einstein deeply believed, has inspired some of the greatest minds in physics today. This approach has yielded both great beauty and profound clarity, although of course following this pathway is not always easy or straightforward. That, for better or worse, is the nature of life.

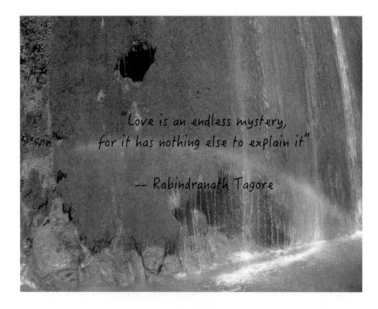

Appendix A: Berry magnetism and the Berry phase

I have had my results for a long time, but I do not yet know how I am to arrive at them.

—Karl Friedrich Gauss

The Berry phase is an example of anholonomy—the phenomenon in which the parameters characterizing a physical system are continuously altered but then in the end most of them return to their initial values (so that one feels that one has "come home"), while one or more other parameters fail to do so. A simple case of classical anholonomy that we have already looked at is the Foucault pendulum. We now briefly outline the essential ingredients of Berry's ideas about geometric phases in quantum-mechanical situations.

Consider a quantum system in an eigenstate ψ_n with energy E_n, described by a time-independent Hamiltonian H:

$$H\psi_n(x) = E_n\psi_n(x). \tag{A.1}$$

In the absence of any additional perturbation, as the system evolves in time, it will remain in the same energy eigenstate, but its wave function will undergo a periodic oscillation of phase with the frequency $E_n/\hbar$. This periodically varying wave function can be expressed more formally as follows:

$$\psi_n(x, t) = e^{iE_n t/\hbar}\psi_n(x). \tag{A.2}$$

Let us now suppose that during the time interval $0 \leq t \leq T$, some parameters $R_1, R_2, \ldots$, which we collectively denote as $\vec{R}(t)$, are slowly changing, with the values of $\vec{R}(t)$ tracing out a curve C in an abstract space. The state of such a system evolves according to the time-dependent Schrödinger equation. However, if the parameters $\vec{R}(t)$ vary sufficiently slowly with time (i.e., adiabatically), then equation (A.1) will remain valid at each instant of time. In the case of *cyclic* evolution, where the system returns to its starting point after a time interval T (that is, $\vec{R}(t) = \vec{R}(t+T)$), we can write:

$$H\psi_n(t) = E_n(t)\psi_n(t) \tag{A.3}$$

$$\psi_n(t=T) = e^{i\gamma}e^{-\frac{i}{\hbar}\oint E_n(t)\, dt}\psi_n. \tag{A.4}$$

In addition to the phase $e^{-\frac{i}{\hbar}\oint E_n(t)\, dt}$, commonly known as a *dynamical phase*, one is allowed to throw in an extra phase factor $e^{i\gamma}$, which depends only on the geometry of the path in the parameter space $\vec{R}$. This phase factor is included to account for what is known as "gauge freedom", which means that one has the freedom to assign a phase to the wave function at each point in parameter space. (The *magnitude* of the wave function, however, cannot be tampered with anywhere.)

If we substitute the wave function in equation (A.4) into the time-dependent Schrödinger equation $H\psi = E_n(t)\psi = i\hbar\partial_t\psi$, it can be shown [7] that:

$$\gamma = i\oint <\psi|\nabla_{\vec{R}}|\psi> \cdot d\vec{R}. \tag{A.5}$$

(An expression such as $<A(x, y)|B(x, y)>$, using Dirac's classic "bra" and "ket" notation, represents a scalar product, and is equal to $\int A^*B\,dx\,dy$.)

Remarkably, one can identify an "effective electromagnetism" lurking in this situation by thinking of $<\psi|\nabla_{\vec{R}}|\psi>$ as an effective magnetic vector potential, which we will denote by $\vec{A}_{\text{eff}}$:

$$\vec{A}_{\text{eff}} = i<\psi|\nabla_{\vec{R}}|\psi>. \tag{A.6}$$

Here, $\nabla_{\vec{R}}$ is the gradient operator in the abstract parameter space. The parameters collected in the vector $\vec{R}$ might, for instance, represent angles θ and ϕ, if the parameter space is, say, the surface of a sphere.

Equation (A.5) then becomes:

$$\gamma = \oint \vec{A}_{\text{eff}} \cdot d\vec{R}. \tag{A.7}$$

A more intuitive understanding of the angle γ may be obtained if we use Stokes' theorem to rewrite the above integral as a surface integral:

$$\gamma = \int_S \nabla_R \times \vec{A}_{\text{eff}} \cdot d\vec{S} = \int_S \vec{B}_{\text{eff}} \cdot d\vec{S}. \tag{A.8}$$

This can be interpreted as saying that we have an effective magnetic field B_{eff}, a gauge-invariant quantity, associated with the gauge field $\vec{A}_{\text{eff}}$, as follows:

$$\vec{B}_{\text{eff}} = \nabla_R \times \vec{A}_{\text{eff}} = i<\nabla_R\psi|\times|\nabla_R\psi>. \tag{A.9}$$

For $\vec{R} = (R_x, R_y)$, we can write the above equation as follows:

$$\vec{B}_{\text{eff}} = i\int_S \left[\partial_{R_x}\psi^*\partial_{R_y}\psi - \partial_{R_x}\psi\partial_{R_y}\psi^*\right] dR_x\,dR_y. \tag{A.10}$$

Why is it reasonable to think of $\vec{B}_{\text{eff}}$ as an effective magnetic field?

If we systematically change the phase of the wave function at all points in space by altering its phase (i.e. if we carry out a gauge transformation, as defined earlier), then to compensate for this alteration, we must simultaneously carry out a change in $\vec{A}_{\text{eff}}$, so that $\vec{B}_{\text{eff}}$ will be unchanged:

$$\psi \to e^{i\beta}\psi; \qquad \vec{A}_{\text{eff}} \to \vec{A}' = \vec{A}_{\text{eff}} - \nabla_R\beta; \qquad \vec{B}_{\text{eff}} \to \vec{B}_{\text{eff}}. \tag{A.11}$$

If one chooses β so as to make $\vec{A}'$ vanish everywhere, then the phase factor γ can be eliminated. However, this will be possible only if $\vec{A}_{\text{eff}} = \nabla_R\beta$ has a solution at all points. That is, the vector potential $\vec{A}_{\text{eff}}$, or the wave function that determines this vector potential, must be smoothly defined everywhere. However, it may happen that this is possible only locally, and there may not exist a well-defined solution at all points in parameter space, because of the singular nature of the wave function. That is, it may not be possible in general to find a global convention that leads to a single-valued, normalizable wave function for all values of the parameters. In such cases, we have a scenario that leads to quantum anholonomy.

Berry curvature and the quantum analogue of the Gauss–Bonnet theorem

One can now assign a *geometrical* meaning to $\vec{B}_{\text{eff}}$, which, unlike $\vec{A}_{\text{eff}}$, is a gauge-invariant quantity. This interpretation of $\vec{B}_{\text{eff}}$ stems from its analogical link to the Gaussian curvature. We can see this from equation (A.8), which can be viewed as the quantum analogue of the Gauss–Bonnet theorem. It defines *quantum curvature* as a sort of angular mismatch, where "angle" more accurately means *phase*. This $\vec{B}_{\text{eff}}$ is called the "Berry curvature" and is denoted by κ_b.

It is this curvature that may lead to non-zero values of γ, as can be seen from equation (A.5). The phase factor γ is a result of quantum anholonomy—a quantum analogue to the angular shift of a classical vector when it has completed a cyclic path in a curved space.

It is important to stress the following points regarding the effective magnetism just defined:

- The effective magnetic field is defined not in ordinary 3-space, but in a more abstract parameter space. In particular, in a two-dimensional electron gas in which quantum Hall states can arise, the parameter space is the space of Bloch vectors (k_x, k_y), which constitutes a two-dimensional torus—namely, the Brillouin zone.
- $\vec{B}_{\text{eff}}$ and the corresponding geometric phase γ depend only on the the geometry of the curve C and the eigenstates of the system.

Appendix B: The Berry phase and 2 × 2 matrices

Michael Berry noted that the mathematics of geometrical phases is, in its simplest form, a subset of the mathematics of 2 × 2 matrices. Therefore, although the concepts and ideas associated with the Berry phase are subtle, there are examples where the underlying mathematics is quite simple—namely, all it takes is understanding the properties of 2 × 2 matrices.

The mathematical framework underlying the Berry phase for the classic textbook example of an electron in a magnetic field in vacuum can be expressed in terms of a 2 × 2 matrix. This is also the case when the two-dimensional electron gas in a square lattice is subjected to a magnetic flux of $\phi = \frac{1}{2}$, and where the fickle electrons can hop from any given nucleus to its nearest neighbors not only along the lattice's two principal axes, but also along the diagonals of the lattice. These two examples were discussed above to illustrate quantum anholonomy.

The appearance of 2 × 2 matrices in quantum physics is associated with cases when the quantum state or wave function is a two-component object. Such states are standardly referred to as *spinors*, and each component corresponds to one of the two possible states of the particle. An example of a two-component wave function is given by the quantum-mechanical states of particles such as electrons. These particles carry intrinsic spin—one of the mysterious aspects of quantum science. Spin is a quantum-mechanical property with no classical analogue. Fermions—the particles that constitute ordinary matter—have *half-integer* spin (meaning that their spin, when measured in units of $\hbar$, is equal to $\frac{1}{2}$ or $\frac{3}{2}$ or $\frac{5}{2}$, etc). Within the set of all fermions, spin-$\frac{1}{2}$ particles constitute by far the most important subset.

With all the mystique surrounding this strange quantum property, there is something very simple, at least in mathematical sense, about the quantum behavior of an isolated spin-$\frac{1}{2}$ particle. It is quite gratifying that such a particle, in the presence of a magnetic field, is completely describable by a 2×2 matrix. The wave function is a spinor, with one component corresponding to spin *parallel* to the magnetic field, and the other component corresponding to spin *antiparallel* to it.

Appendix C: What causes Berry curvature? Dirac strings, vortices, and magnetic monopoles

Although mathematically intriguing, the concepts of effective magnetism and Berry curvature are quite abstract, probably leaving some readers wondering how all this can be expressed in terms of the more familiar quantum-mechanical notion of wave functions. To shed a different and perhaps clearer light on quantum anholonomy, then, we will take a closer look at the wave function for a spin-$\frac{1}{2}$ particle in a magnetic field. The Hamiltonian is as follows:

$$H = -\frac{1}{2}\vec{B} \cdot \vec{\sigma}. \tag{C.1}$$

In quantum physics, the three components of the spin operator $\vec{\sigma}$—namely, σ_x, σ_y, and σ_z—are represented by the 2×2 Pauli spin matrices. The system described by this Hamiltonian can be solved using standard quantum methods. The eigenvalues are given by $E_\pm = \pm B$. It will suffice for us to consider just one of the eigenstates of the system, such as this:

$$\Psi = e^{i\alpha} \begin{pmatrix} e^{-i\phi/2} \sin\frac{\theta}{2} \\ e^{+i\phi/2} \cos\frac{\theta}{2} \end{pmatrix}. \tag{C.2}$$

Here α is an arbitrary phase factor. Setting this phase to a value of one's choice is commonly called the choice of *gauge*. In his famous quantum-mechanics textbook, Leonard Schiff writes:

> We can change the phase of the eigenfunction by an amount [...], which is permissible, since the phases of the eigenfunctions are arbitrary at each instant of time.

One might well think that phases such as α in equation (C.2) are irrelevant. However, it turns out that this is not always the case, as was swiftly realized by the physics community after Michael Berry's discovery of geometric phases, which he argued are experimentally accessible.

The crux of the matter lies in the fact that it is not possible to write an expression for the wave function in such a way that it is nonsingular everywhere in the parameter space $\vec{B}$, as we show below. We will look at three different choices for the phase α, and we will see that there is no "good gauge"—that is, a gauge that works at all points on the Bloch sphere. Otherwise put, there is no expression for α that yields wave functions that are single-valued and normalizable *at all points in the parameter space*. Here are the three cases we'll consider:

1. Suppose we choose $\alpha = 0$. If we fix θ and make a full circle in ϕ, we should come back to the same wave function. However, because of the $\phi/2$ in the above equation, we see that the wave function returns with its phase shifted by π. Thus this choice for α fails to give single-valued wave functions on the Bloch sphere. Still, one might think there could be other ways to accomplish this goal, so that for each couple (θ, ϕ), we have a unique eigenstate.
2. Suppose instead that we choose $\alpha = \phi/2$. There are two tricky points on the sphere now: the north and south poles. At the former, where $\theta = 0$, the eigenstate is $(1, 0)$. At the latter, it is $(0, e^{i\phi})$. That is, at the south pole, the eigenstate depends upon the direction in which we approach the pole. (And by the way, had we chosen $\alpha = -\phi/2$, then the south pole would work just fine, but in that case the *north* pole would have the ambiguous-phase problem.)
3. One can try to overcome this problem of phase ambiguity at the poles by choosing a gauge that is a linear combination of the above two possibilities, each multiplied by a function that vanishes at the singularity:

$$\Psi = \begin{pmatrix} \cos\frac{\theta}{2}\left(\cos\frac{\theta}{2} + \sin\frac{\theta}{2}e^{iB}e^{-i\phi}\right) \\ \sin\frac{\theta}{2}e^{i\phi}\left(\cos\frac{\theta}{2} + \sin\frac{\theta}{2}e^{iB}e^{-i\phi}\right) \end{pmatrix}.$$

Here, B is some arbitrary constant. The good news is that this state is single-valued everywhere, which solves the ambiguity problem at the poles. The bad news, however, is that a new problem arises: at $\theta = \frac{\pi}{2}$, $\phi = B \pm \pi$, the wave function vanishes. This means that the wave function cannot be normalized. That is, if we demand a single-valued function, what we end up with is a vortex somewhere.

We invite readers to try other possibilities and confirm that there is no well-defined gauge that works globally—that is, there is no way to fix α so that we have a single-valued, normalizable wave function everywhere on the Bloch sphere. And thus, if we go back to equation (A.11), we cannot get rid of the geometric phase shift that results from following a cyclic path. This fact results in a singularity in the Berry curvature, and that is what leads to quantum anholonomy.

Dirac strings

Readers can easily verify that in the gauge choices discussed above, the corresponding vector potentials exhibit Dirac string singularity, as displayed below. For instance, using spherical polar coordinates $B_x = B \sin \theta \cos \phi$, $B_y = B \sin \theta \sin \phi$ and $B_z = B \cos \theta$, one can show:

$$\Psi(\alpha = \phi/2) = \begin{pmatrix} \cos \dfrac{\theta}{2} \\ e^{i\phi} \sin \dfrac{\theta}{2} \end{pmatrix}; \quad \vec{A}(\alpha = \phi/2) = -\frac{1}{2B} \frac{1}{(B_z + B)}(-B_y, B_x, 0). \quad (C.3)$$

This wave function is single-valued at all points on the sphere, but it is ill-defined at $\theta = \pi$, and this is reflected in the corresponding vector potential as a singularity along the line $B_z = -B$, which is the predicted Dirac string.

If we flip the sign of α, we get:

$$\Psi(\alpha = -\phi/2) = \begin{pmatrix} e^{-i\phi} \cos \dfrac{\theta}{2} \\ \sin \dfrac{\theta}{2} \end{pmatrix}, \quad \vec{A}(\alpha = -\phi/2) = -\frac{1}{2B} \frac{1}{(B_z - B)}(B_y, B_x, 0). \quad (C.4)$$

This wave function is single-valued at all points on the sphere, but it is ill-defined at $\theta = 0$, and this is reflected as a singularity in the corresponding vector potential along the line $B_z = +B$, which we identify as another Dirac string.

The above two sample calculations give the flavor of why Dirac's singular strings are inevitable.

Appendix D: The two-band lattice model for the quantum Hall effect

We now describe the Berry phase in a quantum Hall system—a square lattice immersed in a magnetic field (the same system as was studied by Hofstadter). As was shown in chapter 7, the situation is described by Harper's equation:

$$\psi_{m+1} + \psi_{m-1} + 2\cos(2\pi\phi m - k_y)\psi_m = E\psi_m. \quad (D.1)$$

Unlike the spin-$\frac{1}{2}$ situation described above, where the parameter space can be realized as a sphere, the parameter space here is the space of Bloch wave-number vectors (k_x, k_y). This space—a Brillouin zone—is not a sphere but a torus, as is shown in figure 9.3. If we are dealing with a *rational* magnetic flux-value p/q, then the edges of the Brillouin zone will be defined by k_y lying within the interval $[-\pi/q, +\pi/q]$ on one axis (as can be seen from the periodicity of the cosine in equation (D.1)), and k_x lying within the interval $[-\pi, +\pi]$ along the perpendicular axis. Such a modified Brillouin zone is called a *magnetic Brillouin zone*.

In the above model, all rational flux-values except for $\phi = 1/2$ support a quantum Hall state. However, for the simple case of flux-value $\frac{1}{2}$, a generalization of Harper's equation in which the crystal electrons can hop along *diagonals* (in addition to nearest-neighbor hopping) turns out to be the ideal model to describe the Berry phase in a quantum Hall system. Such a system maps exactly onto a spin-$\frac{1}{2}$ situation in a magnetic field, as described above. With the diagonal hopping, which we will denote by J_d, the Hamiltonian of this system can be transformed as follows [5]:

$$H(k_x, k_y) = -\vec{h} \cdot \vec{\sigma}, \tag{D.2}$$

$$h_x = 2J_x \cos k_x; \qquad h_y = 2J_y \cos k_y; \qquad h_z = 4J_d \sin k_x \sin k_y. \tag{D.3}$$

Although the problem is thereby mapped to a spin-$\frac{1}{2}$ problem in a magnetic field, we note that there is no point in the Brillouin zone where $h = 0$ (which is the location of the monopole). Therefore, the fictitious monopole in the quantum Hall problem exists *outside the reciprocal space* and, in view of the crossing-point of the two energy eigenvalues of the edge modes, we can pinpoint the monopole inside the gap. We refer readers to the paper by Hatsugai for additional details [5].

Remark: locating the monopole

One way to obtain $\vec{h} = 0$—the point in parameter space where the monopole resides—is to analytically continue k_y in the complex plane. If we write $k_y = k_y^r + ik_y^i$, this yields a Hamiltonian that, in general, is non-Hermitian. Interestingly, it has real eigenvalues when $k_x = \pi/2$ and $\cos k_x^2 + J_d^2 \sin k_x^2 \sin k_y^2 = 0$. The latter equality is possible because k_y is complex, and this fact determines the localization length of the edge modes. This leads to $E = 0$ as a possible eigenstate, where $\vec{h} = 0$, and we have twofold degeneracy.

We close this section with a remark that often makes newcomers trying to understand the quantum Hall effect uncomfortable—namely, although the Chern-number formula involves no edges, thereby implying that the bulk on its own "knows" everything about the nontrivial topology of quantum Hall states, edge states are nonetheless necessary to actually *produce* the quantum Hall effect. In other words, yes, the infinite sample "knows" all about the nontrivial topology associated with the band structure, but to bring these topological aspects to "life" so that we can actually see them, we need to bring in edges. It is the presence of edges that allows us to see the topological aspect of the quantum Hall effect, by providing "shelter" to the magnetic monopole whose fingerprint is "hiding" in the properties of the bulk.

References

[1] Berry M V 1984 Quantal phase factors accompanying adiabatic changes *Proc. R. Soc.* A **392** 45
[2] Simon B 1983 Holonomy, the quantum adiabatic theorem and Berry's phase *Phys. Rev. Lett.* **51** 2167

[3] Shapere A and Wilczek F (ed) 1989 *Geometric Phases in Physics* (Singapore: World Scientific)
[4] Kohmoto M 1989 *Phys. Rev.* B **39** 11943
[5] Hatsugai Y and Kohmoto M 1990 *Phys. Rev.* B **42** 8283
[6] Thouless D, Kohmoto M, Nightingale M P and den Nijs M 1982 *Phys. Rev. Lett.* **49** 405
[7] Griffiths D 1995 *Introduction to Quantum Physics* (Upper Saddle River, NJ: Prentice-Hall) ch 10

IOP Concise Physics

Butterfly in the Quantum World
The story of the most fascinating quantum fractal
Indubala I Satija

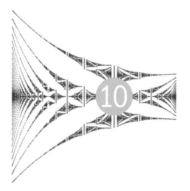

Chapter 10

The kiss precise and precise quantization

> *The sweet, soft freshness that blooms on a baby's limbs—does anyone know where it was hidden so long?*
> *Yes, when the mother was a young girl, it lay pervading her heart in the tender and silent mystery of love—the sweet, soft freshness that has bloomed on the baby's limbs.*
>
> Rabindranath Tagore

Having just swooped through anholonomic loops in curved spaces, we now return to the butterfly landscape—home of the quantum Hall effect. The home we return to is a sweet home and a beautiful home, hosting quantum Hall states, which, as we will soon see, are reincarnations of integral Apollonian gaskets. Nature surprises and delights us again with the revelation that the quantum world, in all its mystery, is a beautiful world when viewed from the proper perspective.

The theme of this chapter is to assign the proper address to each of the inhabitants of the butterfly landscape. Equipped with some knowledge of topology and topological quantum numbers, we will label parts of the butterfly with integers that can be observed in laboratories. In the landscape, these states are nested, and their topological addresses form hierarchical patterns exhibiting self-similarity and universal scaling.

Incidentally, shortly after Douglas Hofstadter found the butterfly's recursive structure, his doctoral advisor Gregory Wannier and the Chilean physicist Francisco Claro proved [1] that any white swath in the butterfly can be labeled by a pair of integers. This important result is known as the "gap-labeling theorem". However, since Claro and Wannier's formulation is somewhat distinct from from the topological approach we are stressing here, we will not go into it.

The correspondence between the butterfly fractal and integer Apollonian gaskets ($\mathcal{IAG}$s) was laid out earlier in this book, in chapter 3. What we there called "$\mathcal{ABC}$" (Apollonian–butterfly connection) will take on a new meaning when viewed from a topological perspective. Each quantum Hall state, characterized by topological quantum numbers, turns out to be an $\mathcal{IAG}$ in disguise, where topological numbers are lurking in the curvatures of the kissing circles. As is explained below, these fractal objects made up of integers encode the topology of the entire family of nested butterflies.

Although the energy axis makes no explicit appearance in this picture, the topological quantum numbers determine the sizes of the energy gaps, and therefore, in a subtle way, the Apollonian viewpoint encodes more than just the partitioning of the butterfly along the magnetic-flux axis. This chapter thus tells, in one fell swoop, the tale of two precisions: "the kiss precise" and "precise quantization".

As was stated way back in chapter 1, the word "gap" in this book refers to a *one-dimensional* vertical white line-segment between two neighboring bands at one specific flux-value. In contrast, to refer to a *two-dimensional* diagonal white area, the word "swath" is used instead. Each two-dimensional swath is the union of an infinite number of one-dimensional gaps belonging to all the values of ϕ lying in an interval. Thus, for example, the four very salient white wings that meet at the butterfly's center—the wings that give the butterfly its name—are swaths in the sense just defined; indeed, *all* swaths are wings of smaller butterflies found at various hierarchical levels inside the large butterfly.

What is remarkable is that every swath in the graph, no matter how large or small it might be, is naturally associated with a pair of topological integers. It is this chapter's purpose to explain this topological labeling of the butterfly graph at every scale, highlighting the topological nature of this self-similarity. Each point in the butterfly graph, whether it belongs to a black band or to a white swath, can be labeled by two integers, which are quantum numbers of topological origin. For a point in a swath, the first of these two integers is the quantum number associated with the Hall conductivity.

We will now give more details of the story by describing how to assign a "topological address" to each of the infinitely many swaths of the butterfly, each one representing a distinct quantum Hall state. Then we will revisit the relationship between the butterfly and Apollonian gaskets, and will show how the topological address of each quantum Hall state is encoded in the integer curvatures of the underlying Apollonian gasket. We will conclude by revealing the existence of

Apollonian gaskets with trefoil symmetry that describe the quantum Hall topology as the sizes of the butterflies with centers at $E = 0$ shrink to zero.

10.1 Diophantus gives us two numbers for each swath in the butterfly

One very general way of determining the topological label for a gap (i.e., the pair of Chern numbers) was sketched in the previous chapter. However, for the square-lattice model that was studied by Hofstadter, there is a very elegant and simple number-theoretical way to determine such labels; in fact, it turns out that the integers in question are the solutions to equations of a sort first studied by Diophantus of Alexandria around 250 AD, and today known as *Diophantine equations*. The Diophantine method yields a complete labeling of all swaths at all scales in the Hofstadter butterfly.

We will now give a Diophantine recipe to determine the Chern numbers attached to all the diagonal swaths of the butterfly. We begin very humbly, looking merely at the (one-dimensional, vertical) gaps in the spectrum of a particular rational flux-value $\phi = \frac{p}{q}$. At this flux-value, there will be q bands, and thus $q - 1$ gaps between them. As we slide up the graph at this fixed flux-value, moving from lowest to highest energy, we can assign to each of these gaps an integer r, running from 1 to $q - 1$. Now let us focus on a particular gap, meaning that we choose some particular value for r.

The next step in our recipe is to write down a very simple Diophantine equation that has the three integers p, q, and r as its coefficients:

$$\sigma p + \tau q = r. \tag{10.1}$$

In this equation, p, q, and r are constants given to us, while σ and τ are unknown integers (positive, negative, or zero) that we wish to solve for. These solutions, σ and τ, will constitute the label that will be assigned to the white swath, of which our chosen one-dimensional white gap is merely one of infinitely many vertical cross-sections. There is, by the way, a most remarkable fact here, which is that this label does not depend on which value of p/q we choose.

To make these ideas very concrete, let us take a maximally simple case. We will calculate the label for the swath that runs diagonally across the whole butterfly, from its lower left-hand corner to its upper right-hand corner. (If you look at figure 10.1, this is the swath labeled "[1, 0]", although identifying it through its label is cheating a little bit, since at this point we are trying to calculate those values. But no matter.) To follow the Diophantine recipe, we need to select some specific flux-value ϕ that has a gap making up part of this swath. That's easy—choose $\phi = 1/3$. Given that $q = 3$, this flux-value's spectrum has three bands and thus just two gaps, of which the lower one belongs to this swath, and that means that $r = 1$. Plugging these three values into equation (10.1) gives us the following Diophantine equation to solve:

$$1\sigma + 3\tau = 1.$$

Nothing could be simpler: $\sigma = 1$ and $\tau = 0$. So these two numbers—[1, 0]— constitute the topological label attached to this very large swath of the butterfly. We

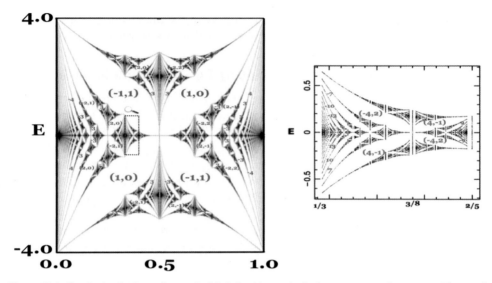

Figure 10.1. Swaths in the butterfly graph, labeled with topological quantum numbers [σ, τ]. The small butterfly on the right (whose φ-axis runs from 1/3 to 2/5) is a blowup of a little region inside the full butterfly just to its left. That region is delimited by a black rectangular box centered at φ = 3/8. In these two butterflies, pairs of integers comprising the topological addresses of swaths have been inserted in swaths that are large enough for the numbers to fit in. In smaller swaths, because of lack of room, we have inserted only the first number of the pair—namely, σ.

will of course want to check that this label, as claimed, doesn't depend on our having reached it through our choice of φ = 1/3. We want it to be *independent* of the flux-value that we chose.

So let us try a different flux-value in the right range—say, φ = 2/5. At this flux-value, there are five bands and thus four gaps, and of these, it's not the lowest this time, but the second-lowest, that lies in the swath in question. (You can see this by locating 2/5 on the φ-axis, then spotting the five bands, and thus the four gaps, located directly above it.) Given that $p = 2$, $q = 5$, and $r = 2$, the Diophantine equation we want to solve is this one:

$$2\sigma + 5\tau = 2.$$

Well, once again, solving this equation is trivial. We need merely pick $\sigma = 1$ and $\tau = 0$, exactly as before, and we see that the label we get this time—[1, 0]—is identical to the label that we calculated earlier.

Although we have shown only two sample cases and found that the labels they gave agreed, this agreement is not a coincidence. It illustrates a general theorem— namely, that a label found through the Diophantine recipe does not depend on the particular flux-value φ that was chosen. One just has to be careful to choose r so that the rth gap in φ's spectrum lies in the swath in question.

Let us take one more example, this time focusing on a smaller two-dimensional gap. The swath we just labeled was a "forward slash"; now we will label a "backward slash". In figure 10.1, our swath runs from the topmost point of the lowest band of $\phi = 1/3$ to the bottommost point of the lower band of $\phi = 1/2$. In the figure, its label is $[-2, 1]$, although once again, strictly speaking, we're not supposed to know those numbers yet, since our purpose is to calculate them.

To calculate this swath's topological label, we need to pick a particular ϕ in between 1/3 and 1/2, so let's once again use $\phi = 2/5$, which has five bands and four gaps between them. This time, however, it's the *lowest* band that belongs to the swath in question, so $r = 1$. Given that $p = 2$ and $q = 5$, our Diophantine equation becomes:

$$2\sigma + 5\tau = 1.$$

One solution to this equation is $\sigma = +3$ and $\tau = -1$, since $3 \times 2 - 1 \times 5 = 1$. Another solution is $\sigma = -2$ and $\tau = +1$, since $-2 \times 2 + 1 \times 5 = 1$. Yet another solution is $\sigma = -7$ and $\tau = +3$, since $-7 \times 2 + 3 \times 5 = 1$. And there are many more.

In fact, there are infinitely many solutions to any such Diophantine equation. Indeed, it is easy to see that if $[\sigma, \tau]$ is a solution of equation (10.1), then so is

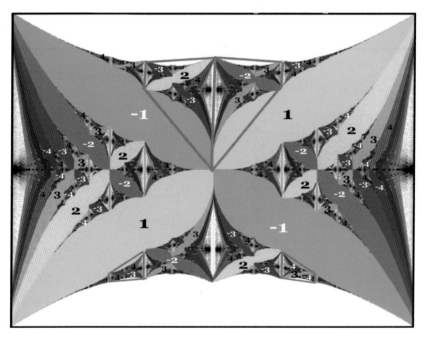

Figure 10.2. In this highly colorful graph, swaths having the same Chern number are shown in the same color. Here we can see that the topological quantum numbers labeling off-center butterflies (such as those framed by pink trapezoids) are determined by those that label the central butterflies, through continuity in ϕ, as is illustrated by the explicit labeling of the swaths labeled by Chern numbers ±2, ±3, and ±4. The fact that it is possible to derive the Chern numbers of all the off-centered butterflies from those that are centered at $E = 0$ highlights the importance of the central butterflies in the butterfly landscape.

$[\sigma + q, \tau - p]$, and so is $[\sigma - q, \tau + p]$. This observation leads us to a formula for the entire family of solutions, given any particular solution $[\sigma, \tau]$:

$$[\sigma + nq, \tau - np], \qquad n = 0, \pm 1, \pm 2, \ldots. \tag{10.2}$$

Which member of this infinite family do we wish to use to label the swath? It turns out that for the rectangular lattice, what we want is the smallest possible σ (in absolute value). In this case, that means $\sigma = -2$, which goes along with $\tau = +1$, and so our topological label for the swath is $[-2, 1]$. We won't bother calculating the label using a different value of ϕ, but the reader is encouraged to do so, in order to confirm that indeed, the swath's label is independent of the flux-value that is used in calculating it.

There are several independent proofs (for specifics, see [2]) of the fact that the minimal-size integer σ that (with an appropriate partner τ) solves the Diophantine equation $\sigma p + \tau q = r$ is the Chern number associated with the Hall conductivity for the rth gap in the spectrum of $\phi = \frac{p}{q}$. The Chern partner of σ—namely, τ—is also a topological number, but at this point in time, the physical significance of τ unfortunately remains obscure. In the colorful butterfly graph shown in figure 10.2, regions with the same Chern number are shown with the same color. Figure 10.3 summarizes some useful identities obeyed by the topological integers.

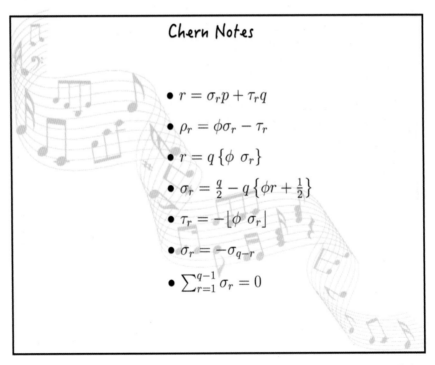

Figure 10.3. A few useful identities related to the topological integers [3]. Here, the notation "$\{x\}$" denotes the fractional part of the real number x, and the notation "$\lfloor x \rfloor$", often called the "floor" function, denotes the integer part of x—that is, the largest integer not greater than x.

10.1.1 Quantum labels for swaths when ϕ is irrational

Equation 10.1 can be rewritten as follows:

$$\sigma \frac{p}{q} + \tau = \frac{r}{q}. \tag{10.3}$$

This equation depends on ϕ being rational, since it makes reference to the numerator p and the denominator q, but one feels tempted to try to extend its meaning to irrational values of ϕ. The first step is of course to replace the expression "p/q" by the real variable ϕ:

$$\sigma\phi + \tau = r/q. \tag{10.4}$$

The remaining question is how to interpret the expression "r/q" in the case of an irrational value of ϕ—and luckily, there is a natural way to do that. The quantity r/q is essentially telling us how far upwards we have climbed in the one-dimensional spectrum belonging to flux-value ϕ; in other words, how many bands have been filled (the filled bands lie below us) and how many bands are empty (the empty bands lie above us). Thus if $r = 1$, we are at the very bottom, with no bands filled, and if $r = q - 1$, we are at the top, with all bands filled. Now since an irrational number is the limit of fractions whose numerator and denominator increase beyond limit, the quantity r/q will approach 0 when we are at the bottom of the spectrum of an irrational ϕ, and it will approach 1 when we are at the top. In short, independently of ϕ, the physical interpretation of 0 is "completely empty", while 1 means "completely full". It now seems very natural to replace the fraction r/q by a new continuous variable ρ—a "filling-factor"—which ranges from 0 to 1 and which tells us how far up we have climbed in the spectrum belonging to ϕ, independently of whether ϕ is rational or irrational. In terms of ρ, our Diophantine equation then becomes the following:

$$\sigma\phi + \tau = \rho. \tag{10.5}$$

Of course, when ϕ and ρ take on irrational values, this is no longer a Diophantine equation, but astonishingly enough, when ρ is picked so that we are at the height of a swath whose label we know from *rational* values of ϕ, this equation has a unique solution for integers σ and τ, and that solution will coincide with the known label for the swath. This fact once again underscores the topological robustness of the labels attached to all the swaths in the butterfly.

10.2 Chern labels not just for swaths but also for bands

Let us go back for a moment to rational values of ϕ, which are the only values whose spectra consist of *bands*, since the spectrum of any irrational ϕ consists of infinitely many isolated *points* making up a Cantor set. So let's assume that ϕ is rational. Now suppose that we would like to assign a quantum label to each of the q bands belonging to ϕ's spectrum, not just to its $q - 1$ gaps. It would seem very likely that such a label would be simply related to the labels of the gaps lying just above and below the band in question. Indeed, that is exactly the case, and the relationship is as simple and natural as one could hope for.

To help us express this relationship, let us attach subscripts to the integers σ and τ making up the labels of the $q-1$ gaps in ϕ's spectrum, counting from the bottom, as always. Thus the lowest gap would have label $[\sigma_1, \tau_1]$, the next-lowest gap would be $[\sigma_2, \tau_2]$, and so forth. We will now invent an analogous notation for bands. Let's write the label for the rth band in ϕ's spectrum (counting from the bottom up, just as for gaps) as follows: $[\sigma_r^b, \tau_r^b]$. With this notation, the label of the band is computed from the labels of its neighboring gaps as follows:

$$\sigma_r^b = \sigma_r - \sigma_{r-1}, \qquad (10.6)$$

$$\tau_r^b = \tau_r - \tau_{r-1}. \qquad (10.7)$$

The band's label is given by the differences between the σ's and the τ's labeling the gap just above and the gap just below.

It can be shown that when $\phi = \frac{p}{q}$, the left side of equation (10.6) can take on only two possible values: either $\sigma_r^b = -p$ or $\sigma_r^b = q - p$. In other words, the Hall conductances associated with the q bands belonging to ϕ form a sequence of integers oscillating between two values. These sequences turn out, in fact, to be what Douglas Hofstadter called η-sequences (see chapter 3).

10.3 A topological map of the butterfly

The topological trajectories lurking in the butterfly can be summarized in a graph of ϕ versus ρ, shown in figure 10.4. Today such graphs are known as *Claro–Wannier diagrams* [4], after Francisco Claro and Gregory Wannier, who revisited the problem of a crystal in a magnetic field shortly after the discovery of the butterfly. A Claro–Wannier diagram is a kind of "skeletal butterfly" that highlights the swaths using straight lines; for each straight line in it, its slope equals the swath's Chern number σ, and the x-intercept determines the swath's other Chern number, τ.

Below we list several important consequences of the relations summarized in what we have called our "Chern notes". In particular, we give the precise rules for filling in the topological map of the butterfly at all scales. In our presentation, we will label every small butterfly with a pair of Chern numbers, writing the pair as "$\langle(\sigma_+, \sigma_-)\rangle$", where the "+" and "−" subscripts symbolize the fact that there is always one positive and one negative Chern number, which together determine the two diagonal swaths of the given butterfly.

1. By consulting the Chern notes, readers can easily verify that any central butterfly whose center is located at a flux-value p_c/q_c whose denominator q_c is even is characterized by the Chern numbers $\langle(\frac{q_c}{2}, -\frac{q_c}{2})\rangle$.
2. According to the Chern notes, the Chern numbers of a hierarchical set of gaps that define the fine structure of the butterfly near any rational flux-value—say, $\phi_0 = p_0/q_0$—are obtained by "tilting" the flux-value and the ρ values a very small amount. Using the relations $\phi = \phi_0 + \delta\phi$ and $\rho = \rho_0 + \delta\rho$, and the

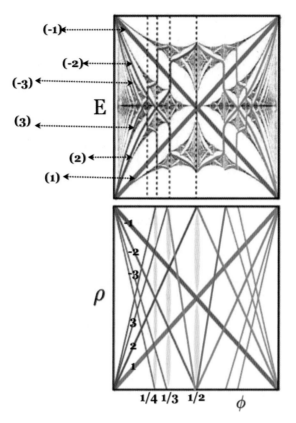

Figure 10.4. The lower graph, made of crisscrossing diagonal lines, is called a *Claro–Wannier diagram* for the butterfly. It is a pictorial representation of the Diophantine equation $\sigma\phi + \tau = \rho$ discussed above. Since this equation is a linear equation with parameters σ and τ, its graph is a straight line with slope σ. This particular Claro–Wannier diagram uses just the values 1, 2, and 3 for σ; these integers can be seen in the figure, labeling the slopes of the lines. A few of the intercepts with the x-axis (actually the ϕ-axis) are also labeled. The upper graph is the usual butterfly with some additional lines shown in color, which outline the swaths that have Chern numbers of 1, 2, and 3. The color-coding in the upper graph matches that used in the Claro–Wannier diagram below it, and this illustrates the deep correlation between the topological structure of the butterfly and the ideas captured in the Claro–Wannier diagram.

corresponding quantum numbers $\sigma = \sigma_0 + \Delta\sigma$ and $\tau = \tau_0 + \Delta\tau$, and then finally taking the limit as $\delta\phi$ and $\delta\rho$ go to zero, we obtain:

$$\phi_0 \Delta\sigma + \Delta\tau = 0; \qquad \frac{\Delta\sigma}{\Delta\tau} = -\frac{q_0}{p_0}. \qquad (10.8)$$

Since both $\Delta\sigma$ and $\Delta\tau$ are integers and p_0 and q_0 are relatively prime, the simplest solutions of equation (10.8) are:

$$\Delta\sigma = \pm n q_0, \qquad \Delta\tau = \mp n p_0, \qquad (10.9)$$

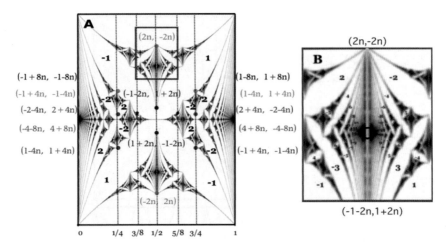

Figure 10.5. In panel (A), we see the Chern-number labeling of some of the major swaths in the butterfly, and their corresponding fine structure. This figure illustrates the fact that the higher-n solutions (that is, with $n > 1$) of the Diophantine equation for a given flux-value ϕ lie very close to that flux-value. Each color-coded dot lies at a rational flux-value (see the numbers at the bottom of the graph) and is located inside a swath whose left and right Chern values are displayed inside parentheses of the same color. Panel (B) is a blowup of the region inside the red box in (A). It shows some of the fine structure near $\phi = 1/2$.

where $n = 0, 1, 2, \ldots$. These solutions describe the fine structure of the butterfly near a flux-value ϕ_0. Those solutions that exist in the immediate vicinity of $\phi = p_0/q_0$ are the "finite n" solutions of equation (10.2), the Diophantine equation that we discussed earlier.

Figure 10.5 is a topological map of the butterfly, showing the distribution of Chern numbers. In addition to the dominant gaps, this plot shows the fine structure of central butterflies, both near their centers and at their edges.

In chapter 3, butterflies were divided into two categories: $\mathcal{K}$-type and C-type (the latter were also called "fountain butterflies"). Our topological labeling reveals that there is a quantitative distinction between these two types. Consider the $n > 0$ solutions of the Diophantine equation for a flux-value p/q. As was stated earlier, Harper's equation does not support these solutions, since only $n = 0$ solutions are realized by this system. Interestingly, these higher-n solutions are located at flux-values very close to $\phi = p/q$, as is shown in figure 10.5. This effect can be seen in that figure, near $\phi = 1/2$.

10.4 Apollonian–butterfly connection: Where are the Chern numbers?

In chapter 3, we showed how integral Apollonian gaskets ($\mathcal{IAG}$s) encode butterfly configurations in the Hofstadter landscape. We now broach the following two key questions:

1. *Do Chern numbers apply in some fashion to Apollonian gaskets? Do they describe some special geometric property of configurations of four kissing circles?*
2. *Given four kissing circles making up an $\mathcal{IAG}$, along with their integer curvatures, what are the Chern numbers of the corresponding butterfly?*

We would ideally like to find an answer to the first question along the lines of our discussion in chapter 8, where we saw how the Gauss–Bonnet theorem relates certain integer invariants computed geometrically from a manifold to the manifold's topological genus (see section 8.2). Unfortunately, however, how to extend the concept of Chern numbers so that it can apply to an $\mathcal{IAG}$ that encodes the butterfly is unknown at this time. We believe a satisfactory answer can be found, perhaps within the mathematical framework of conformal and Möbius transformations. Work along these lines is currently in progress.

Since we cannot answer the first question, we will focus on the second one, following the discussion in chapter 3, which relates an $\mathcal{IAG}$'s four curvature values to the flux-values that define the center and edges of the corresponding butterfly. (See section 3.2). One can then use results from section 3.3 to express Chern numbers in terms of curvatures, as we will describe below.

We recall that three mutually tangent circles with integer curvatures κ_1, κ_2, and κ_3 form an integral Apollonian gasket provided that $\kappa_0(\pm)$—the curvature of the outer bounding or inner circle, as defined below—is also an integer:

$$\kappa_0(\pm) = \kappa_1 + \kappa_2 + \kappa_3 \pm 2\sqrt{\Delta}, \tag{10.10}$$

where

$$\Delta = \kappa_1\kappa_2 + \kappa_2\kappa_3 + \kappa_1\kappa_3. \tag{10.11}$$

Clearly, the Apollonian gasket will have integer curvatures if and only if Δ is a perfect square. Let δ be its square root:

$$\delta = \sqrt{\Delta}. \tag{10.12}$$

As was pointed out in chapter 3, δ is the curvature of the "dual circle"—that is, the circle passing through the tangency points of the three inner circles. It turns out that Δ encodes the Chern numbers of the butterfly at least in the cases where the mathematical framework underlying $\mathcal{ABC}$ is well established.

As was shown in chapter 3, our "Holy Grail" of $\mathcal{ABC}$ is fully realized for central butterflies, since the hoped-for $\mathcal{IAG}$ is the dual of the triplet of Ford circles corresponding to the center and the two edges of the butterfly. In this case, it is an easy exercise to show that $\delta = \sqrt{\Delta}$ is a perfect square. We can then write down an explicit formula for the Chern number of the quantum Hall state in terms of the curvatures of the three inner circles of the corresponding $\mathcal{IAG}$. The Chern numbers for a butterfly centered at flux-value $\phi = \frac{p_c}{q_c}$ are:

$$\sigma_\pm = \pm\frac{\sqrt{\delta}}{2} = \pm\frac{1}{2}(\kappa_1\kappa_2 + \kappa_2\kappa_3 + \kappa_1\kappa_3)^{1/4} = \pm\frac{q_c}{2}. \tag{10.13}$$

As is shown in figure 10.3, central butterflies determine the topological quantum numbers of the entire butterfly landscape, since the topological numbers of off-centered butterflies can be determined by continuity in ϕ across the butterfly gaps. This raises an interesting question as to whether the Apollonian representations of the butterflies centered at $E = 0$ axis play any role in determining the Apollonian gaskets that represent the off-centered butterflies.

Figure 10.6. Chern-2 swaths in the butterfly and the IAGs that correspond to them. The red and blue zones show the K-type and C-type butterflies. The continuity of the swaths having Chern number 2 is reflected in the corresponding IAGs.

Figure 10.6 illustrates the ABC for the fountain butterflies. The relationship between the Chern numbers of these butterflies and the curvatures of the corresponding Apollonian gaskets is a topic still under investigation and will not be discussed here.

10.5 A topological landscape that has trefoil symmetry

We now discuss the ABC idea in a completely different context, where nature seems to want to hand us a big surprise. Suppose we zoom down into the full butterfly following the diamond hierarchy, looking at it on ever smaller scales. It turns out that in so doing, we will encounter a topological landscape possessing trefoil symmetry. This unexpected fact is intimately related to the hidden trefoil symmetry of the butterfly, encountered in chapters 2 and 3, describing the the ϕ-axis scaling of the butterflies constituting the diamond hierarchy.

Let us recall, from chapter 2, that a pair of triplets (p_L, p_c, p_R) and (q_L, q_c, q_R), which determine the coordinates $(\frac{p_L}{q_L}, \frac{p_c}{q_c}, \frac{p_R}{q_R})$ of a generic central butterfly, satisfies the following recursion relation, linking three different generations ($l-1$, l and $l+1$) of butterfly zooms:

$$s_x(l+1) = 4s_x(l) - s_x(l-1), \qquad (10.14)$$

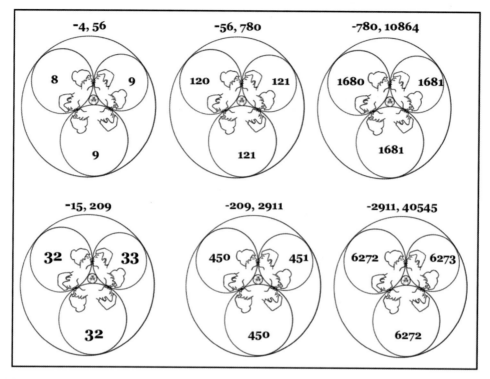

Figure 10.7. A set of Apollonian gaskets having "almost trefoil symmetry". These gaskets encode the sequence (4, 15, 56, 209, 780, 2911, ...), representing Chern numbers associated with the magnetic-flux interval [1/3 → 2/5]. The two numbers at the top of each circle show the curvatures of the outermost and the innermost circles, which are each other's "mirror images" (via circular inversion in a circle that passes through the tangency points of the trio of nearly identical circles).

where $s = p$, q and $x = L$, c, R. In view of equation (10.13), the Chern numbers of the butterflies satisfy the same recursions given by equation (10.14):

$$\sigma(l+1) = 4\sigma(l) - \sigma(l-1). \tag{10.15}$$

Equation (10.15) results in a fixed-point solution of the ratio of integers at two successive levels:

$$\frac{\sigma(l+1)}{\sigma(l)} = R_\sigma(l) \rightarrow 2 + \sqrt{3} = \sqrt{R_\phi}. \tag{10.16}$$

It is easy to show that the topological integer τ has the same scaling behavior as does σ. Therefore, the topological landscape, where the nested quantum Hall states are kaleidoscopic images, exhibits trefoil symmetry. In some special cases, the topological quantum numbers associated with butterflies in the diamond hierarchy are given explicitly by the curvatures of $\mathcal{IAG}$s that asymptotically approach perfect trefoil symmetry, as in figure 10.7. The quest for a deeper understanding of the relationship between (1) the $\mathcal{IAG}$s that encode topology and (2) the dual and symmetric-dual Apollonian gaskets that represent the butterfly is, however, still only in its infancy.

Finally, we note that although it would be premature to speculate on the importance of the hidden trefoil symmetry and the associated topological scaling in the butterfly or the quantum Hall states, results from theoretical physics very often connect to the real world with scaling relations that tend to be "universal". In other words, the validity of the topological scaling may extend beyond the basic model used by Hofstadter to study crystal electrons in a magnetic field. Just as a simple quadratic map can predict the behavior of a dripping faucet (see chapter 1), so topological scaling, emerging from the smallest energy intervals and flux intervals in the butterfly fractal, may have implications that extend beyond the confines of this book. This is an exciting prospect, since the topological integers in this case are the quantum numbers of Hall conductivity states, and hence are observable in laboratory experiments.

10.6 Chern-dressed wave functions

So far, our discussion of topological aspects has been confined to the butterfly spectrum, which depicts the allowed *energies* of an electron in a lattice immersed in a magnetic field. We now wish to point out that the electron's *wave functions* also have topological aspects [5].

As was shown in chapter 6, the wave function for an electron in such a system exhibits the same type of self-similar structure as does the plot of eigen-energies. However, in sharp contrast to bulk matter (which has no edges), the energy gaps in a finite sample (which by definition has edges) contain certain exceptional states that exhibit electrical conductivity. These conducting states are localized at the edges of the sample, and are known as *edge modes*. In fact, the number of edge modes equals the Chern number associated with that gap, and also determines the degree of splitting of the peak in the wave functions for band-edge states. It is curious, however, that the phenomenon of the splitting of the peak arises even in an infinite crystal that has no edges at all. Such edge-modes, residing in the interior of the sample, encode topological quantum numbers associated with Hall conductivity. This "Chern-dressing" of the peaks is shown in figure 10.8.

10.7 Summary and outlook

The fact that quantum Hall states—exotic topological states of matter—are "reincarnations" of integer Apollonian gaskets is a fascinating result. How unexpected that a beautiful and abstract piece of mathematics from well over 2000 years ago would turn up in the midst of "dirty" two-dimensional insulators! Yes, $\mathcal{ABC}$ takes on a special meaning when discussed in the context of topological states of quantum Hall systems. It's fair to say that when we observe the quantum Hall effect in a laboratory, what we are seeing is, in some sense, a reincarnation of an Apollonian gasket from way back in 300 BC!

Although the task of establishing a rigorous mathematical framework relating the butterfly and $\mathcal{IAG}$s is an open problem, all the figures and other types of evidence presented above leave little doubt about the validity of this connection.

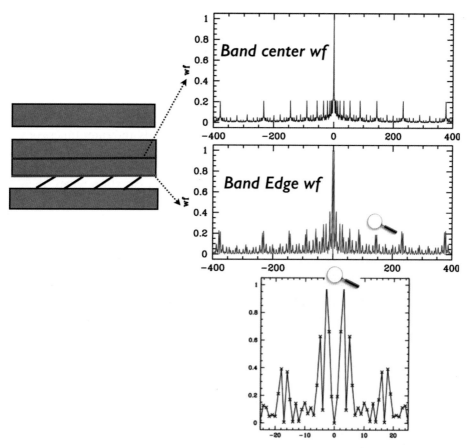

Figure 10.8. In this figure, we are looking at the (theoretically predicted) wave functions for the case where the flux-value ϕ equals the golden mean. The left side of the figure shows three subbands, highlighting a gap that hosts four edge modes, and that therefore represents a quantum Hall state with quantum number 4. On the right side are shown the wave functions belonging to the band's center (blue) and the band's edge (red). Every peak of the band-center's wave function splits into two peaks for the band-edge's wave function. This makes a doublet of size four, encoding the quantum number associated with the quantum Hall state. The figure at the bottom is a blowup that illustrates the doublet structure of the band-edge states.

Circle-packing can be viewed as the art of placing tangent circles on the plane, leaving as little unoccupied space as possible. It is a very attractive field of mathematics, and it is deeply related to the beautiful and rich geometric transformation known as "inversion in a circle". Our discussion above shows that the close-packing of circles holds the secret of precise quantization in the quantum Hall effect.

The butterfly, with its underlying mixture of complexity and order, remains in many ways a profound enigma. To be sure, all fractals, such as the Mandelbrot set, are mysterious. But what is particularly fascinating about the butterfly is how both fractality, which is rooted in two competing periodicities, and topology, which is quintessentially quantum in nature, are interwoven in it. Furthermore, each point in the empty swaths of the butterfly graph describes, to a close approximation, a macroscopic quantum effect of astonishing precision, which is immune to all the standard confounding factors in condensed-matter systems, such as impurities and interactions.

The butterfly graph as a whole describes all possible phases of a two-dimensional electron gas that arise as one varies the filling-factor ρ and the magnetic field ϕ. With a remarkable mix of fragmentation of bands and smoothness of gaps, all of these infinitely many phases, each one characterized by an integer, not only coexist, but form a fractal made entirely of integers. The order and the complexity of the butterfly show how nature reacts to a quantum situation where there are two competing periodicities. Experimentalists believe that the study of the butterfly offers the possibility of discovering materials with novel exotic properties that are beyond our present imagination. Who knows how many more mysteries and hidden treasures are yet to be discovered in the butterfly, and in related landscapes?

Quantum Hall states are the simplest examples of topological insulators that provide a topological classification of states of matter. It is conceivable that there may emerge a corresponding classification of butterfly-type structures, based on symmetry and topology, as one considers graphs analogous to the butterfly that arise in other topological insulators.

The fact that integer quantum Hall states are reincarnations of integral Apollonian gaskets opens new doors for understanding other topological states in terms of Apollonian gaskets. In fact, there are examples of "rational" Apollonian gaskets, where the rational curvatures of the first four mutually tangent circles give all circles with rational curvatures... This might be a potential route to the understanding of fractional quantum Hall states, but as of yet it is unexplored.

As we think of the butterfly and its relation to the quantum Hall effect, it is important to remember how the underlying theme of the quantum Hall effect has cropped up in a wide variety of seemingly unrelated problems in physics; thus, there are publications with such titles as "Black holes and quantum Hall effects", "Quantum Hall quarks", "Quantum computation in quantum Hall systems", and "Higher-dimensional quantum Hall effect in stringtheory". This suggests that there exists a much broader perspective from which to view the butterfly fractal, home of the quantum Hall effect, and host of all possible quantum Hall states of non-interacting fermions.

We conclude our discussion with a another piece of poetic mathematics, by Thorold Gosset[1] with a wishful speculation that nature has perhaps found ways to use the following mathematics in some physically interesting ways.

---

[1] Thorold Gosset sent a copy of the poem to Donald Coxeter (whose area of expertise was higher dimensions) on the occasion of his wedding, in the Round Church in Cambridge. It was be published in *Nature*, January 9, 1937.

The Kiss Precise (Generalized)
by Thorold Gosset

And let us not confine our cares
To simple circles, planes and spheres,
But rise to hyper flats and bends
Where kissing multiply appears.
In n-ic space the kissing pairs
Are hyperspheres, and Truth declares—
As n + 2 such osculate
Each with an n + 1 fold mate.
The square of the sum of all the bends
Is n times the sum of their squares.

References

[1] Claro F H and Wannier G H 1979 *Phys. Rev.* B **19** 6068
[2] Dana I, Avron Y and Zak J 1985 *J. Phys. C: Solid State Phys.* **18** L679
[3] Sajita I I 2014 *Topology and self-similarity of Hofstadter butterfly* arXiv:1408.1006 [cond-mat.dis-nn] (unpublished)
[4] Wannier G H 1978 A result not dependent on rationality for Bloch electrons in a magnetic field *Phys. Status Solidi* **88** 757
[5] Satija I and Naumis G 2013 *Phys. Rev.* B **88** 054204

Part IV

Catching the butterfly

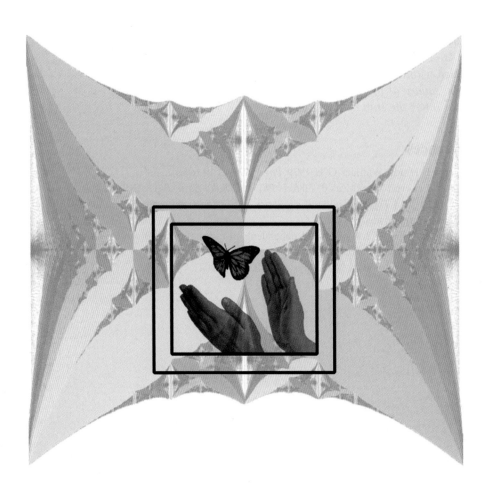

IOP Concise Physics

Butterfly in the Quantum World
The story of the most fascinating quantum fractal
Indubala I Satija

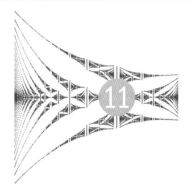

Chapter 11

The art of tinkering

Receiving emails is like receiving artificial flowers. Send me real ones!
—Indradev[1]

This short chapter is devoted to the glory of physics laboratories, where theoretical ideas are tested and new phenomena are uncovered.

Experimentation plays many roles in science. One of its key roles is to test proposed theories. Another is to reveal the need for new theories, either by showing that an accepted theory is incorrect or by exhibiting an unexpected new phenomenon that is in need of explanation. The quantum Hall effect is an example of the latter.

Beyond the beautiful mathematics and the abstract concepts, the ultimate fate of any theory rests on what transpires when concrete experiments are carried out.

---

[1] The author's father. To understand the essence of this quote, one needs only to compare the tingling that one feels upon looking at the sentence that Dirac himself wrote out on a Moscow blackboard (see section 8.5) with the more mundane feeling one has when reading the same sentence simply typed out on a page.

Theory and experiment go hand in hand, and neither can survive without the other. If theoreticians are composers, then experimentalists are performers—the people who bring the music to life. And experimental science is every bit as much about the art of tinkering as it is about the pursuit of scientific knowledge.

The most basic rule of science is that if you have a theory about some aspect of how the world works, you must absolutely test it through experiments. If the experiments confirm your theory, that is terrific news; if not, then you must revise your theory, and further experimentation must be carried out. As Richard Feynman once put it:

> *It doesn't matter how beautiful your theory is;*
> *it doesn't matter how smart you are.*
> *If it doesn't agree with experiment, it's wrong.*

On the surface, Nature seems to behave in an extremely complicated way. And yet, very often, underlying this complexity are found simple and beautiful patterns, when the phenomena are described in the language of mathematics. For example, we do not know why $F = ma$, yet we still have full faith in this marvelous equation, because so many observations are consistent with it. Physics, like natural science in general, is a reasonable enterprise based on carefully acquired experimental evidence, constructive criticism, rational discussion, and a sense of aesthetics. It provides us with knowledge of the physical world, and it is experiments that provide the evidence that underpins this knowledge.

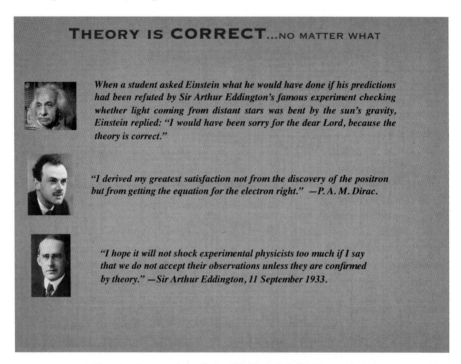

(Einstein photograph credit: Bangkokhappiness/Shutterstock.com)

It is important to know that revolutionary theories like Einstein's theory of general relativity, de Broglie's theory of wave–particle duality, and Dirac's prediction of antiparticles won approval—and in some cases, Nobel prizes—only after their predictions were confirmed by experiment. Of course, some of the greatest of all physicists have believed in their theories no matter what, simply because they were convinced that their theories possessed an inevitable kind of mathematical beauty. These scientists, having a deep sense of faith in the profound elegance of nature, were persuaded that nature ought to obey the simple, symmetrical, mathematical rules that they had come up with [2].

11.1 The most beautiful physics experiments

The history of science is, in many ways, a history of experimentation. Major milestones in how we understand the world have been marked by experiments so ingenious, so simple, and so Earth-shaking that they can take your breath away.

Robert P Crease, a member of the Philosophy Department at the State University of New York at Stony Brook and the official historian at Brookhaven National Laboratory, recently asked physicists to nominate the most beautiful experiment of all time. The ten most frequently named experiments have some features in common, including simplicity and beauty. Most of the experiments took place on tabletops, and none required more computational power than that of a slide rule or calculator. The following list, taken from *Physics World*, was ranked according to popularity, the top prize going to an experiment that vividly demonstrated the quantum nature of the physical world [1]. The winners provide a bird's-eye view of more than 2000 years of discovery.

- Young's double-slit experiment applied to the interference of single electrons.
- Galileo's experiment on falling bodies (1600s).
- Millikan's oil-drop experiment (1910s).
- Newton's decomposition of sunlight with a prism (1665–6).
- Young's light-interference experiment (1801).
- Cavendish's torsion-bar experiment (1798).
- Eratosthenes' measurement of the Earth's circumference (3rd century BC).
- Galileo's experiments with balls rolling down inclined planes (1600s).
- Rutherford's discovery of the nucleus (1911).
- Foucault's pendulum (1851).

Other experiments that were nominated by physicists included:
- Archimedes' experiment on hydrostatics.
- Roemer's observations of the speed of light.
- Joule's paddle-wheel heat experiments.
- Reynolds' pipe-flow experiment.
- Mach and Salcher's acoustic shock wave.
- Michelson and Morley's measurement of the null effect of the ether.
- Röntgen's detection of Maxwell's displacement current.
- Oersted's discovery of electromagnetism.

- The Braggs' x-ray diffraction of salt crystals.
- Eddington's measurement of the bending of starlight.
- Stern and Gerlach's demonstration of space quantization.
- Schrödinger's cat thought-experiment.
- Trinity test of a nuclear chain reaction.
- Wu *et al*'s measurement of parity violation.
- Goldhaber's study of neutrino helicity.
- Feynman's dipping of an O-ring into a glass of ice water.

This last example showed that the post-launch disintegration of the space shuttle Challenger on January 28, 1986 was due to the fact that the primary O-ring was not properly sealed in unusually cold weather at Cape Canaveral. Its inclusion in the above list is particularly gratifying to those who are deeply frustrated by bureaucracy. While NASA officials were explaining in detail why it would have cost billions of dollars to measure the rate at which frozen O-rings would recover their shape (and therefore why NASA was not negligent in the death of the seven astronauts aboard the Challenger that fateful day), Richard Feynman, listening in his capacity as a presidential committee member, clamped a section of O-ring with a clamp he had bought for $1.43 the day before at a local hardware store, and dunked it in a glass of ice water that had been provided to him as a committee member. At the height of the bureaucratic doubletalk, Feynman pressed his button to ask a question. With the TV cameras focused on him, he removed the O-ring from the water and released the clamp. The world watched while the O-ring regained its original form a thousand times too slowly to guarantee a safe shuttle launch.

Perhaps each one of us has our own personal favorite. For some, it could be the discovery of x-rays by Wilhelm Röntgen (for which the first Nobel Prize in Physics was awarded, in 1901), simply because of the enormous importance of x-rays from then on; for others, it could be the Casimir effect (conjectured in 1948 and experimentally confirmed in 1996), which reveals a mind-boggling roiling and boiling taking place ceaselessly and ubiquitously in the supposedly "empty" vacuum. And then for others, it could be the discovery of the Higgs boson, whose yet-to-be-measured properties may encode the next level of secrets of this universe.

As will be shown in the next, and concluding, chapter of this book, the challenge of detecting traces of the Hofstadter butterfly in a physics laboratory has been a long journey, filled with many creative ideas and thoughts. Many dedicated scientists are pursuing this journey today, but despite a number of encouraging first steps, they are still far from arriving at the hoped-for destination.

References

[1] http://www.explainthatstuff.com/great-physics-experiments.html
[2] Dirac P A M 1939 The relation between mathematics and physics Published in *Proc. R. Soc. Edinburgh* **59** 122–9 (lecture delivered upon presentation of the James Scott Prize, February 6, 1939)

IOP Concise Physics

Butterfly in the Quantum World
The story of the most fascinating quantum fractal
Indubala I Satija

Chapter 12

The butterfly in the laboratory

Measure what is measurable, and make measurable what is not so.
—Galileo Galilei

An experiment is a question that science poses to Nature, and a measurement is the recording of Nature's answer.
—Max Planck

Ever since it was first revealed in print, some 40 years ago, the eye-catching butterfly plot has tantalized experimental physicists. They would deeply love to observe its fractality in their laboratories, but finding traces of it has proved remarkably difficult. Some people believe that laboratory observations of the fractal nature of the butterfly could well pave the way for new kinds of materials with exotic properties whose potential is yet to be imagined. But others are skeptical. Does nature truly behave in the curious way that Douglas Hofstadter's "Gplot" predicts? Why should anyone believe in the highly counterintuitive properties of a bizarre-looking graph that was calculated using an extremely stripped-down model of a crystal?

Among the drastically idealized assumptions underlying the model that gave rise to the fractal butterfly are the following: (1) each electron in the crystal is very tightly bound to its nucleus; (2) the electrons in the crystal move in only two dimensions; (3) the crystal electrons, despite their electrical charges, do not interact with each other; (4) there is just one Bloch band in the crystal; (5) the sole Bloch band's defining equation is maximally simple—and there are numerous other assumptions as well. And so one might well wonder: do the weird mathematical properties of the butterfly, all flowing out of this unrealistically simple model, have any chance of ever being seen in a real laboratory setup?

Hunting the butterfly: a new frontier in exploring properties of matter

As was discussed in earlier chapters, the achievement of a theoretical understanding of the quantum behavior of an electron in a crystal lattice under the influence of a magnetic field was an important milestone in condensed-matter physics, and it took almost 40 years of research. The collective effort involved some of the pioneers of 20th-century physics, including Lev Landau, Rudolf Peierls, Lars Onsager, Gregory Wannier, and Mark Azbel'. However, the full richness of this quantum system started to come into view only after the contributions made by Douglas Hofstadter in 1976.

The butterfly is a pictorial representation of the fractal energy spectrum of a two-dimensional crystal immersed in a perpendicular magnetic field B. It can be seen as the inevitable outcome of a "competition" between two characteristic areas that define the physical situation: (1) the area intercepting one natural flux quantum $\Phi_0 = \frac{h}{e}$, and (2) the area of one unit cell of the given crystal (a^2). (As was pointed out in chapter 7, the situation can also be described in terms of two competing temporal periodicities, or two competing amounts of magnetic flux.) This competition gives rise to a spectrum having many properties that are unprecedented in any area of physics, and this is of course most provocative.

Unfortunately, however, the experimental conditions for the salient features of the Hofstadter butterfly to emerge clearly from hiding are quite stringent. The central problem is that for those telltale traces to be detectable, the magnetic flux passing through a unit cell of the crystal has to be on the order of one flux quantum, which for a normal crystal would require an enormously intense magnetic field. More quantitatively, the condition is this:

$$\phi = \frac{\Phi}{\Phi_0} = \frac{Ba^2}{(h/e)} \approx 1, \tag{12.1}$$

which, in the case of typical semiconductor lattices, amounts to magnetic fields B of strength greater than 60 000 tesla (see figure 12.1). Such intense magnetic fields were unimaginable when Hofstadter did his work, and even with today's technology they remain far out of range.

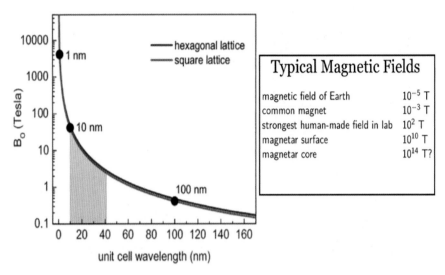

Figure 12.1. This graph shows, as a function of the size of a lattice's unit cell, the strength of the magnetic field that would be needed for the dimensionless flux-value ϕ to be on the order of 1 (where fractal effects become clearly visible). If we are dealing with normal crystals with typical lattice spacings—about 1 nanometer—then the technical challenge of creating a sufficiently strong magnetic field is huge. Moreover, for there to be any hope of the setup yielding observable traces of fractality, the field strength must also be controlled very carefully and measured with great accuracy. The three labeled dots on the curve in the graph correspond to, respectively: (1) the unit cell of a typical crystal; (2) the unit cell in the graphene experiment described in the text; and (3) the unit cell of an artificial crystal lattice, as will be discussed below.

When Hofstadter published his *Physical Review* article in 1976 [1], there was no chance of actually observing the fractal properties that he described in it. Nonetheless, in section 10 of that paper, Hofstadter very briefly commented on the possibility of experimentally observing traces of the butterfly, as is shown below:

> **X. POSSIBLE EXPERIMENTAL TEST**
>
> Finally, I would like to comment on the possibility of looking for the features predicted by this model experimentally. At first glance, the idea seems totally out of the range of possibility, since a value of $\alpha = 1$ in a crystal with the rather generous lattice spacing of $a = 2$ Å demands a magnetic field of roughly 10^9 G. It has been suggested, however (by Lowndes among others), that one could manufacture a synthetic two-dimensional lattice of considerably greater spacing than that which characterizes real crystals. The technique involves applying an electric field across a field-effect transistor (without leads). The effect of

Hofstadter (1976)

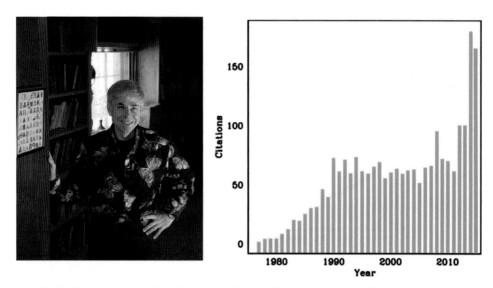

Figure 12.2. Citations by year of Hofstadter's 1976 *Physical Review* B paper [1] (from Google Scholar). The recent upward surge is due primarily to the experimental confirmation of some aspects of the butterfly graph. The photograph of Douglas Hofstadter wearing a butterfly shirt is reprinted here with his permission.

We live in a special time in the history of science, when during a single professional lifetime, experimental techniques can move from being "absolutely impossible" to being "routinely realizable". This turns out to be the case for the strange spectra of crystal electrons in magnetic fields, and it is reflected in the recent upsurge in the number of citations of Hofstadter's original paper, as one can see in figure 12.2. The reason behind the upsurge is that ingenious new experimental techniques have allowed the fractal behavior of real physical systems to begin to emerge from the woodwork. This is a most exciting thing to see happen!

For years, experimentalists dreamed up innovative ways to try to catch glimpses of the elusive butterfly, which came out of its chrysalis (Douglas Hofstadter's Regensburg notebooks) in the mid-1970s. However, it took almost 40 years before actual traces of a butterfly were found in the quantum world. (Coincidentally, it took almost the same number of years, starting with Lev Landau's 1933 paper, to work out the essential theory of the quantum behavior of Bloch electrons in a magnetic field!) To conjure up the Hofstadter butterfly in a laboratory requires sophisticated tinkering with pieces of experimental apparatus—arguably a high form of art—combined with deep originality, patience, and perseverance. For that reason, this achievement is justly considered to be a huge leap in materials-science research, a leap that will impact both the fundamental and technological frontiers of science.

Below we will give a brief overview of some of most striking recent advances in nanotechnology, and how they have led to laboratory confirmation of one of the most intriguing theoretical predictions in condensed-matter physics. Readers will get a taste of what went on behind the scenes in order to realize the extremely challenging laboratory conditions that allowed Hofstadter's Gplot fractal graph to emerge, at least a bit, from hiding. We note that in addition to the various approximations alluded to above, all the experiments described below were carried out at non-zero temperatures, and therefore it may appear almost magical that real-world laboratory setups have now fairly accurately reproduced strange phenomena formerly belonging only to the highly theoretical and highly idealized world explored by Hofstadter.

The experimental efforts to catch the butterfly (or if not to catch it, then at least to sight it) can be grouped into two distinct categories, and stem from two different stages of technological development:
(1) ones that use artificial crystals, with a cell size of approximately 100 nanometers and a magnetic-field strength B of roughly 1 tesla (beginning around 1990);
(2) ones that use moiré patterns made with graphene, with a cell size of approximately 10 nanometers and a magnetic-field strength B of roughly 50 tesla (beginning around 2013).

It is important to point out that laboratory setups do not *directly* measure the allowed and forbidden electron energies that make up the butterfly graph. Experimental verification of the butterfly involves measuring other quantities, and it is only from them that certain aspects of the butterfly graph can be indirectly inferred.

What do the laboratory experiments actually measure?

The experiments to be described below measure the longitudinal resistance R_{xx} and the transverse (or Hall) resistance R_{xy}, as discussed in chapter 7. The theoretical variables of interest are the reciprocals of these measured quantities—specifically, the longitudinal conductivity $\sigma_{xx} = 1/R_{xx}$ and the transverse conductivity $\sigma_{xy} = 1/R_{xy}$. To make a comparison with theoretical predictions, what one wants to look at is regions where σ_{xx} equals zero and where σ_{xy} does *not* equal zero. Such regions constitute plateaus in the quantum Hall regime that correspond, in the butterfly graph, to gaps. In short, making a plot of the experimentally measured longitudinal and transverse conductivities as a function of magnetic field strength ϕ reveals gaps, which hopefully should agree with the gaps predicted by the butterfly. The process of constructing a *theoretical* plot of σ_{xy} versus ϕ, given the butterfly graph, is illustrated in figure 12.3. (This kind of calculational process can also be run backwards, meaning that an experimentally measured spectrum can be constructed from an *experimental* plot of of σ_{xy} versus ϕ.)

In addition, experiments can also measure ρ, the density of charge carriers. The filling-factor ρ, measured as a function of ϕ, gives the Chern number σ, the quantum number associated with Hall conductivity. In particular, σ is the slope of the various ρ-versus-ϕ lines (recall the Diophantine equation $\rho = \sigma\phi + \tau$, discussed in chapter 10). A theoretical set of graphs of ρ versus ϕ, known as a Claro–Wannier diagram, is displayed on the left side of figure 12.4, for a handful of values of σ. Such a diagram is a kind of skeletal butterfly graph. In it, curves of constant σ are straight lines having integer slopes (see chapter 10, figures 10.3 and 10.5). As we will see below, experiments revealing traces of such a Claro–Wannier diagram provide some of the most convincing arguments for the detection, in the laboratory, of the butterfly fractal.

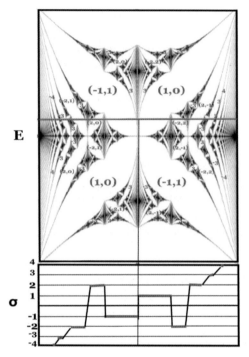

Figure 12.3. This graph illustrates the construction of theoretical Hall plateaus from the butterfly graph, as a function of the magnetic field strength ϕ, while keeping the Fermi energy constant. These theoretical Hall plateaus can then be compared to the actual ones measured in the laboratory. Or it can be done the other way around—that is, the experimental measurements of the Hall conductivity can be used to construct an empirical energy-spectrum graph, and that graph can then be compared to the butterfly graph. We note that Hall conductivity is a nonmonotonic function of ϕ and also changes sign—two telltale features of the butterfly graph that are absent in a quantum Hall system without an underlying lattice.

In the real world, where various idealizations used in the derivation of the Hofstadter spectrum may hold only partially, the following two features can be taken as quite strong indications of the presence of the butterfly:

- The Hall conductance can vary nonmonotonically and can even fluctuate in sign as the magnetic flux is varied. This is illustrated in figure 12.3.
- The Hall conductance plateaus remain quantized in integral multiples of e^2/h. The experimentally observed quantized levels coincide with the theoretically predicted slopes of the gap trajectories in the ρ-versus-ϕ Claro–Wannier diagram, in accordance with the Diophantine equation $\rho = \sigma\phi + \tau$ described above.

Not just tinkering: numerical simulations of experimental systems

It is important to note that experimental studies are often accompanied by numerical simulation of the experimental system—a theoretical kind of modeling that helps in understanding the system and in guiding the interpretation of what is observed in the laboratory. In other words, *computational* experiments and *laboratory* experiments go hand in hand, as one tries to extract information from any type of experimental data.

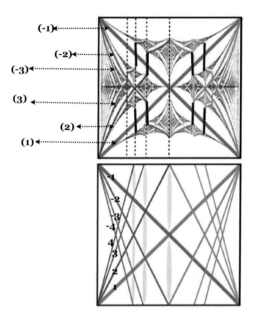

Figure 12.4. This figure shows a Claro–Wannier diagram—a "skeletal butterfly"—where the energy gaps of the Hofstadter spectrum are simplified down to linear trajectories parametrized as $\rho = \sigma\phi + \tau$. In the actual butterfly diagram, the linear trajectories become discontinuous, as is shown by the black vertical lines in between the colored lines. The integers shown in the graph are Chern numbers—that is, the quantum numbers associated with Hall conductivity. As is explained in the text, the Claro–Wannier diagram is crucial for interpreting the experimental results.

In order to observe the butterfly spectrum experimentally, the single-band picture described in earlier chapters is not always suitable. In the fervent chase after the Hofstadter butterfly, some experiments are done in a low-magnetic-field regime, where the spectrum is dominated by Landau levels that gradually widen into bands as the magnetic field is increased. Figure 12.5 displays the results of a numerical simulation of such a situation. In it, one sees the so-called "Landau fan" regime exhibiting discrete Landau levels characterized by the single quantum number σ. Here, the graph of electron density versus flux-value ϕ (actually, the reciprocal of ϕ, in this case) passes through the origin and corresponds to $\tau = 0$ in a Claro–Wannier diagram. In the high-magnetic-field regime, the Landau-fan structure is preserved except that the straight lines are now characterized by *two integers* (σ, τ), which is consistent with the Claro–Wannier diagram. In this case, computational simulations reveal many butterflies, with each Landau level mutating into a butterfly as lattice effects come into play.

12.1 Two-dimensional electron gases, superlattices, and the butterfly revealed

In order to try to observe the Hofstadter butterfly in a laboratory, one needs to have a two-dimensional electron gas in a crystal lattice subjected to a perpendicular magnetic field. Moreover, the dimensions of the lattice's unit cell and the strength of the producible magnetic fields should be such that the magnetic flux intercepted by a

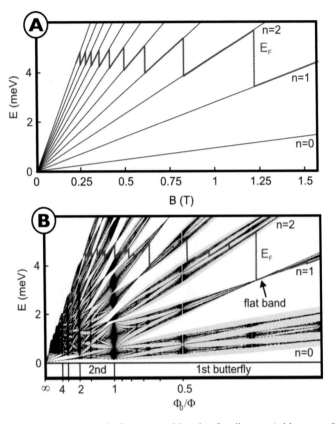

Figure 12.5. Panel (A) shows the theoretically computed Landau-fan diagram (with energy E proportional to magnetic field B, as discussed in chapter 7), often referred to as the conventional quantum Hall effect—that is, the quantum Hall effect without a crystal lattice. The zigzagging red line shows the Fermi energy for a fixed electron density as a function of the magnetic-field strength. Panel (B) shows how the discrete Landau levels are modified by the introduction of a two-dimensional periodic potential, neglecting the coupling between the Landau levels, where each Landau level splits into a band. (Adapted from [2] reproduced with permission.)

unit cell of the lattice (as measured in units of the magnetic flux quantum hc/e) should be on the order of unity, as is shown in figure 12.2.

To capture the fractal aspects of the butterfly experimentally, one needs to be able to detect not just the largest gaps, but also minigaps inside the spectrum. It is therefore essential to have the possibility of tuning (1) the charge carrier density, (2) the Fermi energy, and most importantly (3) the size of the lattice and the strength of the magnetic field, so that the magnetic length, the lattice spacing, and the gap sizes are all comparable. Finally, the purity of the sample is important, and experiments have to be done at low temperature. The discussion below describes how some of these challenges are met in laboratories.

Experimental realization of a two-dimensional electron gas

Two-dimensional electron gases—key players in the hunt for the butterfly—were first realized experimentally around 1960 [3]. It was shown that electrons can be made to accumulate on the surface of a semiconducting silicon crystal by applying a voltage to the crystal. In earlier studies, electrons had been confined between silicon (a semiconductor) and silicon oxide (an insulator). Later studies used GaAs and AlGaAs, both semiconductors, the combination being commonly known as a "GaAs–AlGaAs hetero-junction".

Artificial crystals

Experimentalists have, in their chase for the butterfly, devised setups with a two-dimensional superlattice located directly above a two-dimensional electron gas. This imposes a superstructure having a spatial period that is far larger than that of a typical crystal, thus resembling the idea proposed by Hofstadter—namely, that of making a crystal with such a large unit cell that one sidesteps the need to create a magnetic field of some wildly inconceivable strength. The period of the lattice and the amplitude of the periodic modulation are *adjustable parameters*, which allow the experimentalist to manipulate the widths of minibands and minigaps, in order to try to observe telltale traces of fractal aspects of the butterfly.

The minigaps in the fractal energy spectrum become observable only if the *magnetic length* $l_B = \sqrt{h/eB}$, which characterizes cyclotron motion, is of the same order as the wavelength of the periodic potential, which characterizes the Bloch waves. Therefore, the ability to tune lattice parameters in order to resolve tiny gaps in the butterfly is essential if one is striving to reveal fractal aspects of the spectrum.

When one is using an ordinary crystal lattice, where the interatomic spacing is a few ångströms, attaining sufficiently large magnetic fields is impossible, as the field strengths would have to be in excess of 10 000 tesla. The main experimental effort therefore has been to lithographically define artificial superlattices with unit-cell dimensions on the order of tens of nanometers; the idea is that with such large unit cells, magnetic fields achievable in the lab can yield a magnetic flux ϕ that is sufficiently large to "net" the butterfly.

One way to create an artificial two-dimensional periodic potential that mimics a two-dimensional crystal with a huge unit cell is to manufacture a metallic layer perforated by periodically spaced tiny holes whose diameter is approximately half the lattice period a. Figure 12.6 shows a schematic diagram of such an artificial crystal lattice placed above a two-dimensional electron gas confined between two layers of semiconductors. Without going into detail, we will merely note that there are quite a few ways to create an artificial crystal in which all spatial degrees of freedom are modulated in a periodic fashion, just as in a real crystal, but with the added benefit that experimenters have full control over the periodicity and geometry of this crystal. We refer readers to Martin Geisler's doctoral thesis [2] for various references and details.

Some of the early pieces of research that exploited this type of methodology to try to demonstrate the butterfly spectrum are:
1. "Magnetoresistance oscillations in a grid potential: indication of a Hofstadter-type energy spectrum" (1991) [4].
2. "Landau subbands generated by a lateral electrostatic superlattice: chasing the Hofstadter butterfly" (1996) [5].

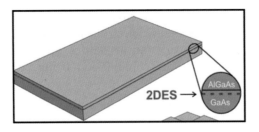
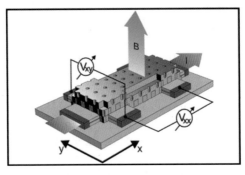

Figure 12.6. A cartoon version of a system—a modulated sample (a metallic layer with holes)—placed directly over a two-dimensional electron gas. The system is subjected to an applied current I and a perpendicular magnetic field B. Both the longitudinal voltage V_{xx} and the transverse (or Hall) voltage V_{xy} can be measured, thus determining the longitudinal resistance R_{xx} and the transverse (or Hall) resistance R_{xy}—or their reciprocals, the longitudinal conductance σ_{xx} and the transverse (or Hall) conductance σ_{xy}. (Adapted from [2] reproduced with permission.)

3. "Evidence of Hofstadter's fractal energy spectrum in the quantized Hall conductance" (2001) [6].
4. "Laterally modulated 2D electron system in the extreme quantum limit" (2004) [7].
5. "Detection of a Landau band-coupling-induced rearrangement of the Hofstadter butterfly" (2004) [8].

Figure 12.7 shows results from one such experiment [8]—one of the first experiments that provided some evidence for the reality of the butterfly fractal. It exhibits the measured Hall conductance, along with a comparison of these empirical data with theoretical results that, using gaps of the butterfly, predict the sizes of the Hall plateaus. As was discussed in section 7.3, the weak-lattice limit (or the strong-field limit) gives a different topological map of the butterfly, since its primary gaps are now labeled with quantum numbers $(0, 1)$ instead of $(1, -1)$. This is due to the fact that if we begin with the Landau-level description of the quantum Hall effect and perturb it with a lattice, then the energy diagram is parametrized by $1/\phi = q/p$ (instead of $\phi = p/q$), and this implies that each Landau level splits into p subbands. We note that in this case, the Hall conductivity is quantized with quantum number $(\sigma + N - 1)$, where N is the index of the Landau band.

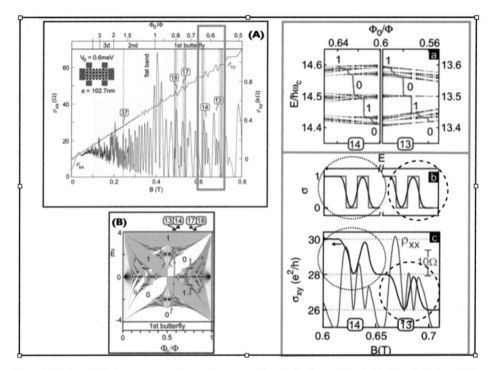

Figure 12.7. Panel (A) shows the *experimentally measured* longitudinal resistivity (in black) and Hall resistivity (in red), using an artificial crystal having square lattice cells of size $a = 102.7 \pm 0.5$ nm. (Here the resistivities are denoted by the Greek letter "ρ", in contrast to the usual roman "R".) Panel (B) shows the theoretical butterfly spectrum, where the integers in the graph are values of the topological quantum number σ, which determines the Hall conductivity. On the right side of this figure, panels (a–c) highlight the green region of (A), where the magnetic field varies between 0.6 and 0.7 tesla. In this region, Landau bands $N = 13$ and $N = 14$ are filled, and the inverse magnetic flux ϕ^{-1} (denoted in the figure by $\frac{\Phi_0}{\Phi}$) is roughly 2/3. This leads to a three-fold splitting of the band. Panel (a) shows a theoretical blowup of butterflies, along with a sequence of σ values as the Fermi energy enters various gaps. Specifically, the sequence (0, 1, 0, 1) for the Hall-conductance quantum number σ (written in black, next to the descending staircase-like red lines) comes out of the Diophantine equation. To make it possible to compare this theoretical plot with experimental findings, panel (b) displays this same theoretically predicted sequence of σ-values using the sizes of gaps as plateau widths (shown in red) obtained from the theoretical butterfly. To simulate disorder and finite temperature, this curve is convoluted (black)—a procedure that is necessary, since it is not easy to calculate the butterfly spectrum for a dirty system at finite temperature. Panel (c)— a blowup of part of the green region of part (A) of this figure—shows the experimentally observed longitudinal conductivity and Hall conductivity. The expected nonmonotonic behavior, where the Hall conductivity nearly takes on quantized values, confirms the theoretical prediction. The dotted circles in (b) and (c) help bring out the agreement between theory and experiment. (Adapted from [2] reproduced with permission.)

These early results, although encouraging, did not go far enough to reveal any fine-grained structure hinting at fractality. One limitation was that it was not possible to tune the magnetic field or the carrier density. Consequently, these early experiments did not observe quantized minigaps in the spectrum, which are a crucial aspect of the butterfly fractal. One feature of this experimental system—and a feature not present in the highly idealized model studied by Hofstadter—was the coupling between the Landau bands. It turns out that this coupling cannot be ignored, particularly as one tries to resolve higher-order gaps. Therefore, although the experiment we have just described confirmed some of the minigaps (we refer readers to the original paper for further details), the original Hofstadter butterfly spectrum does not apply to this setup, because it did not take into account any kind of coupling between Landau bands.

12.2 Magical carbon: A new net for the Hofstadter butterfly

The year 2013 witnessed a major breakthrough in the laboratory verification of the Hofstadter butterfly spectrum. At the heart of this remarkable achievement lay a wonder material—graphene—a completely new kind of two-dimensional crystal, and a rapidly rising star on the horizon of materials science and condensed-matter physics.

At the present time, graphene is the ultimate two-dimensional conducting system. It can be thought of as a single layer of carbon atoms that has been isolated from a graphite crystal. Appendix A summarizes some of the properties of this astonishing material, whose discovery was rewarded by the 2010 Nobel Prize in Physics [9].

This natural two-dimensional crystal soon emerged as a highly suitable candidate for catching the butterfly in a laboratory. The key breakthrough was the idea of using moiré patterns, which are interference patterns that become visible when two lattices are superimposed, as shown in figure 12.8. This elegant geometrical trick, carried out by superimposing two different hexagonal lattices—graphene and boron nitride—created an emergent hexagonal lattice with a far larger unit cell, whose linear dimensions were on the order of 10–14 nanometers. The experimenters were also able to use very powerful magnetic fields, having strengths of up to 45 tesla. These parameter values in the B-versus-a space depicted in figure 12.2 lie well inside the critical gray region in that figure—the only zone in which it would be theoretically possible to detect some of the fractal aspects of the butterfly graph. We note that as long as the point (B,a) falls in the desirable gray zone in figure 12.2, it is advantageous to have a relatively small lattice cell and a relatively large magnetic field, because that combination of parameters gives the best chance for detecting minigaps in the spectrum. Numerous laboratories substantiated this idea, and the reader is referred to the original papers.

As Cory Dean and his collaborators explained in their 2010 paper [10] in *Nature Nanotechnology*, superimposing graphene on the standard substrate of silicon dioxide yields a highly disordered superlattice. By contrast, superimposing graphene on the hexagonal substrate of boron nitride is a more promising idea, because boron nitride has a lattice constant very close to that of graphene, and it also has an atomically smooth surface that is relatively free of dangling bonds and charge traps. And indeed, the moiré pattern that arises as a result of superimposing a layer of graphene on a layer of hexagonal boron nitride and then rotating it forms a hexagonal superlattice (see figure 12.8) whose length scale is on the order of 10 nanometers (roughly 40 times greater than the length scale of the lattices of graphene and boron nitride). This moiré superlattice, in combination with very powerful magnetic fields, opens up unprecedented experimental avenues for testing the fractal spectrum.

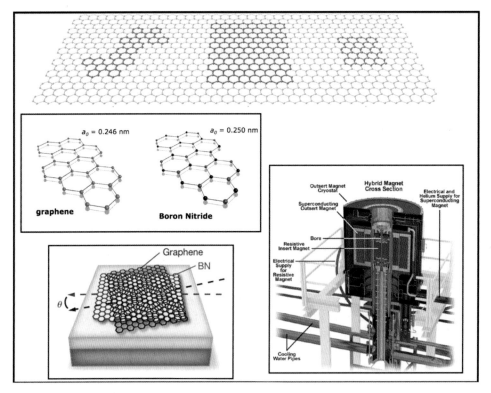

Figure 12.8. At the top is shown the hexagonal lattice structure of graphene. The drawing just below it shows the very similar hexagonal lattice structures of graphene and boron nitride. The drawing just below that depicts the key idea, showing what happens when a layer of graphene is superimposed on a layer of boron nitride and then slightly rotated; the result is a moiré pattern that still possesses hexagonal symmetry, but on a considerably larger scale. Indeed, the linear size of the unit cell of the emergent moiré lattice is on the order of 10 nanometers, which is much greater than that of the two substrates, which is about 0.25 nanometers. Finally, the drawing on the lower right shows the powerful magnet at the National High Magnetic Field Laboratory that was used in the experiment. The cross-section reveals the complexity of the equipment, and thus affords some sense of the remarkable recent advances in generating very high magnetic fields. This national laboratory houses several world-record magnets, including a 45-tesla hybrid magnet, which combines resistive and superconducting magnets to create the strongest steady magnetic field available anywhere. The lab's 35-tesla resistive magnet is the strongest such magnet in the world, and its 25-tesla Keck magnet boasts the highest homogeneity of any resistive magnet. (Adapted from [11] with permission and the permission of the National High Magnetic Field Laboratory.)

Comparing GaAs/ALGaAs and graphene-hexagonal boron nitrate heterostructures

GaAs/ALGaAs superlattices, although they have a very large lattice spacing (about 100 nanometers), unfortunately have a very limited ability to tune the carrier density and other parameters, which is crucial if one wishes to map out the entire Hofstadter butterfly spectrum. In order to be able to observe the miniband structure, it is necessary to reduce the period of the superlattice to length scales comparable to the Fermi wavelength of the electrons, which is about 50 nanometers in GaAs heterojunctions. Furthermore, fabricating such lattices while maintaining coherent registry and without introducing disorder is a daunting task.

In contrast, superimposing graphene and hexagonal boron nitride (hBN), whose crystal lattices are structurally almost identical to each other, results in a periodic moiré pattern. The size of the unit cell of the moiré superlattice is a function of the angle between the two substrate lattices, and it can be tuned so that the superlattice runs through all the different desired length scales. Moreover, hBN provides an ideal substrate for achieving graphene devices with high mobility, which is crucial for the purposes of high-resolution quantum Hall measurements. Finally, gating in graphene allows the Fermi energy to be continuously varied through the entire Bloch band of the moiré lattice. An experimental setup with lattice sizes on the order of 10 nanometers and with high magnetic fields (up to 30–40 tesla) has the possibility of giving high enough magnetic flux-values to allow many minigaps of the butterfly spectrum to be observed. In summary, a moiré superlattice produced by superimposing graphene with hBN yields an ideal platform for experimentally capturing the fractal structure of the Hofstadter butterfly.

Three research groups—one based in Manchester, England, another at Columbia University, and a third at MIT—worked independently at the National High Magnetic Field Laboratory in Florida and observed signatures of the butterfly fractal. Each of these groups used moiré superlattices that had relatively large unit cells (their dimensions were on the order of tens of nanometers), but which differed in other aspects. The Manchester and MIT groups used single-layer graphene, while the Columbia-led experiment used two layers of graphene. By trying out different relative orientations of the two lattices, the teams were able to find orientations that yielded superlattices with appropriately large spacings. The teams then determined the energy spectra of the electrons in their superlattices by measuring the electrical conductivity in very strong magnetic fields—up to about 35 tesla for the Columbia group, 17 tesla for the Manchester group, and 43 tesla for the MIT group. These experimental findings were received with great excitement by the condensed-matter physics community. Here are some of the main articles:

1. "Hofstadter's butterfly and the fractal quantum Hall effect in moiré superlattices" (2013) [11].
2. "Cloning of Dirac fermions in graphene superlattices" (2013) [13].
3. "Massive Dirac fermions and Hofstadter butterfly in a van der Waals heterostructure" (2013) [12].

Figures 12.9, 12.10 and 12.11 provide a small sampler of experimental results. The graphs in them, all reflecting real laboratory studies, give a taste of the kind of measurements done in the laboratory and how they seem to confirm at least some aspects of the butterfly's intricate structure. The quantum Hall states that were observed were found to correspond well with the predicted spectral gaps in a Hofstadter-type energy spectrum. In addition to Hall conductivity, the experimenters also measured the filling-fraction ρ, and from the graph of ρ versus magnetic flux ϕ they were able to obtain the topological quantum numbers σ and τ. The references given below contain much more information, and show results from many different measurements. Taken all together, they make a strong collective case that the real energies of real electrons in real crystals immersed in real magnetic fields do in fact

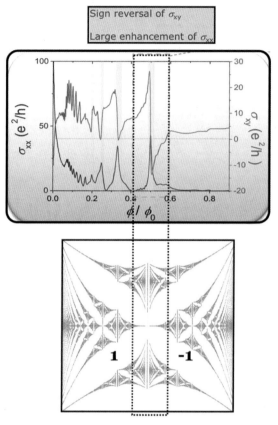

Figure 12.9. The upper half shows two graphs from the experiment of Dean *et al*, showing both longitudinal and Hall conductivity plotted as functions of the flux-value. The sudden sign-reversal of σ_{xy} (the Hall conductivity), accompanied by an enhancement of the longitudinal conductivity near flux-value $\phi = 1/2$, is in complete agreement with the butterfly graph displayed underneath. The gapless spectrum belonging to $\phi = 1/2$ (recall that since the denominator is even, it has two bands that kiss in the middle) leads to a large value of σ_{xx}, and the sign-reversal of the Chern number is reflected in the sign-reversal of σ_{xy}. (Adapted from [11] reproduced with permission.)

possess some of the key features of the highly theoretical spectrum discovered by Douglas Hofstadter in the mid-1970s.

We emphasize that such experimental measurements do not constitute a direct detection of the fractal spectrum, but they strongly suggest the existence of a Hofstadter butterfly that has not yet quite flown into the net. It goes without saying that no experiment could ever see the butterfly in its entirety, since it is a fractal with structure on infinitely many levels, and of course nothing in nature goes down infinitely far like that, with infinite amounts of detail. Nonetheless, when one gathers together many different observations, and especially when one recalls that these data come from three independent groups, one can have little doubt that many of the essential aspects of the exotic nesting behavior first spotted by Hofstadter have actually been detected in real laboratories.

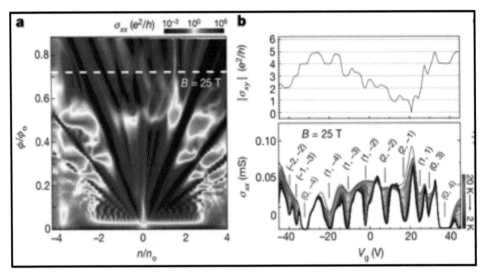

Figure 12.10. A sampler of experimental results of Dean and collaborators from their 2013 *Nature* paper, showing what happens with a smaller lattice cell. Panel (a) shows a Landau fan diagram, which reveals some of the fine structure of the butterfly, lurking in the color-coded graph of longitudinal conductance. Panel (b) shows the Hall conductivity for a fixed value of the magnetic field, corresponding to the dashed white line near the top of panel (a). The panel on the lower right shows the results for various temperatures ranging from 2 K to 20 K, where the integers in parentheses show the quantum numbers (σ, τ). The upper right panel shows the Hall conductance at 2 K. (Adapted from [11] reproduced with permission.)

Below we quote some recent statements from physicists involved in this research as they shared their results with reporters, who of course are always eager to learn about exotic new scientific discoveries:

> *This is a very good example of a fundamental discovery that opens doors that we do not even know about yet. Why go to a distant planet? We go there to discover what's out there. We do not yet know what this new world will result in and what will emerge out of this.*
> —Cory Dean, experimental physicist at the City College of New York.

> *We found a cocoon. No one doubts that there is a butterfly inside.*
> —Pablo Jarillo-Herrero, experimental physicist at MIT.

> *Using the 45 tesla hybrid magnet, researchers at the MagLab observed the long-predicted but never-before-seen fractal known as the Hofstadter butterfly. This work enriches our understanding of the basic physics of electrons in a magnetic field and opens a new route for exploring the role of topology in condensed matter systems.*
> —A post on the website of the National High Magnetic Field Laboratory.

12.3 A potentially sizzling hot topic in ultracold atom laboratories

Ultracold atoms at micro-kelvin to nano-kelvin temperatures are now emerging as a highly versatile new tool for exploring many condensed-matter phenomena, such as

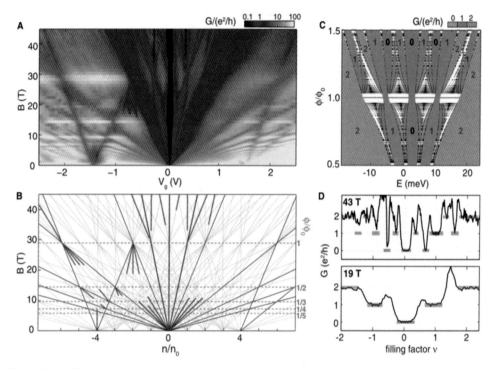

Figure 12.11. These graphs are taken from a 2013 *Science* paper by Hunt and collaborators ([19] reprinted with permission). Panel (A) shows the Hall conductance (here denoted by the letter "G") that was measured for magnetic fields up to 45 tesla. The lower graph (B) (reproduced with permission of Professor Greiner, Harvard University) shows a theoretical Claro–Wannier diagram, where energy gaps in the Hofstadter spectrum are represented as linear trajectories $\rho = \sigma\phi + \tau$. The gray lines indicate gaps for $-4 \leqslant \sigma \leqslant 4$. The red lines superimposed on the theoretical Claro–Wannier diagram were taken directly from the upper graph, and thus constitute experimental data. Note the considerable agreement between the gray theoretical lines and the red experimental ones. Panel (C) shows the theoretical Hofstadter-like energy spectrum for the lowest Landau level of a system incorporating both spin and symmetry-breaking terms. The black points indicate regions of dense energy bands; intervening spectral gaps are color-coded to the associated Hall conductance. Panel (D) shows the conductance traces within the Landau level at magnetic-field values of 43 tesla (top) and 19 tesla (bottom). The emergence of Hofstadter-butterfly minigaps, recognizable as nonmonotonic sequences of quantized conductance plateaus, is evident in the 43-tesla data. The monotonic sequence observed in the 19-tesla data shows the quantum Hall effect in the total absence of a lattice potential, or for a very weak lattice potential. (Adapted from [12] © 2013 American Association for the Advancement of Science, reproduced with permission.)

the quantum Hall effect. Here, highly disciplined neutral cold atoms are "fooled" into behaving like charged particles in a magnetic field. This trickery is achieved by using laser-induced hopping, where a controlled phase can be imposed upon particles moving around a closed loop, thus creating a synthetic magnetic field. This artificial field, mimicking a real magnetic field, makes totally neutral fermionic atoms (i.e. atoms that are fermions—that is, atoms with half-integer spin), such as ^{6}Li and ^{40}K, act in much the same way as charged electrons act in a real magnetic field.

A two-dimensional crystalline environment for these cold neutral atoms can be manufactured by making "optical lattices"—artificial lattices formed by the

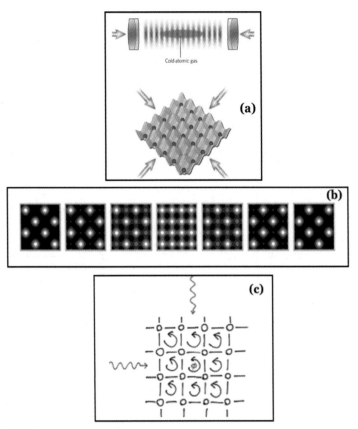

Figure 12.12. The top panel (a) shows an optical standing wave generated by superimposing two laser beams (Figure reproduced from [14] with permission, copyright 2008 Nature Publishing Group.). The antinodes (or nodes) of the standing wave form a periodic array of microscopic laser traps for the atoms. Such a "crystal of light", in which cold atoms can move and can be stored, is called an *optical lattice.* If several standing waves are made to overlap, higher-dimensional lattice structures can be formed, such as the two-dimensional optical lattices shown in panel (b) (Figure reproduced with prermission of [15], Harvard University.). Panel (c) is a cartoon (Figure reproduced from [16] with permission. Copyright 2016 American Physical Society.) showing ultracold atoms interacting with two perpendicular laser beams (red) that, being coherent with each other, can interfere and produce path-dependent phases. The atoms, even though they are electrically neutral, behave just like charged particles in a two-dimensional lattice in a magnetic field. They can tunnel between lattice sites, where tunneling involves absorbing and emitting photons in the laser beams.

interference of laser beams, producing a spatially periodic pattern (see figure 12.12). When an atom interacts with light (an electromagnetic field), the energy of its internal states depends on the intensity of the field. Therefore, a spatially varying field intensity induces a spatially varying potential energy that acts as a potential-energy landscape that can be used to trap ultracold atoms in a periodic fashion in space. The resulting arrangement of trapped atoms has much in common with an ordinary crystal lattice [17].

Two independent research groups have created optical lattices with ultracold neutral atoms moving in them, and these atoms mimic the behavior of electrons in an ordinary crystal lattice moving under the influence of a strong magnetic field. The result is a novel realization of the idealized Hamiltonian used by Hofstadter [18].

In one such setup, rubidium atoms tunnel between sites in a two-dimensional square lattice in a regime where tunneling involves absorbing and emitting photons produced by lasers. Interaction with these lasers produces path-dependent phases for the atoms, and this mimics a magnetic field acting on the lattice-bound electrons. It turns out that in this fashion, one can make sufficiently intense synthetic magnetic fields that the system enters the desirable regime in which it is possible to observe aspects of the fractal spectrum of a two-dimensional electron gas in an ordinary lattice in a real magnetic field.

As of this writing, cold-atom experiments have not yet observed the fractal pattern that is associated with the model explored by Hofstadter, but they have provided compelling evidence that the model's Hamiltonian has been physically realized. Extensive efforts are currently underway in various laboratories to capture, for once and for all, the celebrated fractal beast. However, it still remains to be seen whether ultracold electrically neutral atoms—a very orderly species acting like cloned soldiers—can mimic the behavior of noninteracting charged particles. But there is great promise in the optical lattices produced by crisscrossing laser beams and the controlled motion of neutral atoms in such lattices. Perhaps this technique will provide just the right landscape that will allow empirical confirmation of the theoretical model studied by Hofstadter.

Meanwhile, the obsession with spotting this beloved butterfly somewhere in the real world continues to occupy center stage in our scientific adventures. As we commemorate the four decades of the Hofstadter butterfly, we hope that this journey will lead us into brand-new territories—and along the way revealing yet other unknown and

unimagined secrets of this wonderful and endlessly intriguing quantum world that we macroscopic classical creatures have the good fortune to inhabit.

Appendix: Excerpts from the 2010 Physics Nobel Prize press release [9]

Graphene—the perfect atomic lattice

A thin flake of ordinary carbon, just one atom thick, lies behind this year's Nobel Prize in Physics. Andre Geim and Konstantin Novoselov have shown that carbon in such a flat form has exceptional properties that originate from the remarkable world of quantum physics. Graphene is a form of carbon. As a material, it is completely new— not only the thinnest ever, but also the strongest. As a conductor of electricity, it performs as well as copper. As a conductor of heat, it outperforms all other known materials. It is almost completely transparent, yet so dense that not even helium, the smallest gas atom, can pass through it. Carbon, the basis of all known life on Earth, has surprised us once again.

Geim and Novoselov extracted the graphene from a piece of graphite such as is found in ordinary pencils. Using regular adhesive tape they managed to obtain a flake of carbon with a thickness of just one atom—this at a time when many believed it was impossible for such thin crystalline materials to be stable. However, with graphene, physicists can now study a new class of two-dimensional materials with unique properties. Graphene makes experiments possible that give new twists to the phenomena in quantum physics. Also a vast variety of practical applications now appear possible, including the creation of new materials and the manufacture of innovative electronics. Graphene transistors are predicted to be substantially faster than today's silicon transistors and result in more efficient computers. Since it is practically transparent and a good conductor, graphene is suitable for producing transparent touch screens, light panels, and maybe even solar cells. When mixed into plastics, graphene can turn them into conductors of electricity while making them more heat-resistant and mechanically robust. This resilience can be utilized in new super-strong materials, which are also thin, elastic, and lightweight. In the future, satellites, airplanes, and cars could be manufactured out of the new composite materials.

References

[1] Hofstadter D R 1976 Energy levels and wave functions of Bloch electrons in rational and irrational magnetic fields *Phys. Rev.* B **14** 2239
[2] Geisler M 2005 The Hofstadter butterfly and quantum interferences in modulated 2-dimensional electron systems *PhD thesis* University of Stuttgart doi:10.18419/opus-6610
[3] Fowler A B, Fang F F, Howard W E and Stiles P J 1966 Magneto-oscillatory conductance in silicon surfaces *Phys. Rev. Lett.* **16** 901
[4] Gerhardts R, Weiss D and Wulf U 1991 Magnetoresistance oscillations in a grid potential: indication of a Hofstadter-type energy spectrum *Phys. Rev.* B **43** 5192–5

[5] Schlosser T, Ensslin K, Kotthaus J P and Holland M 1996 Landau subbands generated by a lateral electrostatic superlattice: chasing the Hofstadter butterfly *Semicond. Sci. Technol.* **11** 1582–5

[6] Albrecht C, Smet J H, von Klitzing K, Weiss D, Umansky V and Schweizer H 2001 Evidence of Hofstadter fractal energy spectrum in the quantized Hall conductance *Phys. Rev. Lett.* **86** 147

[7] Melinte S *et al* 2004 Laterally modulated 2D electron system in the extreme quantum limit *Phys. Rev. Lett.* **92** 036802

[8] Geisler M C, Smet J H, Umansky V, von Klitzing K, Naundorf B, Ketzmerick R and Schweizer H 2004 Detection of a Landau band-coupling-induced rearrangement of the Hofstadter butterfly *Phys. Rev. Lett.* **92** 256801

[9] http://www.nobelprize.org/nobel_prizes/physics/laureates/2010/press.html

[10] Dean C R *et al* 2010 *Nature Nanotech.* **5** 722–6

[11] Dean C R *et al* 2013 Hofstadter's butterfly and the fractal quantum Hall effect in moiré superlattices *Nature* **497** 598

[12] Hunt B *et al* 2013 Massive Dirac fermions and Hofstadter butterfly in a van der Waals heterostructure *Science* **340** 1427

[13] Ponomarenko L A *et al* 2013 Cloning of Dirac fermions in graphene superlattices *Nature* **497** 594

[14] Bloch I 2008 Quantum coherence and entanglement with ultracold atoms in optical lattices *Nature* **45** 7198

[15] Greiner M 2003 Ultracold quantum gases in three-dimensional optical attice potentials *PhD thesis* Ludwig-Maximilians-Universität München (http://greiner.physics.harvard.edu/Theses/PhD_greiner.pdf)

[16] Ching C and Mueller E J 2013 *Physics* **6** 118

[17] For a general review, see; Bloch I 2005 Ultracold quantum gases in optical lattices *Nat. Phys.* **1** 23–30

[18] Aidelsburger M, Atala M, Lohse M, Barreiro J T, Paredes B and Bloch I 2013 Realization of the Hofstadter Hamiltonian with ultracold atoms in optical lattices *Phys. Rev. Lett.* **111** 185301

Miyake H, Siviloglou G A, Kennedy C J, Burton W C and Ketterle W 2013 Realizing the Harper Hamiltonian with laser-assisted tunneling in optical lattices *Phys. Rev. Lett.* **111** 185302

[19] Bloch I 2008 Quantum coherence and entanglement with ultrcold atomos in optical lattices *Nature* **453** 1016

IOP Concise Physics

Butterfly in the Quantum World
The story of the most fascinating quantum fractal
Indubala I Satija

The butterfly gallery: Variations on a theme of Philip G Harper

Large Divided Oval Butterfly. This bronze sculpture by Henry Moore, weighing in at over eight tons, seems to hover weightlessly over the surface of the reflecting pool at Berlin's House of World Cultures.

It is always fascinating to hear what kinds of variations a great composer can come up with when creatively exploring the possibilities lurking implicitly within a given theme—for example, Sergei Rachmaninoff's *Rhapsody on a Theme of Paganini*, J S Bach's *Goldberg Variations*, and so forth—and when the composer is Nature herself, one's curiosity will be particularly piqued.

In this mini-gallery, the theme is Harper's equation and all that comes out of it. On this theme, nature has composed some beautiful variations that reflect the immense richness, diversity, and universality of the patterns inherent in the theme. This gallery of various reincarnations of the Hofstadter butterfly features Bloch electrons that reside in various types of non-square lattices (including the quasi-periodic Penrose lattice), or in lattices that are subject to additional complications, such as electron–electron interactions or a Zeeman splitting or spin–orbit coupling.

The first display in the Gallery is from the 1969 paper by Dieter Langbein. Although today readers can immediately recognize the butterfly in this image, Langbein, who did not know the 1964 paper by Azbel' and probably had no background in number theory, did not realize his graph's recursive aspect—the heart and soul of this graph. Otherwise, as Douglas Hofstadter said in the Prologue, this image might have been known the world 'round as the "Langbein butterfly".

The last display in the gallery is from the world of particle physics—indeed, from the colorful world of quarks and antiquarks, described by the laws of quantum chromodynamics (QCD). This rare beast, here dubbed the "quantum-chromodynamical butterfly", was first sighted, by a lovely coincidence, in the University of Regensburg's Physics Department—the very place where Gplot was first sighted, in November of 1974. Thus the venue where the Hofstadter butterfly was born some 40 years ago has now given birth to a sequel—perhaps encoding new secrets of nature.

Without further ado, we now embark on our *Variations on a theme of P G Harper*.

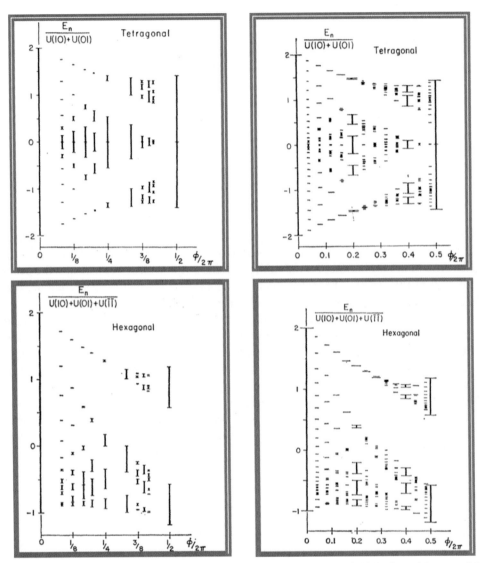

Bloch electrons in tetragonal and hexagonal lattices, by Dieter Langbein, from his 1969 *Physical Review* article "The tight-binding and the nearly-free-electron approach to lattice electrons in external magnetic fields" [1] (image reproduced with permission). The upper two panels show results for a tetragonal (square) lattice, while the lower two are for a hexagonal lattice. Also, the left panels and the right panels, respectively, come from analytic and numerical work.

Gplot, by Douglas Hofstadter, from his 1975 University of Oregon doctoral thesis "The energy levels of Bloch electrons in a magnetic field" [2] (image reproduced with permission).

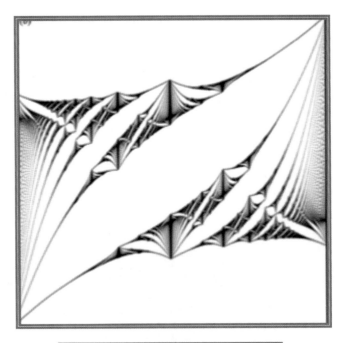

Top panel: *Electrons in a hexagonal lattice* by Francisco Claro and Gregory Wannier, from their 1979 *Physical Review* B article "Magnetic subband structure of electrons in hexagonal lattices" [3] (copyright American Physical Society). Bottom panel: *Electrons in a honeycomb lattice*, by R Rammal, from his 1985 *Journal de Physique* article "Landau level spectrum of Bloch electrons in a honeycomb lattice" [4].

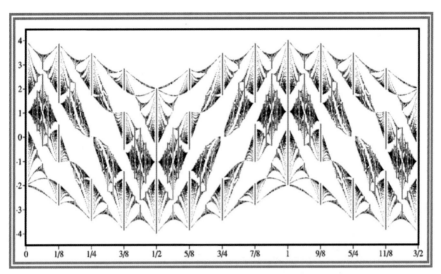

Bloch electrons cavort in kagome lattices, by Yi Xiao, Vincent Pelletier, Paul M Chaikin, and David A Huse. So-called "kagome lattices" look like the woven pattern of a bamboo basket. The term was coined by Japanese physicist K Fushimi, who spliced together the roots "kago" (bamboo basket), and "me" (woven pattern). This graph is taken from Xiao *et al*'s 2003 *Physical Review* article "Landau levels in the case of two degenerate coupled bands: kagome lattice tight-binding spectrum" [5] (Copyright American Physical Society).

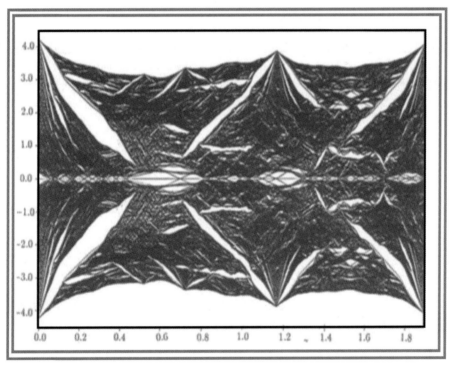

Harper meets Penrose, by Tetsuo Hatakeyama and Hiroshi Kamimura, from their 1987 *Solid State Communications* article "Electronic properties of a Penrose tiling lattice in a magnetic field" [6] (reproduced with kind permission of Elsevier).

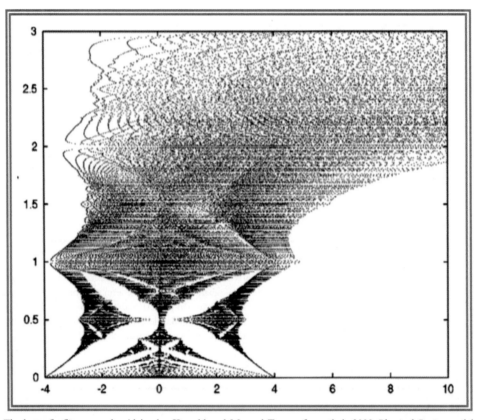

The butterfly flies away, by Alejandro Kunold and Manuel Torres, from their 2000 *Physical Review* article "Bloch electrons in electric and magnetic fields" [7] (copyright American Physical Society).

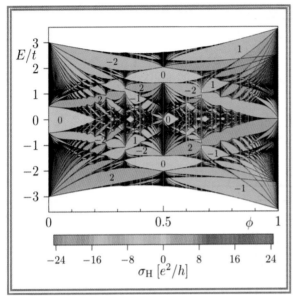
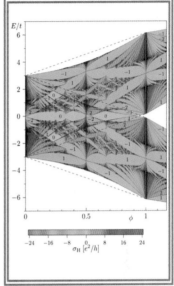

Electrons remember their spin, by Wouter Beugeling, featuring spin-orbit coupling (left) and Zeeman splitting (right) for the honeycomb lattice, from Beugeling's 2012 University of Utrecht doctoral thesis "Topological states of matter in two-dimensional fermionic systems" [8] (copyright American Physical Society, reproduced with permission).

Braid structure when electrons interact at long distance, by Jean Bellissard and Armelle Barelli, from their 1993 *Journal de Physique* article "Semiclassical methods in solid-state physics: two examples" [9].

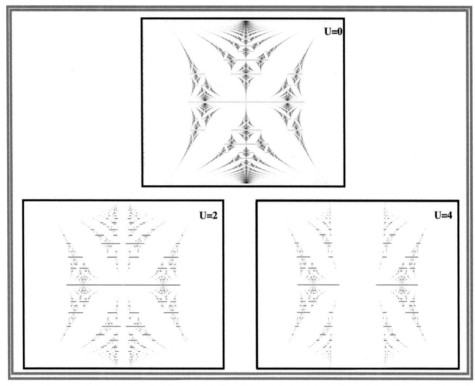

Electrons interact with each other, by Hyeonjin Doh and Sung-Ho Suck Salk, from their 1998 *Physical Review* article "Effects of electron correlations on the Hofstadter spectrum" [10] (copyright American Physical Society).

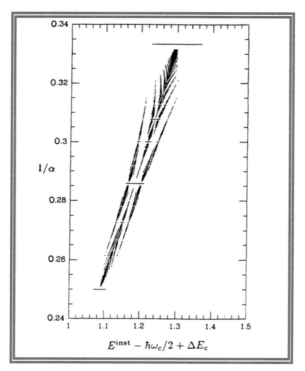

Instantons carry traces of the Hofstadter butterfly, by Denise Freed and Jeffrey A Harvey, from their 1990 *Physical Review* article "Instantons and the spectrum of Bloch electrons in a magnetic field" [11] (copyright American Physical Society).

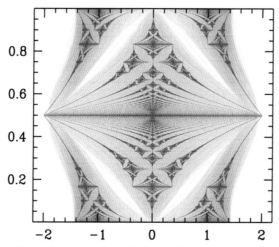

Quantum-chromodynamical butterfly, by Gergely Endrödi, from his 2014 paper "QCD in magnetic fields: from Hofstadter's butterfly to the phase diagram", given at the 32nd International Symposium on Lattice Field Theory, held at Columbia University [12]. (Reproduced with permission of Dr Endrödi, University of Frankfurt.)

References

[1] Langbein D 1969 The tight-binding and the nearly-free-electron approach to lattice electrons in external magnetic fields *Phys. Rev.* **180** 633
[2] Hofstadter D R 1975 The energy levels of Bloch electrons in a magnetic field *PhD thesis* (Eugene, OR: University of Oregon)
[3] Claro F H and Wannier G H 1979 Magnetic subband structure of electrons in hexagonal lattices *Phys. Rev.* B **19** 6068
[4] Rammal R 1985 Electrons in a honeycomb lattice *J. Physique* **46** 1345
[5] Xiao Y, Pelletier V, Chaikin P M and Huse D A 2003 Landau levels in the case of two degenerate coupled bands: kagome lattice tight-binding spectrum *Phys. Rev.* B **67** 104505
[6] Hatakeyama T and Kamimura H 1987 Electronic properties of a Penrose tiling lattice in a magnetic field *Solid State Commun.* **62** 79–83
[7] Kunold A and Torres M 2000 Bloch electrons in electric and magnetic fields *Phys. Rev.* B **61** 9847
[8] Beugeling W 2000 Topological states of matter in two-dimensional fermionic systems *PhD thesis* University of Utrecht
[9] Bellissard J and Barelli B 1992 Semiclassical methods in solid-state physics: two examples *J. Physique* **3** 471–99
[10] Doh H and Salk S-H S 1998 Effects of electron correlations on the Hofstadter spectrum *Phys. Rev.* B **57** 1312
[11] Freed D and Harvey J A 1990 Instantons and the spectrum of Bloch electrons in a magnetic field *Phys. Rev.* B **41** 328
[12] Endrödi G 2014 QCD in magnetic fields: from Hofstadter's butterfly to the phase diagram *32nd Int. Symp. on Lattice Field Theory* (New York: Columbia University)

IOP Concise Physics

Butterfly in the Quantum World
The story of the most fascinating quantum fractal
Indubala I Satija

Divertimento

While writing this book, I had the privilege of communicating by email with some very special people to whom I am extremely grateful. Below I share a few of these emails, since they tell interesting tales that are now part of the history of this book-writing project.

1. Who came up with the name "butterfly"?
Here are some replies to the queries I sent out asking who coined the term "butterfly" for Hofstadter's graph of the spectrum of Harper's equation.

Douglas Hofstadter
  In answer to your query, I myself did not come up with the name "butterfly" for Gplot. It never occurred to me to see it that way (or if it did at some point, it would just have been a random private thought, and I wouldn't ever have suggested that term as a way to refer to the graph—that's not my style, or at least it wasn't my style back then). It's a charming term, no doubt, but it didn't come from me! I have no idea who first called Gplot a "butterfly" in print. Sorry not to be able to help you out on that! Yours, Doug.

Francisco Claro
Emeritus Professor of Physics at the Universidad Católica de Chile in Santiago, Chile
  Having no idea who first named Hofstadter's Butterfly, I wrote Jean Bellissard to find out. Am still waiting for his reply and am also keeping an eye on my files to see if the butterfly's birth certificate appears somewhere. Francisco

Jean Bellissard
Professor of Mathematics and Physics at Georgia Tech
  I have been thinking about your question and my conclusion is: I don't know. I thought Doug Hofstadter was at the origin. I have scanned my own archives to figure

out when I started using the term. I feel like it was after the end of the eighties. I gave a talk at the Seminar Bourbaki in 1991, which appeared in Astérisque in 1992. "Le Papillon de Hofstadter, d'après B. Helffer et J. Sjöstrand", Séminaire Bourbaki, 44e année, 1991–92, #745 (novembre 1991). Published in Astérisque, 206, (1992), pp. 7–40.

Apparently that was the first time I was using the term "Hofstadter Butterfly". I did not invent it, but it was in the air for a long time and I just chose to use this term as an advertisement for the problem. However, scanning my own work prior to that date (with Rammal, Barelli, Seiler, Simon) or the work of others during the eighties about this problem (Aubry, Wilkinson, Helffer-Sjöstrand), I did not see this term being used before 1991. I wonder whether the term did not come from people using it in connection with the high T_c superconductors. I remember P. W. Anderson discussing the Hofstadter spectrum in connection with "flux phase" and the RVB theory. But I could not point to this reference explicitly. I suppose you already asked Yosi Avron, since the picture of the butterfly which is now very famous came from one of his former students.

Well, if you find out, please let me know. I have buried this period into my memory, since I am now surrounded by people much younger who have no clue about what I am talking about concerning anything before 1995.

Finally, the answer to my query was found... The word "butterfly" first occurred in none other than Douglas Hofstadter's very own *Physical Review* article [1], on page 2241, where he writes:

The large gaps form a very striking pattern somewhat resembling a butterfly...

Yes, in the end it was Douglas Hofstadter himself who first used the term "butterfly" in his *Physical Review* article, but over the years he forgot all about having done so!

When asked quite recently what he thought about this unexpected revelation, Hofstadter wrote: "I still think that this kind of concrete visual association was not really my own personal style. On the other hand, I vividly remember that back in those days, my mother always used to see animal forms in my abstract line drawings—drawings in which I never intended to depict any animals at all—and so I surmise that one day when I showed her Gplot, she probably said to me, in an offhand way, 'It looks like a butterfly!' and that was most likely the origin of the term."

2. The Butterfly meets Dick Feynman

Doug Hofstadter shared with me the following tale, which triggered endless speculations in my mind about what the butterfly graph might have meant to Richard Feynman, one of the greatest minds of the last century.

In the summer of 1975, when I was in the throes of working out the details of Gplot's structure and proving that the "smeared" versions of Gplot were always perfectly continuous, I had the good fortune of visiting Feynman in his Caltech office and spending a few hours in conversation with him. That afternoon we talked about many things, but when I showed him Gplot and explained to him what it was, he was absolutely thunderstruck, and instantly asked his secretary to photocopy it not just for himself but even more for his teen-aged son Carl, who, he told me, would love it.

When I told Feynman about the smeared versions of Gplot, he spontaneously used the verb "jiggle" instead of "smear", and so, in my prologue, as a tip of the hat to him, I used that verb a couple of times. I also recall that almost the very moment that I wrote Harper's equation on his blackboard, Feynman casually commented that it looked like it must be its own Fourier transform, a fact that I myself had recently realized (but only after months of thinking about it). Needless to say, the lightning-quickness of his insight bowled me over.

In my Physical Review article and in my PhD thesis, I very briefly thanked Feynman, though I didn't explain in either place why I did so. What I was actually thanking him for was his great enthusiasm for the grace of Gplot, which was so instant and so infectious that it gave me a big boost of self-confidence.

After I read this tale, my curiosity and sheer hunger to know what transpired between father and son as they stared together at the butterfly graph were mixed with delicious thoughts, drowning me in a wild ocean of imagination. In my attempts to contact Carl Feynman, I sent emails to several members of the faculty and the administrative staff at Caltech, to Perseus Books (the publisher of a collection of Feynman's letters, edited by Feynman's daughter Michelle), and to Feynman's old friend and colleague Freeman Dyson. I received only two responses: one from Freeman Dyson, who said that he did not know Carl Feynman (I was touched by his immediate response to my email), and one from Alan Rice, Caltech's Division Administrator for Physics, Mathematics, and Astronomy, who wrote the following note:

Dear Dr Satija,

Richard Feynman was a striking man, and many individuals recall their interactions with him, even after several decades. (Dick and I first met in 1980 while waiting to cross a busy street. His response to my question was so passionate that he simply stopped walking halfway across, stranding us in traffic!)

Thanks for your interest in expanding Feynman's scientific legacy, it's a very rich fabric.

Carl does not often join in the publicity. We will, however, pass along your question and leave him to respond if he wishes.

Alan Rice

And yes, I finally did hear back from Carl Feynman, and here is his response:

Dear Prof Satija,

I would have been 13 or 14 in 1975. I later became fascinated by fractals, like many mathematically-inclined people of my generation, but at the time I would of course not have heard of them. The image of the Hofstadter butterfly is familiar to me, in the sense that when I see the red-and-blue presentation of it, I think to myself, "Ah, that thing again." Alas, I have no memory of seeing—or talking about—the Hofstadter butterfly in 1975. Presumably I saw it sometime between that day and this.

I'm sorry to be such a lousy source. If you have more questions, I am of course happy to cooperate.

–Carl F

And so, in the end, did father and son ever talk about the butterfly? We may never know...

Robert Boeninger, Richard Feynman, and Doug Hofstadter
Sometimes it seems that we live in an incredibly small world, as the following tale shows. While my book was being copy-edited, I found out about a new book of quotations from Richard Feynman, lovingly compiled by his daughter Michelle, and so, being a great devotee of Feynman, I immediately bought it. In riffling through it I came across a surprising claim by Feynman to the effect that number theory plays no role in physics (a fact for which he said he knew no reason). Since this Feynman quote was so close in spirit to the Michio Kaku remark I'd already put in chapter 2's conclusion, I felt I should definitely include it there as well. Doug Hofstadter, when he saw this small new addition, was very intrigued, and asked me if the compilation of Feynman quotes told when or where Feynman had made this provocative comment. I accordingly looked it up in the book and wrote back to him, saying that it was apparently taken from a letter that Feynman wrote in May, 1969 to someone named Robert Boeninger, a name completely unknown to me. When Doug received my email he instantly wrote back as follows:

Figure 14.1. Robert Boeninger and Douglas Hofstadter meeting in Göttingen, early February 1975.

Wow! This is a totally unexpected revelation, and is quite incredible! You would have had no reason to suspect this, Indu, but Robert Boeninger is one of my closest, dearest, oldest friends. We met way back in fourth grade (early 1954), when we were both just 9 years old, at Stanford Elementary School. Ever since then, we've been super-close friends, sharing thoughts about everything under the sun, always deeply fascinated by number theory. Our passions for mathematics grew together and were intricately intertwined for many years. In the early 1960s, Robert and I talked all the time about eta-sequences, INT, and related matters, and for years we explored mathematics together. In fact, my book "Gödel, Escher, Bach" grew out of a 32-page letter I wrote (in longhand!) to Robert in the summer of 1972.

Some years later, in the mid-1970s, Robert was living in Hamburg, Germany, far north of Regensburg, where I was living, and a couple of times during my struggles during that forlorn period, I badly needed a break, so I took the eight-hour train ride and stayed for a day or two with Robert and his family. The most intense such get-together, however, was when Robert and I both took trains, he heading south and I north, and met at the midpoint: on a platform of the train station in Göttingen, the fabled university town where Gauss, Dirichlet, Riemann, Hilbert, Planck, Born, and countless other stellar math and physics figures once lived. (See Figure 1.) During that unforgettable weekend, Robert and I talked nonstop about my trials and tribulations with my Doktorvater, and together we tried to dream up ways to prove that Gplot was an infinitely recursive structure. A beautiful long walk we took through the Göttingen forests, unexpectedly winding up at an exquisite teahouse atop a high hill, later inspired the dialogue "Magnificat, Indeed" in "Gödel, Escher, Bach".

In short, Robert was on the front lines with me all the way from my exuberant eta-sequence days to my despairing doctoral days. Toward the end of the Acknowledgments in my thesis, I wrote this paragraph: "Perhaps the most needed moral support of all was given by my old friend Robert Boeninger. My conversations with him in Stanford, Hamburg, and Göttingen were highly valuable. They deeply influenced my patterns of thought, and my state of mind, for which I owe special thanks." And so, quite marvelously, the quote from Feynman that you stumbled upon twists Robert Boeninger, Richard Feynman, and Doug Hofstadter magically together in a tight little loop. I am smiling with amazement and delight!

3. Three mementos from Doug Hofstadter's filing cabinets

Douglas Hofstadter was recently rummaging around in his old files connected with his doctoral research, and he chanced upon a number of intriguing documents that provide a glimpse into what the Hofstadter butterfly and the science underlying it meant to some of the pioneers of condensed-matter physics. Here are three of those items.

3.1. Gregory Wannier writes to Lars Onsager

The first item is a letter written by Gregory Wannier to Lars Onsager in August of 1975. (Hofstadter had no memory of how this letter had come to be in his files, nor of why it was in such bad shape and on pink paper.) Onsager was a Norwegian-born

American physical chemist and theoretical physicist who won the Chemistry Nobel Prize in 1968. It was Onsager's work, along with that of Rudolf Peierls, that opened the way for Peierls' student P G Harper to come up with his now-famous equation, the equation whose eigenvalues form the butterfly graph.

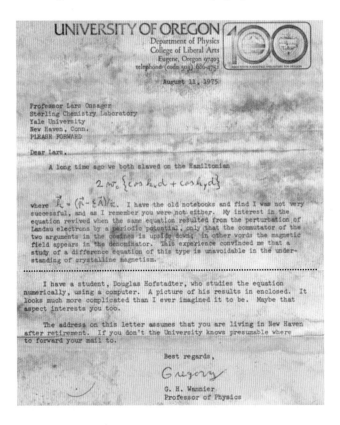

Incidentally, Gregory Wannier does not refer to the equation as "Harper's equation". He himself had independently found exactly the same equation in the mid-1950s and had struggled with it for a long time, though he never published any of his findings, which apparently were sparse. Then in the mid-1970s, Wannier, Gustav Obermair, and Alexander Rauh—none of them aware of Harper's work—rediscovered the equation and started working hard on it together. Mysteriously, it was only in the late 1970s that they finally learned of Harper's work.

In his letter, Wannier describes some progress he has recently made on the problem, and then, toward the end, he mentions Hofstadter's computer plot, saying: "It looks much more complicated than I ever imagined it to be." It doesn't seem at all likely that if Wannier had understood the graph's recursive structure at that time, he would have completely left out all mention of that structure. After all, the ideas about infinite nesting were just too surprising and too radical to blithely skip over. We can thus infer that even at that late date, Wannier hadn't yet fully understood or accepted the graph's recursive nature. (Of course, by the time Hofstadter had his doctoral defense, only four months later, Wannier was on board with, and even excited about, the new ideas.) Onsager died in 1976 and Wannier in 1983, and we do not know whether Wannier ever heard back from Onsager.

3.2. Leo Falicov savors the butterfly

The second item is a referee's report, dated March 1976, concerning Douglas Hofstadter's article "Energy levels and wave functions of Bloch electrons in rational and irrational magnetic fields" [1], which Hofstadter had submitted a month earlier to the *Physical Review*. In it, the anonymous referee waxes effusive about the article, likening it to a Mozart *divertimento*.

```
                    REFEREE'S REPORT ON
        "Energy Levels and Wave Functions of Bloch Electrons
                In Rational and Irrational Magnetic Fields"
                              by
                      Douglas R. Hofstadter


        I find this paper delightfully written, almost like
    a Mozart divertimento. It is evident that the author has
    done a very conscientious job and has found pleasure in
    discovering the mathematical intricacies of the problem
    under study.
        I would personally like to see the paper published as
    is, but the following objections could be raised:  (1)
    probably the Journal of Mathematical Physics is a much more
    suitable journal than the Physical Review, in order to fit
    the contents and spirit of the paper. Even though the
    author makes some exotic suggestions for experiments at
    the end, the paper is fundamentally a mathematical paper.
    (2) notwithstanding the pleasure I had in reading the
    paper in its present form, I should recognize that for its
    contents, the paper is inordinately long.  I would hate to
    see it cut to less than one-half of its present size, but
    this can be accomplished, albeit sacrificing the delight-
    ful style.


    EDITORIAL NOTE                                    #BP632
      Please attend to items marked with red arrow on attached form.
```

Sixteen years later, the theoretical physicist Leo Falicov met Hofstadter and revealed to him that he had been the anonymous voice behind that report. Falicov, who was born in Argentina in 1933 and died in the United States in 1995, not only was a world-renowned condensed-matter theorist, but also loved poetry and was an accomplished pianist. A curious twist in the story is that one of Falicov's own coworkers—probably a post-doc of his—was Dieter Langbein, who in 1969 had come within a hair's breadth of discovering the recursive secrets of the Harper spectrum.

3.3. Gregory Wannier embraces the butterfly

The final item is a short note written to Hofstadter at Indiana University in early 1980 by his old *Doktorvater*. Wannier presumes that since his ex-student has left physics for other pastures, he probably no longer wishes to keep copies of his *Physical Review* article. It is touching to read Wannier's words: "For me, on the other hand, the article is like one of my own"; they reveal that in 1980, he had come to embrace with deep affection the ideas that several years earlier he had dismissed as "numerology". Wannier finally realized the beauty of the ideas that had been under his nose for a while without his understanding them, and in the end he was very gracious about acknowledging his less than ideal behavior in the

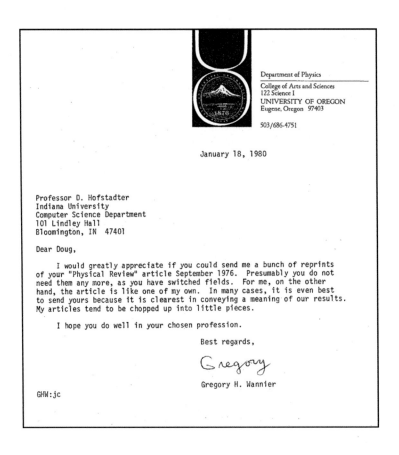

years 1974–1975 (this was in a 1979 letter that Hofstadter came across but decided is too personal to publish here). Thus was rekindled all the prior warmth between *Doktorant* and *Doktorvater*.

Reference

[1] Hofstadter D R 1976 Energy levels and wave functions of Bloch electrons in rational and irrational magnetic fields *Phys. Rev.* B **14** 2239

IOP Concise Physics

Butterfly in the Quantum World
The story of the most fascinating quantum fractal
Indubala I Satija

Gratitude

I am blessed to have a very warm, talented, and highly dedicated group of friends and colleagues who have helped me shape this book.

First, I feel tremendous gratitude to Douglas Hofstadter for sharing his memories of when he first glimpsed Gplot, and for writing an inspiring history of the butterfly graph. My small book about the Hofstadter butterfly would not have been complete without Doug's prologue, "The grace of Gplot", for which I just do not have enough words to thank him. When I first sent him an out-of-the-blue email with a nearly complete draft of this book, I waited for his response for a few days, which seemed like an eternity. I was in a state of sheer panic, with "butterflies in my stomach", imagining that my book's draft had disappeared—even from his trash!—and I could almost hear the deafening sound of the dreaded e-shredding. At last, though, I received his reply, and it was not only with huge relief but with sheer joy that I read it. His encouraging remarks and kind words to this complete stranger about her book added new sparks and colors to my confidence, and in addition, the friendly way he talked about poetry and outdoor adventures brought new life to the last phase of my writing. Here is his reply, which he wrote on March 9, 2015 (and at 3.35 AM, on top of it all!):

Dear Prof. Satija,

Thank you so much for your kind note a few days ago. I was astonished and hugely flattered to learn that someone had written a book in which Gplot plays a starring role. I'm very gratified to know that you have done this.

This evening I downloaded your draft and looked carefully through it, from start to finish. It is an amazing tour de force! I never realized that there was any connection between Gplot and Apollonian gaskets (marvelous, magical mathematical objects that they are!), or between Gplot and the Berry phase, or between Gplot and many other things that you mention. I would have to read your book carefully from cover to cover in order to understand the many connections that you describe, but even without having done so, it is thrilling to me to see the great excitement and the deeply sincere, almost childlike sense of wonder that you exude in describing all these beautiful, interrelated scientific phenomena. That's a wonderful gift!

You asked if I could describe what I felt when I first glimpsed Gplot (which was back in the fall of 1974 in Regensburg, Germany). Well, your request comes at a very timely moment, since just this evening I wrote a long letter to a friend in which I did exactly what you wished. I'll attach that letter to this email. It will give you a pretty accurate picture of how I felt when I first glimpsed Gplot's hidden essence, sitting all alone in my quiet Regensburg office one November day in 1974.

I wish to thank you once again for the magnificent labor of love that you have produced. I will be very proud once it comes out, and I'll certainly purchase numerous copies to give to my more scientifically inclined friends (and even to a few others!).

I send you my warmest wishes, and I hope that your sparklingly lively book will enjoy a truly grand success!

*Sincerely,
Douglas Hofstadter.*

P.S.—Before sending this letter off, I looked you up on the Web and broke into a broad smile when I saw a photo of you standing at the edge of the Grand Canyon. That was quite a coincidence, since just this past November I hiked down the Canyon and then back up with my son Danny and his friend Charles.

I also saw your tribute to your father[1], in which you describe, after he had lost the ability to speak, his wordless way of mouthing the words to poetry that he loved and had memorized, and how he expired while filling in the final words to a favorite poem that you couldn't fully remember. That is an amazingly touching image, and between the lines you convey many powerful emotions and intangible ideas. Poetry, like music, is one of humanity's great collective creations, and savoring beautiful poems is a deeply rewarding way to spend some of one's precious time on earth.

As Doug described in the prologue, when he met me in person a few weeks later, he volunteered to be the copy editor of the book. Being a perfectionist of the highest order, perhaps he could not condone any writeup about his beloved butterfly that did not live up to his standards. In any case, Doug made numerous revisions that brought immense clarity and logical flow to almost all parts of the book. He was extremely critical at times, raising important questions that eventually led to huge improvements.

So the book has been cleansed, both inside and outside. It has been given a long dip in the holy water of the Ganges, purifying its soul. And Doug, in his humble way, mentioned in one email to me that he was simply trying to polish the "diamond in the rough". Well, he adjusted almost every paragraph of the book, yet without changing its DNA. In spite of many revisions, I could always recognize my "Cinderella". By now, her rags have almost disappeared, replaced by beautiful

---

[1] See also http://www.washingtonpost.com/opinions/poetry-and-my-father/2012/01/19/gIQA9KSLJQ_story.html

clothes and ornaments, but she still remains the same old Cinderella that I created before handing Doug my manuscript.

For Doug, proofreading chapter 6 was a kind of homecoming, bringing him back to the "home" he had left behind 40 years ago. And so chapter 6 evolved into a "guest lecture" by him. His passion and intimacy with the subject of this book is evident in almost every sentence he wrote in that chapter. Perhaps it reflects his nostalgia for physics or his love for his beloved Gplot, to whom he is still attached like a child.

I should also mention that Doug was not just the copy editor of my book, he was also my mentor and friend. His detailed lively emails filled with humor, in-depth analysis, and comforting words were always source of great joy and inspiration. They kept me moving in this project—a task that at times felt like it had no end in sight. Doug was so open and forthcoming that we communicated without the slightest hesitation to express our likes and dislikes about various aspects of the task of perfecting this book. Yes, it was a great pleasure and a great privilege to work with him. Some parts of me may be exhausted from this ultramarathon, but deep inside, I feel rested and rewarded from this incredible, somewhat magical journey that I did not envision when I started writing this book. I am truly blessed and fortunate to have had this opportunity.

So this book is at last ready to come out of its hiding. Thank you, Doug, thank you so much for being so gracious, so kind—I am touched.

From the bottom of my heart, my very special gratitude goes to my two wonderful colleagues Erhai Zhao and Predrag Nikolic. Erhai was always there for me when I needed suggestions or clarifications on various scientific issues to fine-tune my writing. In my numerous moments of confusion, he was my savior whom I could call on at any time. Having Erhai, particularly during the last stages of this journey, was absolutely critical, and I cannot imagine crossing the finish line without him. I also owe many thanks to Predrag Nikolic, who brought clarity to my understanding of various condensed-matter concepts. Special thanks to Francisco Claro, for his various comments that added some new colors to the book. Various discussions and comments from Greg Huber were also very helpful.

My unbounded gratitude goes to John Blackwell, Sarah DeBauge, and Daniel Dakin (see figure 15.1), without whom this book would have never seen its completion. Both John and Sarah were students of mine, and they radiated such passion for quantum physics that we had an immediate resonance. They were such a joy to work with, and lots of fun to be around. During my most intense periods of writing, we three would often meet up for lunch or a drink, and our conversations would often digress from the book, gravitating mostly towards Feynman, whom we all adored. It was a sheer delight to hear John tell us some incredible stories about Feynman.

During these conversations, it occurred to me that John must have been quite an anomaly in his business career, since for a period of ten years, during his business trips, he was constantly reading Feynman's "red books", and he would spend all his evenings in hotel rooms thinking about what it all meant.

A mathematics graduate from University College London, John contributed a few paragraphs to various parts of the book. I am specially thankful to him for his

Figure 15.1. From left to right: Dan, John, Indu, Doug, and Sarah at George Mason University, 2015.

suggestion that I start the chapter on fractal geometry with the Mandelbrot set, as it opened up new ways for me to see the butterfly graph. How right John is in his statement: "Physics professors seem to wonder if they should have been math professors and vice versa. This book illustrates that the real fun is at the boundary between the two." John also tirelessly proofread the book several times, helping me to overcome my seemingly infinite deficiency regarding the word "the", and correcting countless other errors.

Sarah was an art-history major at Stanford University, and these days, at George Mason University, she is taking physics courses. In addition to her lovely rendition of "The kiss precise" in chapter 0, Sarah's artistry is reflected in many images of this book. She, too, proofread the book again and again, and she graciously tended to my frequent requests to fix old images as well as creating new ones. I admire her patience, as very often my instructions were very imprecise, since I myself didn't know what I wanted, and so figuring it out required a lot of experimentation. Having Sarah, someone so well-versed in both art and scientific culture, was a real treasure during the course of writing this book.

I am indebted to Dan for his thorough and comprehensive reading of the book, for checking every equation and figure, and for his various comments and suggestions about what fit in and what did not. His instincts, his insights, and his ability to spot errors with microscopic precision were all highly impressive. Dan and I collaborated about a decade ago before he left academia and moved on to challenges in industry. His love for physics has nonetheless stayed on course—the butterfly plot has been attached to the pinboard in his office ever since 2006

Figure 15.2. The pinboard in Dan Dakin's office. The butterfly picture has been posted on this pinboard since 2006, when Dan moved to his office at Optical Air Data Systems.

Figure 15.3. Top: four generations of my family in India. Bottom: my family in USA.

(see figure 15.2). Also, his seven-year-old daughter has taken to signing her name *Saraħ*. Perhaps the Hofstadter butterfly has a nickname, too—the *ħ-butterfly*.

I deeply want to thank my family for their support and enthusiasm (figure 15.3), and for having put up with my butterfly-mania during the writing of this book.

There are other people whose presence in the background I constantly felt and who definitely had a role in shaping this book. In particular, I want to acknowledge two of my teachers. At Bombay University, Abbas Rangwala had such a friendly and lively way of teaching quantum mechanics and introducing abstract concepts that I fell in love with those things almost instantaneously. And at Columbia University, my teacher and PhD mentor Richard Friedberg had the gift of a nearly magical mathematical intuition, and his extraordinary visual way of approaching abstract concepts is a gift I got from him. It made theoretical physics so much fun and joyful. This style has stayed with me ever since my student days, becoming part of me, and in my own teaching I continue to strive to pass it on to my students.

Photograph of Richard Friedberg, reproduced here with his permission.

This book is a tribute to my teacher Richard Friedberg, one of the most brilliant physicists I have encountered in my life. It was a great privilege for me to do my PhD thesis with him. In the process of writing this book, I had many sudden awakenings, reminding me how those days of graduate studies at Columbia University transformed me and made me the person and physicist I am today. One of the most important things that my work with Richard cultivated in me was to be absolutely in love with what I do, and to remain almost immune to the highly competitive culture that pervades the scientific community today. Among the top scorers in the famed Putnam exam in mathematics in 1956, Richard was unconventional and absent-minded, and he cared little about what was going on around him. Many adored him just the way he was. I love the following quote from his book, *Adventurer's Guide to Number Theory*: "The difference between the theory of numbers and arithmetic is like the difference between poetry and grammar", which reveals a side of Richard that many may have missed. I should also mention that one of the most memorable moments of my life was the day of my PhD exam (scheduled at noon) when he did not show up. When we called his home, he was still sleeping, but he quickly arose and rushed to join the other members of my PhD committee (namely, Joaquin Luttinger, Pierre Hohenberg, and Alfred Muller), who patiently waited over an hour until he showed up.

I am at a complete loss for words to thank my dearest friend Radha Balakrishnan—a very special person and collaborator who has brought immense joy and magic to my scientific adventures. After reading a draft of my book, Radha wrote, "When I saw your mention of poetic mathematics, I was so reminded of the legends of Lilavati and her father Bhaskara, who gave her math problems in the form of poetry."

How naïve it was of me to have forgotten *Lilavati*, a beautiful and highly inspiring Indian tale from the 12th century! Bhaskara II, who lived from 1114 to 1185, was one of India's greatest mathematicians and astronomers. In addition, he wrote *Lilavati*, which consists of 279 verses in Sanskrit, each of which sets up a mathematics problem. Much like the legends of Archimedes' bathtub and Isaac Newton's apple, there is a legend that claims that Bhaskara composed that book for his daughter Lilavati. Readers can easily access further details of the intriguing tale of Lilavati by searching for it on-line, but with a bit of caution, as there are numerous misquotations on the web about *Lilavati*. (I am very grateful to Professor M S Sriram, a scholar on the history of the subject, for pointing some of them out.)

Bhaskara concludes *Lilavati* by stating: "Joy and happiness are indeed ever-increasing in this world for those who have *Lilavati* clasped to their throats, decorated as the members are with neat reductions of fractions, multiplications, and involutions, pure and perfect as are the solutions, and tasteful as is the speech which is exemplified." Coincidentally, 2014, the year in which I started writing this book, was the 900th anniversary of the birth of Bhaskara—an event commemorated by several academic conferences across India—and it almost feels as if behind my intense drive to write this book there was an angel, reminding me to honor this great mathematician and to salute his poetic mathematics. In fact, I decided to close my book with a coda that is a short excerpt from *Lilavati*—a little gem of poetic verse where love, beauty, and mathematics are intertwined, causing a special and unparalleled glow. Just before the coda, I included special poems that were inspired by the Hofstadter butterfly. They were written in response to the "Lilavati Competition in Poetic Math and Science", which I launched. The first of these poems is one that was composed jointly by myself and Doug Hofstadter. I sent Doug a first attempt, he revised it, then it went back and forth via email for a while, and we argued like little kids about a few words here and there. The final result pleases us both, though.

Many, many thanks to the staff members of IOP Concise—Jacky Mucklow, Chris Benson, Steve Trevett, Andrew Giaquinto, Karen Donnison, Joel Claypool, and Jeanine Burke for their dedicated efforts in bringing this book to fruition.

In conclusion, I want to pay tribute to two great physicists who continue to guide me and to inspire me in all the ups and downs of my quest to fathom the mysterious quantum world.

First, I wish to pay deep tribute to Paul Adrien Maurice Dirac, who coded my DNA with his principle of mathematical beauty: "We do not really know what the basic equations of physics are, but they have to have a great mathematical beauty. *We must insist on that...*" There's nothing deeper or truer than that! And yet I feel compelled to mention that I have one small disagreement with Dirac, for he once said: "I do not see how a man can work on the frontiers of physics and write poetry at the same time. They are in opposition." To me, however, the two activities are not incompatible at all.

Secondly, I want to say how grateful I am to Richard Feynman—a great physicist, great teacher, and great human being, with an incredible zest for life—for all the joy he continues to bring to our lives, inspiring us in our scientific adventures and struggles, and showing us all how to be creative in science while enjoying every minute of it. I am reminded of a quote from a student of Michelangelo: "The great ruler of Heaven looked down and, seeing these artists' attempts, resolved to send to Earth a genius. He further endowed the genius with a true moral philosophy and a sweet, poetic spirit, so that the world would marvel."

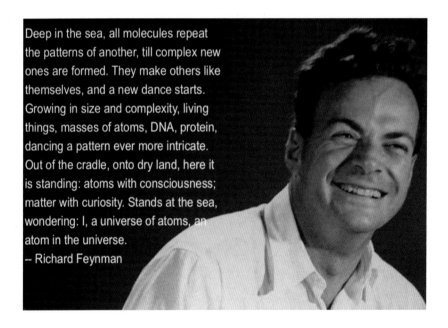

Deep in the sea, all molecules repeat the patterns of another, till complex new ones are formed. They make others like themselves, and a new dance starts. Growing in size and complexity, living things, masses of atoms, DNA, protein, dancing a pattern ever more intricate. Out of the cradle, onto dry land, here it is standing: atoms with consciousness; matter with curiosity. Stands at the sea, wondering: I, a universe of atoms, an atom in the universe.
– Richard Feynman

IOP Concise Physics

Butterfly in the Quantum World
The story of the most fascinating quantum fractal
Indubala I Satija

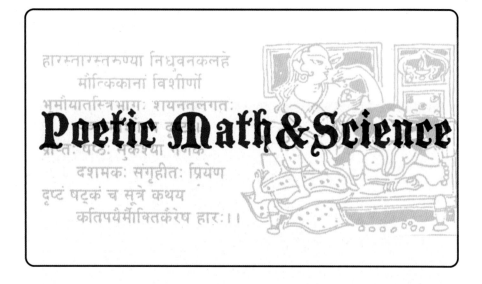

Bulletin of Sciences, 1984, Vol 1, p21–23

Before concluding my book with a coda—a short excerpt from *Lilavati*, a little gem of poetic verse where love, beauty, and mathematics are braided together—I briefly intervene, to a handful of poems inspired by the Hofstadter butterfly.

The first one is a "Salute" composed by Douglas Hofstadter and myself, our love letter to this divine entity. In addition, I include four other poems that I received in response to a challenge I launched on the Web, called the "Lilavati Competition in Poetic Math and Science". As I stated in the announcement of this competition:

Scientists and mathematicians often strive for mathematical beauty with the same depth and eagerness as painters, sculptors, and other artists. The Lilavati competition strives to interlace the abstract poetry of science and mathematics embedded in its great equations and intriguing concepts with the poetry of natural beauty and language, adding a new dimension to one of the highest forms of artistry created by humans...

I am most grateful to the contributors—Bala, Carlos, Julia, and Sachin—for having lent their own types of verbal joy to this celebration of four decades of the Hofstadter butterfly. For me, the gesture of concluding the book with short pieces of verse is also a special way to honor Bhaskara, whose 900th birthday was celebrated last year, and also a very special way to pay homage to my late father, who loved poetry.

Salute

by Indu Satija and Douglas Hofstadter

O crystal in magnetic field,
Pervaded by de Broglie waves,
Within your depths there lay concealed
A butterfly of mystic form.
In your cocoon its wingtips furled,
Long hidden from the outer world.

A shell sat on the beach to pluck,
For someone chancing by with luck.
Professors, gazing skywards, missed it.
A student, peering downwards, found it:
The butterfly's immortal coils
Revealed by Rumpelstiltskin's toils.

Destructive quantum interference
Produced a graph of strange appearance.
Because two periods are contending,
Bands split and split; it's never-ending!
Though paradoxical and weird,
Gplot turns sensible when smeared.

At first it seemed irrational,
As crazy as a Cantor set.
Who'd have dreamt that such would be
Found floating in the Fermi sea?
Yet thus has Nature rendered mates
The bands of Bloch and Landau states.

Pearls from Peierls, Harper's Fareys,
Berry phases, tags from Chern,
Quantum anholonomy,
The Hall effect — so much to learn!
Even ancient Grecian gaskets
Decorate its nestedness.

It's one, yet it's infinity,
Eternity, sublimity,
Divinity and mystery;
It's raga, yet it's poetry.
We hail thee, Nature, for your art.
You showed us secrets from your heart.

On the Hofstadter Butterfly

The young enthusiastic Douglas Hofstadter
Wondered how planar electrons did scatter
So he programmed his computer
To the chagrin of his tutor
Opening a new chapter in condensed matter.

Hofstadter's butterfly is a wondrous beast
Its fractal wings a veritable visual feast
Weirder still, without a doubt
When you actually print it out
Is its awesome spectrum, to say the least!

V. Balakrishnan
Dept. of Physics, IIT Madras
Chennai, India

Butterflies in the fields

By Carlos A. R. Sa de Melo (August 31st, 2015)

Soon, the morning sun in the horizon will rise...

The lovely colorful butterfly first seen last night

Will flap its wings to emerge from its nocturnal disguise

And depart for the meadows in rapid flight.

It will join the Hofstadter's butterflies in the fields

The magnetic ones... with flux threading their wings...

Oh! How different this butterfly is from typical yields,

Since it spins and orbits in voluptuous swings...

From the Zeeman and spin-orbit fields further emerge

Quadri-winged insects with topological patterns,

That when integrated to Berry phases converge.

Like Hofstadter's, they too exhibit vigorous fractality,

Resembling geometric drawings in ancient caverns,

And bringing to the physical world the sense of immortality!

(Oh! How beautiful are butterflies in all types of fields!)

Electron Liberated
By Julia Savich

Step, step, stepping,
ascending into the intricate mystery above.

Branching, reaching, growing,
smoothly sweeping the infinite
open space,
yet uncontrollably fumbling
over imperceptible obstacles along the way.

Turning, only to find sharp corners blocking the visible path.
No gradual transitions to be made,
only unwieldy walls preventing fluid forward motion.

Nowhere to go but out.
Out and away from the weight of the crushing emptiness around.
Deserted by comrades,
completely surrounded by symmetric nothingness.
Barren voids glare all-knowingly,
hiding the coveted escape from their
blind prisoner.

Spinning, ambling about, trying to find a break.

Suddenly, the limitless arching world falls into place around.
Strong lines guide the path to freedom.
Only boundless expanse obscures the
distant passage.

Climb, climb, climbing,
In every direction, gliding
from isolated point to isolated point.
The captive, floating out of the
restricting entrapments,
halts to take rest before flying again.
He lifts up, unhindered, to the levels beyond,
away from the controlling walls that once bound him.

Rising, falling, searching,
hunting in every direction for the next place to go.

"My Beautiful Quantum World"
Sachin Kumar, Delhi University

Sometimes I feel same as Einstein
Imagination works better than I do

Feynman's fantasy world where he used to live
And same like him I defined my dream

Just like Kekule on a sleepy night
I fell in a hole and there was no light

Was shrunk by a field It wasn,'t a fall
Quantum states were handy I was that small

I was made of either energy or strings
Two units of a sense to see 3-D things

Was sensing the world just like a bat
I was throwing something and receiving them back

Yes, these are my personalized photons
They check things
Without disturbing them at all my photons were reflected

By field energies and also deviated
On probability basis

This magnetic Earth also shrunk with me
Conserved circular momentum generated big field

The field became huge as I saw big cyclotron
In my case no problem Coz I had no electron

Tunneling was fun was eating energy
And one day I reached an electronic colony

I had a memory of beautiful big world
But quantum is noisy and uncertain as well

Witth open two eyes view wasn't that great
When I closed one became 2-D surface

My dream fantasy when I closed one eye
Chaos turned to garden with several butterflies

Some near some far all flying to my face
And where ever I see it looks like the same

Noise turned to beauty I thank you Hofstadter
You made a way for quantum look better

IOP Concise Physics

Butterfly in the Quantum World
The story of the most fascinating quantum fractal
Indubala I Satija

Coda

*The square root of half the number of a swarm of bees
is gone to a shrub of jasmine;
and so are eight-ninths of the whole swarm:
a female is buzzing to one remaining male
that is humming within a lotus flower in which he is confined,
having been allured to it by its fragrance at night.
Say, lovely woman, the number of bees.*

From the twelfth-century book *Lilavati*, a volume of poetic mathematics consisting of 279 verses written in Sanskrit by Indian mathematician Bhaskara II (1114–1185 A.D.). It is believed that Bhaskara composed this volume for his daughter Lilavati. Translated by Henry Thomas Colebrooke.

IOP Concise Physics

Butterfly in the Quantum World
The story of the most fascinating quantum fractal
Indubala I Satija

Selected bibliography

Below, in the interests of providing a useful historical overview, we provide a chronological list of selected references concerning the theory of Harper's equation and its spectrum (the Hofstadter butterfly), grouped in four categories: (1) papers going up till the discovery of the butterfly (these include some classic papers from pioneers in the field); (2) papers published after the discovery of the butterfly; (3) popular or less technical articles; and (4) review articles. The second category includes certain papers that are of historic importance, as they appeared right after the discovery of the butterfly, as well as papers reporting important breakthroughs or results that, in our view, revealed new facets of this rich subject. (Experimental papers are not included in this list.) Some of the papers listed below have also been cited in various chapters in the book, of course. Given that as of the current writing, Hofstadter's paper has been cited roughly 2500 times (according to Google Scholar), there will inevitably be many valuable papers that are not included in this short and necessarily highly subjective list.

1. Historical development of the subject of electrons in a magnetic field: 1930–76

   Landau L D 1930 Paramagnetism of metals *Z. Phys.* **64** 629
   Peierls R E 1933 On the theory of diamagnetism of conduction electrons *Z. Phys.* **80** 763
   Luttinger J M 1951 The effect of a magnetic field on electrons in a periodic potential *Phys. Rev.* **84** 814
   Adams E N II 1952 Motion of an electron in a perturbed periodic potential *Phys. Rev.* **85** 41–50
   Onsager L 1952 Interpretation of the de Haas–van Alphen effect *Phil. Mag.* **43** 1006
   Harper P G 1955 Single band motion of conduction electrons in a uniform magnetic field *Proc. Phys. Soc.* A **68** 874
   Luttinger J M and Kohn W 1955 Motion of electrons and holes in perturbed periodic fields *Phys. Rev.* **97** 869–83
   Brailsford A D 1957 The magnetic energy levels of electrons in metals *Proc. Phys. Soc.* A **A70** 275

Zil'berman G E 1957 Behaviour of an electron in a periodic electric field and a uniform magnetic field *Sov. Phys.—JETP* **5** 208

Kohn W 1959 Theory of Bloch electrons in a magnetic field: the effective Hamiltonian *Phys. Rev.* **115** 1460

Wannier G H 1962 Dynamics of band electrons in electric and magnetic fields *Rev. Mod. Phys.* **34** 645

Blount E I 1962 Bloch electrons in a magnetic field *Phys. Rev.* **126** 1636

Roth L M 1962 Theory of Bloch electrons in a magnetic field *J. Phys. Chem. Solids* **23** 433–46

Wannier G H and Fredkin D R 1962 Decoupling of Bloch bands in the presence of homogeneous fields *Phys. Rev.* **125** 1910–5

Azbel' M Ya 1964 Energy spectrum of a conduction electron in a magnetic field *Sov. Phys.—JETP* **19** 634

Azbel' M Ya 1964 On the spectrum of difference equations with periodic coefficients *Sov. Math.—Dokl.* **5** 1549

Zak J 1963 Magnetic translation group *Phys. Rev.* A **134** 1602

Zak J 1964 Magnetic translation group II. Irreducible representations *Phys. Rev.* A **134** 1607

Pippard A B 1964 Quantization of coupled orbits in metals II. The two-dimensional network, with special reference to the properties of zinc *Phil. Trans. R. Soc. Lond.* A **265** 317

Chambers W G 1965 Linear-network model for magnetic breakdown in two dimensions *Phys. Rev.* A **140** A135

Gerlach E and Langbein D 1966 Tight-binding approach to electrons in a crystal potential and an external magnetic field *Phys. Rev.* **145** 449

Butler F A and Brown E 1968 Model calculations of magnetic band structure *Phys. Rev.* **166** 630

Langbein D 1969 The tight-binding and the nearly-free-electron approach to lattice electrons in external magnetic fields *Phys. Rev.* **180** 633

Rauh A, Wannier G H and Obermair G 1974 Bloch electrons in irrational magnetic fields *Phys. Status Solidi* B **63** 215

Rauh A 1974 Degeneracy of Landau levels in crystals *Phys. Status Solidi* B **65** K131

Rauh A 1975 On the broadening of Landau levels in crystals *Phys. Status Solidi* B **69** K9

Hofstadter D R 1976 Energy levels and wave functions of Bloch electrons in rational and irrational magnetic fields *Phys. Rev.* B **14** 2239

2. A selection of relevant references from 1976 onwards

Aubry S 1978 The new concept of transition by breaking of analyticity in crystallographic models *Solid State Sci.* **8** 264–78

Wannier G H 1978 A result not dependent on rationality for Bloch electrons in a magnetic field *Phys. Status Solidi* B **88** 757

Claro F and Wannier G H 1979 Magnetic subband structure of electrons in hexagonal lattices *Phys. Rev.* B **19** 19

Wannier G 1980 Bloch electrons in a magnetic field: reduction to one dimension *J. Math. Phys.* **21** 2844

Aubry S and André G 1980 Analyticity breaking and Anderson localization in incommensurate lattices *Ann. Israel Phys. Soc.* **3** 133

Thouless D J, Kohmoto M, Nightingale M P and den Nijs M 1982 Quantized Hall conductance in a two-dimensional periodic potential *Phys. Rev. Lett.* **49** 405

Bellissard J and Simon B 1982 Cantor spectrum for the almost Mathieu equation *J. Funct. Anal.* **48** 408

MacDonald A H 1983 Landau-level subband structure of electrons on a square lattice *Phys. Rev. B* **28** 6713

Thouless D J 1983 Bandwidth for a quasiperiodic tight binding model *Phys. Rev. B* **28** 187

Ostlund S and Pandit R 1984 Renormalization-group analysis of the discrete quasiperiodic Schrödinger equation *Phys. Rev. B* **29** 1394

Wilkinson M 1984 An example of phase holonomy in WKB theory *J. Phys. A: Math. Gen.* **17** 3459

Wilkinson M 1984 Critical properties of electron eigenstates in incommensurate systems *Proc. R. Soc. Lond. A* **391** 305

Rammal R 1985 Landau level spectrum of Bloch electrons in a honeycomb lattice *J. Physique* **46** 1354

Wilkinson M 1987 An exact renormalisation group for Bloch electrons in a magnetic field *J. Phys. A: Math. Gen.* **20** 4337

Helffer B and Sjöstrand J 1989 Semi-classical analysis for Harper's equation. III: Cantor structure of the spectrum *Mémoires de la Société Mathématique de France* **39** 1–124 http://arxiv.org/pdf/1602.05111.pdf

Bell S C and Stinchcombe R B 1989 Hierarchical clustering in the spectrum of incommensurate systems *J. Phys. A: Math. Gen.* **22** 717

Anderson P W 1990 Theories on high-temperature superconductivity *Int. J. Mod. Phys. B* **4** 181

Barelli A, Bellissard J and Rammal R 1990 Spectrum of 2D Bloch electrons in a periodic magnetic field: algebraic approach *J. Physique* **51** 2167

Freed D and Harvey J A 1990 Instantons and the spectrum of Bloch electrons in a magnetic field *Phys. Rev. B* **41** 328

Avron J, van Mouche P H M and Simon B 1990 On the measure of the spectrum for the almost Mathieu operator *Commun. Math. Phys.* **132** 103

Last Y and Wilkinson M 1992 A sum rule for the dispersion relation of the rational Harper equation *J. Phys. A: Math. Gen.* **25** 6123

Callan Jr C G and Freed D 1992 Phase diagram of the dissipative Hofstadter model *Nucl. Phys. B* **374** 543

Han J H, Thouless D J, Hiramoto H and Kohmoto M 1994 Critical and bicritical properties of Harper's equation with next-nearest neighbor coupling *Phys. Rev. B* **50** 11365

Wiegmann P B and Zabroding A V 1994 Bethe-ansatz for the Bloch electron in magnetic field *Phys. Rev. Lett.* **72** 1890

Wilkinson M 1994 Generalized Wannier function and renormalization of Harper's equation *J. Phys. A: Math. Gen.* **21** 8123

Last Y 1994 Zero measure spectrum for the almost Mathieu operator *Commun. Math. Phys.* **164** 421

Ketoja J A and Satija I I 1995 Self-similarity and localization *Phys. Rev. Lett.* **75** 2762

Ketoja J A and Satija I I 1997 The re-entrant phase diagram of the generalized Harper equation *J. Phys. Condens. Matter* **9** 1123–32

Rüdinger A and Piéchon F 1997 Hofstadter rules and generalized dimensions of the spectrum of Harper's equation *J. Phys. A: Math. Gen.* **30** 117

Puig J 2004 Cantor spectrum for the almost Mathieu operator *Commun. Math. Phys.* **244** 297

Ávila A and Jitomirskaya S 2009 The ten martini problem *Ann. Math.* **170** 303

Janecek S, Aichinger M and Hernańdez E R 2013 Two-dimensional Bloch electrons in perpendicular magnetic fields: an exact calculation of the Hofstadter butterfly spectrum *Phys. Rev.* B **87** 235429

Kimura T 2014 Hofstadter problem in higher dimensions *Prog. Theor. Exp. Phys.* 2014**103** B05

3. Popular or less technical articles

Hofstadter D R 1984 A nose for depth: Gregory Wannier's style in physics *Phys. Rep.* **110** 273

Avron J E, Osadchy D and Seiler R 2003 A topological look at the quantum Hall effect *Phys. Today* **39** 38

Hofstadter D R 2004 Of Rumpelstilzchen and of Gplot *Schmetterlinge im Festkörper: Zur Geschichte des Hofstadter-Butterflys* ed C Forstner (Lehrstuhl für Wissenschaftsgeschichte: Universität Regensburg)

Hofstadter D R 2013 Butterflies in the stomach *Soc. Phys. Stud. Obs.* Winter p 6

4. Review articles

Simon B 1982 Almost periodic Schrödinger operators: A review *Adv. Appl. Math.* **3** 463–90

Sokoloff J B 1985 Unusual band structure, wave functions and electrical conductance in crystals with incommensurate periodic potentials *Phys. Rep.* **126** 189

Guillement J P, Helffer B and Treton P 1989 Walk inside Hofstadter's butterfly *J. Physique* **50** 2019–58

Last Y 1995 Almost everything about the almost Mathieu operator, I *XIth Int. Congress Math. Phys. (Paris)* (Cambridge, MA: International Press) pp 366–72

Rössler U and Suhrke M 2000 Bloch electrons in a magnetic field: Hofstadter's butterfly *Adv. Solid State Phys.* **40** 35

Satija I I 2016 A Tale of Two Fractals: The Hofstadter butterfly and integral Apollonian gaskets *Eur. Phys. J.* at press (http://arxiv.org/abs/1606.09119)

We conclude with a short summary of some of the most seminal papers in the field, as provided by Jean Bellissard, who is one of the field's key players.

I want to emphasize the following contributions, which I consider to be seminal for this problem.

Original papers: Landau 1930 (diamagnetism of metals), Peierls 1933 (effective approach to periodicity and magnetic field).

Onsager 1952: magnetic oscillations. Seminal paper, used every day today in magnetic field experiments, together with the book of I M Lifshitz and A M Kosevich, "Theory of magnetic susceptibility in metals at low temperatures", which is THE reference on magnetic oscillation and semiclassical analysis. Experimentalists at the

High Magnetic Field facilities (Tallahassee, Grenoble, Toulouse...) use this book as their main reference.

Among the earliest contributions, Luttinger (1951), Adams (1952), Luttinger and Kohn (1955) are certainly important ones. Brailsford (1957), Blount (1962), Roth (1962), Wannier and Fredkin (1962), Pippard (1964) were also significant. The work of Zak (1964) was also a breakthrough. It was eventually used by Alex Grossmann to prove that the C^*-algebra generated by the magnetic translations was type II, which, at the time, was a tremendous result. (I believe he published this article in 1968 in an obscure Israeli journal. Grossmann was a close friend of Zak.)

But I would put an emphasis on the work of Chambers published in 1965, since he proposed a formula that has been used over and over by mathematicians since I brought it to the attention of Barry Simon in 1982 (see our joint paper). Needless to say, Chambers already understood the renormalization scheme, but he explicitly claims that it is useless in practice since physical effects will wash out the small gaps (which is true, but so what? the usual prejudice of physicists against math). I am not sure if he knew about Azbel"s work, but I did not check.

You should also note the work of Langbein (1969), Rauh (1974), and Rauh, Wannier, and Obermair (1974) at the time when Hofstadter was working on his PhD thesis under Wannier's guidance.

All this accumulation of works culminated with the PhD thesis of Douglas R Hofstadter in 1975. It is a very early example of fractal, at a time when Benoît Mandelbrot had not yet coined the word "fractal". Hofstadter was a student of Wannier, together with Francisco Claro. Claro and Wannier published a paper in 1978 in which they introduce a gap labeling, which was eventually interpreted as the corresponding Chern numbers of the part of the spectrum with energy smaller than the gap.

In 1978, Aubry and André made a breakthrough for the Harper model. First they introduced a duality (due actually to Bernard Derrida from Saclay) leading to an argument in favor of an Anderson metal–insulator transition (this was really new at the time). In addition, they numerically computed the Lebesgue measure of the spectrum as a function of the coupling constant v in the almost Mathieu operator, and found the wonderful formula $4|v-1|$. This formula was later proved by using an argument due to Thouless. They also surmised that the Lyapunov exponent was zero for $v < 1$ and $Ln(v)$ otherwise.

The rigorous mathematical studies really started in 1981. I proposed a gap-labeling theorem based on K-theory (Bellissard and Simon 1982), then the use of Aubry duality and of the KAM theorem led to a proof of the existence of a metal–insulator transition (1982). Then came my paper with Barry Simon in which we proved in a generic way that all gaps but the central one were open for the almost Mathieu operator. This was proved rigorously for the Harper model by Choi, Elliott, and Yui (1990) in a beautiful paper, demonstrating that many gaps were actually tiny.

Michel Herman gave a very simple argument to prove that the Lyapunov exponent (in the transfer-matrix approach) was bounded from below by the Aubry–André formula (published at the end of a very long paper sometime in 1984).

Do not overlook the seminal series of papers by Helffer and Sjöstrand during the eighties. Using a modern version of semiclassical analysis (Hörmander's microlocality), *they proved a great number of results only hinted at by physicists before. They used the work of Azbel' and some of his formulae to analyze the spectrum more carefully. I contributed to this program in a different way, using the C*-algebraic approach.*

My work with Rammal (1990) gives explicit formulas for the gap edges at rational magnetic fluxes, proving that the derivative was discontinuous. I proved in 1994 that the gap edges were Lipshitz continuous for a much broader class of models than just the Harper one.

Most of these mathematical works in the 1980s were incomplete: a large class of irrational numbers (with zero Lebesgue measure) for the magnetic flux was excluded for a long time. In addition, in the one-dimensional version of the almost Mathieu model, there is a phase which might be responsible for some resonance effect and a subset of zero Lebesgue measure was excluded from the results.

The Ten Martini problem was more challenging, though. Not only did it ask about whether the spectrum was a Cantor set for ALL irrational flux-values, but also it was addressing implicitly the question of the nature of the spectral measure. The main new tool of study came during the last 15 years with the development of the theory of cocycles. The earliest hint in this direction came from Michel Herman before he died, followed by Raphael Krikorian and Hagan Eliasson. Another breakthrough came with the work of Yoccoz (another former student of Michel Herman) on Siegel disks; Yoccoz introduced a technique, due to Brjuno, for including all possible irrational numbers that were inaccessible before. Artur Ávila changed the game by developing the theory of cocycles. Svetlana Jitomirskaya, who worked for a long time with Yoram Last, then with Bourgain, jumped on this wagon and was able to finish the job with Ávila. Today, there are still tiny corners left over for which we do not know the nature of the spectral measure, but it is almost tight.

What is remarkable is that this problem has been worked on by a very large number of scientists, both in the physics community and in the mathematical community as well. I once listed 200 seminal papers from physicists that could be counted as important, and I then realized that most of the leading figures in solid-state physics had contributed to the problem. The mathematical community dealing with the problem used techniques coming from dynamical systems, from C*-algebras, and from PDE's, to fill up the multiple holes that remained over time. It is a remarkable topic. And the consequences will last for a very long time.

"So what do you think it's all about? Life, I mean.
What's the purpose? What are we doing here?"

"To work hard ... to love someone ... and to have some fun ....
And if you are lucky ... somebody loves you back."

Katherine Hepburn's reply to her biographer, from *Kate Remembered* by A Scott Berg

◎ 编辑手记

    本书是我们工作室引进的一部高级科普著作英文影印版.中文书名可译为:《量子世界中的蝴蝶:最迷人的量子分形故事》.

    印度 LiveMint 的记者兼作家迪普利·德索萨的推荐语是:"很少有人能读完这一整本书,但这本书丰富而广阔的想法让我惊讶."

    本书的两大亮点是:

    1.由蝴蝶图形的发现者道格拉斯·霍夫施塔特(Douglas Hofstadter)倾情作序!

    2.有史以来第一本讲述"霍夫施塔特蝴蝶(Hofstadter's Butterfly)"的书籍!

    分形成为显学的标志是 1975 年在法国巴黎出版了一本由美籍法国数学家、经济学家 B. B. 曼德尔布罗特(Benoît B. Mandelbrot)所著的《分形对象——形、机遇与维数》(Les Objeis Fractuls:Forn, Hasardet Dimension).

    1976 年,也就是分形理论广为人知的最初几年,道格拉斯·霍夫施塔特当时还是俄勒冈大学(University of Oregon)的一名物理学研究生,正在试图理解磁场存在下晶体中电子的量子行为.当他把电子的容许能量画成磁场的函数图来进行探索时,他发现这幅图就像一只蝴蝶,有着谁也没有预料到的高度复杂的递归结构.此图只是由它自己的无数的复制品组成,无限深地相互嵌套着.这幅图最初被称为"上帝的图像",现在被物理学家和数学家亲切地称为"霍夫施塔特蝴蝶".

E-1

本书由道格拉斯·霍夫施塔特作序,是有史以来第一本讲述"霍夫施塔特蝴蝶"的书."霍夫施塔特蝴蝶"是一个美丽而迷人的图形,位于物质量子理论的核心.本书讲述了蝴蝶的故事和它与量子霍尔效应的联系.它揭示了霍尔电阻惊人的精确量子化背后的秘密是如何被编码在数学的一个叫作拓扑的分支中的.拓扑学揭示了一个球体和一个立方体之间隐藏的数字量,同时将它们与甜甜圈和咖啡杯区分开来.量子霍尔效应背后的深层拓扑现象是物理学的一个抽象版本,它构成了傅科摆日常活动的基础;它可以被认为是这种进动量子的表亲,被称为贝里相.

本书的一开始提到了古希腊数学家阿波罗尼乌斯(Apollonius),他在公元前300年创造了"椭圆"和"双曲线"这两个术语,并探索了相切圆的性质,最终促使了阿波罗尼垫圈的出现.法国哲学家勒内·笛卡儿(René Descartes)重新发现了这个问题,然后又由诺贝尔化学奖获得者弗雷德里克·索迪(Frederick Soddy)发现,他在一首名为《精确吻》(the Kiss Precise)的诗中赞美了这个问题.本书使用了量子力学的一些概念,主要采用了几何方法,最终将阿波罗尼乌斯、笛卡儿和索迪的《精确吻》联系起来.

"蝴蝶"图形与许多其他重要的现象密切相关,包括傅科摆、准晶体、量子霍尔效应等.在本书中,作者以完美的个人风格讲述了这个故事,采用了大量丰富而生动的历史轶事、照片、美丽的图像甚至诗歌,将她的书变成一场饕餮盛宴,让你的眼睛、心灵和灵魂尽享科学的魅力.

本书从图书分类上分可算做科普类图书,但它是一本高级科普书.

写科普书是中国人的短板,除了早期的高士其,叶永烈(刚刚离世,享年80岁)等就没有太广为人知的了.大专家不愿写(因其没名没利),小人物写不了(因其无法深入浅出),而且中国后来的教育文理分家,所以文理兼备者甚少.如果再奢求点艺术修养那就更寥若晨星了,所以目前还是以引进为主流.

下面对本书作者做一点介绍.本书作者是一位女性,叫金杜·萨蒂亚(Indubala I. Satija).她出生于印度,在孟买长大,从孟买大学(University of Mumbai)获得物理学硕士学位后,她来到纽约,在哥伦比亚大学(Columbia University in the City of New York)获得了理论物理学博士学位.目前,她是弗吉尼亚州费尔法克斯的乔治梅森大学(George Mason University)的物理学教授.她最近的研究领域包括拓扑绝缘体、玻色-爱因斯坦凝聚体和孤子.她发表了许多科学论文,这是她的第一本书.

物理学是金杜的初恋,而户外活动则是她的第二个恋人.她和丈夫苏希尔(Sushil)住在华盛顿特区的郊区的波托马克(Potomac),她是美国国家标准与技术研究所的物理学家.金杜和苏希尔都是马拉松选手,他们也喜欢徒步旅行和骑自行车.他们有两个孩子:拉胡尔(Rahul)是生物学家,而尼娜(Neena)是一名调查记者.

道格拉斯·霍夫施塔特为本书的序作者,更是位名人,他是"霍夫施塔特蝴蝶"分形的发现者,著名认知科学家,出生于纽约,其父亲是诺贝尔奖获得者、物理学家罗伯特·霍夫施塔特(Robert Hofstadter).他曾担任科普杂志《科学美国人》的专栏作家(1981—1983).道格拉斯·霍夫施塔特的首部著作《GEB——一条永恒的金带》获得了1980年的普利策奖和美国图书奖,那本书已被译作多种文字流传世界.

　　1958—1959年他就读于日内瓦国际学校;

　　1965年,他以优异的成绩从斯坦福大学(Stanford University)毕业;

　　1975年获得俄勒冈大学物理学博士学位,在那里他研究了磁场中布洛赫电子的能量水平,从而发现了被称为"霍夫施塔特蝴蝶"的分形.

　　本书的内容十分引人入胜.单从目录中就可以看出:

1. 分形族
2. 几何、数论和蝴蝶:友好的数字和密切联系的圈子
3. 阿波罗与蝴蝶之间的联系
4. 准周期的模式与蝴蝶
5. 量子世界
6. 量子力学式的婚姻和不听话的孩子
7. 另一种量子化:量子霍尔效应
8. 拓扑和拓扑不变量:量子霍尔效应的拓扑序言
9. 贝里相和量子霍尔效应
10. 精确的吻和精确的量化
11. 铸补的艺术
12. 实验室里的蝴蝶
13. 蝴蝶画廊:菲利普·G.哈帕主题变奏曲
14. 游戏曲
15. 感激之情
16. 诗意的数学和科学
17. 终曲
18. 相关检索

　　关于为什么要引入本书版权,不想说那些高大上的谎话,引用一个名人典故来说明.

　　奥威尔在谈到为何写作时,曾经调侃,写作可以报复一下少年时欺负自己的那些成年人.借用此调侃笔者也可以说:编辑出版一些高大上的科学著作是借此消除一下青年时被它们难住的心理阴影.笔者求学的时代恰逢20世纪80

年代那个全民爱科学的年代,像大多数国人一样,笔者也对数理科学的著作非常迷恋.但许多时候都是半途而废,原因是预备知识不够,所以这些没能卒读的阴影多年之后还在.今天与当年比起来,知识储备似乎多了一些,看看有没有一点改变呢?

在自媒体时代传统出版业的垄断被打破,科学著作发表的平台一下子变得很多.这个变化的优点是使每一个有发表欲的人都得到了满足,但这又同时带来了一个更大的危机,那就是对人们有限的注意力的分流.当泥沙俱下之时,真正的珍珠往往会被遗失.这时出版机构及编辑的本质就会被重新认识,那就是选择与鉴赏.一个编辑的核心能力绝不是简单的技术性操作,而是鉴赏.发现的眼光远胜于貌似努力的勤奋.现在数理的公众号很多,许多人都发布了他(她)以为好的选题内容,但优秀的真的不多.笔者有把握将本书推荐给读者,因为它从任何角度看都是优秀的.

本书即将付梓之际,赶上了百年不遇的疫情.在空荡荡的办公大楼中笔者想起早年间顾城有一首著名的诗:

> 我在幻想着
> 幻想在破灭着
> 幻想总把破灭宽恕
> 破灭却从不曾把幻想放过

不知是否适合此时的心境.

刘培杰
2020 年 4 月 23 日
于哈工大

刘培杰物理工作室
已出版(即将出版)图书目录

序号	书　名	出版时间	定　价
1	物理学中的几何方法	2017—06	88.00
2	量子力学原理.上	2016—01	38.00
3	时标动力学方程的指数型二分性与周期解	2016—04	48.00
4	重刚体绕不动点运动方程的积分法	2016—05	68.00
5	水轮机水力稳定性	2016—05	48.00
6	Lévy噪音驱动的传染病模型的动力学行为	2016—05	48.00
7	铣加工动力学系统稳定性研究的数学方法	2016—11	28.00
8	粒子图像测速仪实用指南:第二版	2017—08	78.00
9	锥形波入射粗糙表面反散射问题理论与算法	2018—03	68.00
10	混沌动力学:分形、平铺、代换	2019—09	48.00
11	从开普勒到阿诺德——三体问题的历史	2014—05	298.00
12	数学物理大百科全书.第1卷	2016—01	418.00
13	数学物理大百科全书.第2卷	2016—01	408.00
14	数学物理大百科全书.第3卷	2016—01	396.00
15	数学物理大百科全书.第4卷	2016—01	408.00
16	数学物理大百科全书.第5卷	2016—01	368.00
17	量子机器学习中数据挖掘的量子计算方法	2016—01	98.00
18	量子物理的非常规方法	2016—01	118.00
19	运输过程的统一非局部理论:广义波尔兹曼物理动力学,第2版	2016—01	198.00
20	量子力学与经典力学之间的联系在原子、分子及电动力学系统建模中的应用	2016—01	58.00
21	动力系统与统计力学:英文	2018—09	118.00
22	表示论与动力系统:英文	2018—09	118.00
23	工程师与科学家微分方程用书:第4版	2019—07	58.00
24	工程师与科学家统计学:第4版	2019—06	58.00
25	通往天文学的途径:第5版	2019—05	58.00
26	量子世界中的蝴蝶:最迷人的量子分形故事	2020—06	118.00
27	走进量子力学	2020—06	118.00
28	计算物理学概论	2020—06	48.00
29	物质,空间和时间的理论:量子理论	即将出版	
30	物质,空间和时间的理论:经典理论	即将出版	
31	量子场论:解释世界的神秘背景	即将出版	
32	计算物理学概论	即将出版	
33	行星状星云	即将出版	

刘培杰物理工作室
已出版（即将出版）图书目录

序号	书 名	出版时间	定 价
34	基本宇宙学：从亚里士多德的宇宙到大爆炸	即将出版	
35	数学磁流体力学	即将出版	
36	高考物理解题金典（第2版）	2019—05	68.00
37	高考物理压轴题全解	2017—04	48.00
38	高中物理经典问题25讲	2017—05	28.00
39	高中物理教学讲义	2018—01	48.00
40	1000个国外中学物理好题	2012—04	48.00
41	数学解题中的物理方法	2011—06	28.00
42	力学在几何中的一些应用	2013—01	38.00
43	物理奥林匹克竞赛大题典——力学卷	2014—11	48.00
44	物理奥林匹克竞赛大题典——热学卷	2014—04	28.00
45	物理奥林匹克竞赛大题典——电磁学卷	2015—07	48.00
46	物理奥林匹克竞赛大题典——光学与近代物理卷	2014—06	28.00

联系地址：哈尔滨市南岗区复华四道街10号 哈尔滨工业大学出版社刘培杰物理工作室
网　　址：http://lpj.hit.edu.cn/
邮　　编：150006
联系电话：0451—86281378　　13904613167
E-mail：lpj1378@163.com